Spatial Cognition

The GeoJournal Library

Volume 39

Spatial Cognition

Geographic Environments

Spatial Cognition

Geographic Environments

by

ROBERT LLOYD
Department of Geography
and the Center of Excellence in Geographic Education,
University of South Carolina, Columbia, U.S.A.

SPRINGER-SCIENCE+BUSINESS MEDIA, B.V.

A C.I.P. Catalogue record for this book is available from the Library of Congress

ISBN 978-90-481-4783-0 ISBN 978-94-017-3044-0 (eBook)
DOI 10.1007/978-94-017-3044-0

Printed on acid-free paper

For

Lena and Rich
Beth and Mark

Table of Contents

LIST OF TABLES

LIST OF FIGURES

Preface

This book was born out of a number of seminars that I have been teaching in the Department of Geography at the University of South Carolina. I found my original seminars on behavioral geography and cartography were gradually adding topics on spatial cognition over time as I became interested in the various ideas that became chapters in this book. The seminars would discuss the literature on some cognitive science topic and discuss how the ideas related to geography. Many of the discussions turned into interesting research projects for me and theses or dissertations for the students. The final inspiration to begin the book came when I taught a seminar on neural networks. It became clear to me that the connectionist view of the world had something to offer geography.

The book should appeal to researchers and students interested in relating cognitive processes to spatial problems. It could be used as a text in a general course on spatial cognition. Since many of the example relate to maps and map reading , it would also be an appropriate text in a course on cognitive cartography . Although my intention was not to write a book on geographic methods, the various discussions of neural networks should provide the interested reader with an initial understanding of these models and how they can be used to do geographic research.

I would like to acknowledge the support of the National Science Foundation for grants to the author that supported various research projects discussed in the book. I would also like to acknowledge the students in my seminars, particularly Rex Cammack, David Patton and Elisabeth Nelson, for contributing their ideas.

CHAPTER 1

GEOGRAPHY AND COGNITIVE SCIENCE

1.1. Introduction

The current interest that geographic researchers have in spatial cognition has directly evolved from the behavioral paradigm established in geography in the late sixties (Cox and Golledge 1969) and early seventies (Downs and Stea 1973). Early landmark studies in this era by researches, such as Lynch (1960) on images of cities, Gould (1966) on mental maps, Lowenthal (1961) on environmental images, and Wolpert (1964) on decision-making processes, had sown the seeds of some new ideas that took root. These roots established a cognitive perspective that attracted the attention of geographers interested in a variety of topics. Geographers had always been interested in exploring environments and representing them on maps and this new interest in spatial cognition was a natural extension of this tradition. Because the basic concepts of spatial cognition focus on the environmental information we all have stored in our memories, these concepts have a broad range of applications in many subareas of geography.

Spatial cognition is an important research area in both geography and psychology (Figure 1.1) with the two disciplines sharing a fundamental theoretical interest in (1) the processes used to encode spatial information into memory, (2) the nature of the internal representations, and (3) the decoding processes used with internal representations for making decisions (Garling and Golledge 1993). In addition to theoretical concerns, geographers almost always have a distinct interest in the particular environment being represented and in the spatial behavior of persons who have the environment represented in their memory. In addition, because geographers have invariably desired to represent environments on cartographic maps, they also have been concerned with how effectively their maps communicated information about environments. It is the need to communicate about spatial information and to understand behaviors taking place in environments that has caused geographers to be interested in the cognitive processes people use when interacting with cartographic maps and cognitive maps. Certain environments have unique characteristics that make their cognitive representations especially interesting. Some types of spatial behavior are uniquely tied to characteristics of the environment through these cognitive representations.

An interesting example of an interaction between cartographic and cognitive maps was discussed by Lloyd (1993). The map of New Orleans represented in Figure 1.2 is similar to a map that accompanied the classified advertisements in the June 30,1991, edition of *The Times-Picayune*. The newspaper's cartographer had constructed the map following the usual conventions and had oriented the map with north at the top of the page. This uncomplicated map had an unidentified lake in the North and an unidentified river that flows west to east across the map. The lake (Pontchartrain) and the river (Mississippi) were, perhaps, too well known to require an explicit label, but some areas in the city and streets had been identified on the map. This

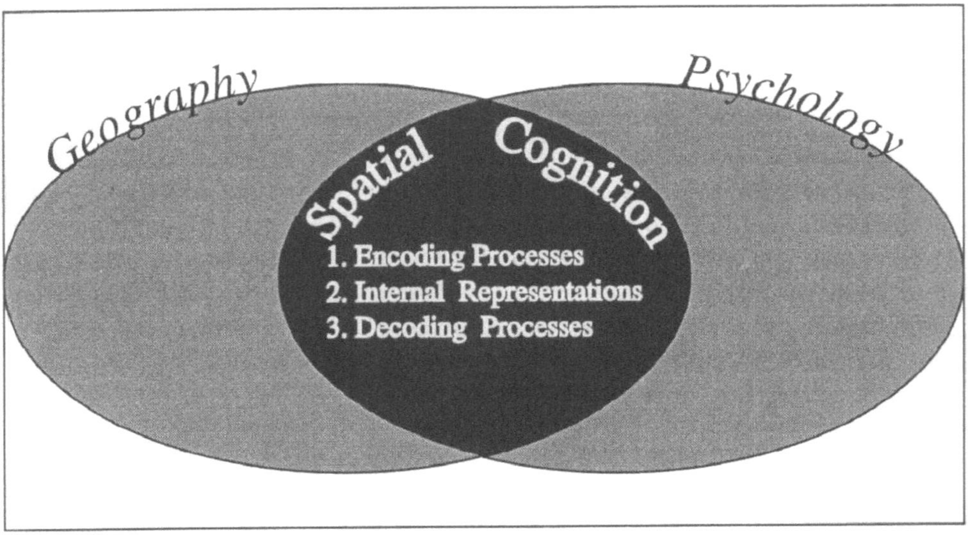

Figure 1.1: Spatial cognition is a research area of interest for both geography and psychology. Both disciplines are interested in fundamental ideas related to encoding processes, internal representations, and decoding processes.

suggested the cartographer made the map primarily for readers who were already familiar with the local environment. Additional strong evidence of this assumed familiarity can be found in labels selected for the map. The map was unusual because it presented propositional information used by local residents ,who were experienced navigators, on a cartographic map. If an uninitiated map reader assumed north was at the top of the map, the verbal coding on the map would have appeared to be at odds with the frame that enclosed it. Because the map had no north arrow, a stranger might have incorrectly concluded that east was at the top of the map. The map communicated with the nomenclature a local person would use to navigate through the city. The river's influence as the primary reference axis for this navigation process was obvious. The section of the city between the river and the lake was located on the same side of the river as the Eastern United States. Once this notion was accepted as reasonable, one could extend the same logic to the West Bank being located in the South. Other local navigation conventions may have been equally confusing to a stranger. The convention of presenting virtually all maps with north at the top has made it standard for the word *up* to be substituted for north and *down* for south (Shepard and Hurwitz 1984). A stranger might have assumed *above* and *below* were to be treated as equivalent terms for *up* (north) and *down* (south). This assumption also would have been incorrect. The direction of flow for the river, which was not necessarily obvious from just looking at the map, was the crucial information. Keeping in mind that water flows downhill, *above* translated to upstream from a street that intersects the river and *below* translated to downstream from a street. Any location in the environment then could be communicated by a coordinate system defined by the side and flow of the river. This meant that east (of the river) was in the north and west (of the river) was in the South. Ignoring the bends in the river, this generally also meant *above* was in the West and *below* was in the East. See Lewis (1976) for a more detailed explanation of how the residents of New Orleans describe their urban environment.

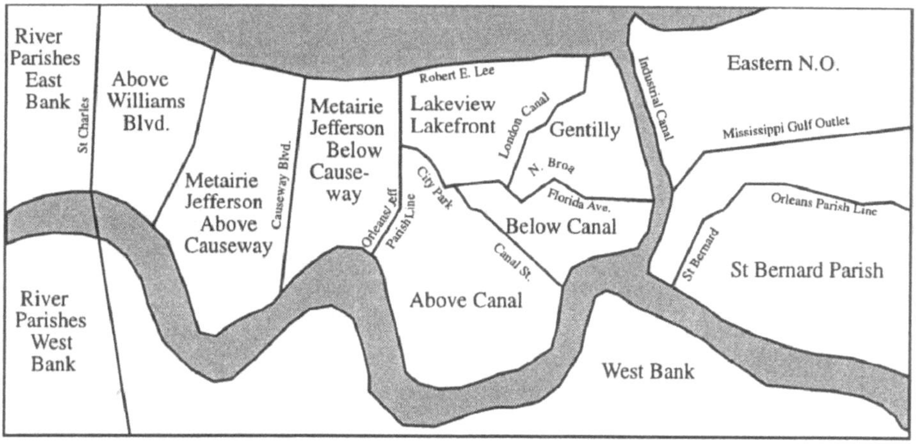

Figure 1.2: The place names on this map of New Orleans depict the propositions used for navigation by local residents. A similar map appeared in the June 30, 1991, edition of *The Times-Picayune*. Redrawn here iwth permission.

We learn spatial information about our environment all the time. Just like the residents of New Orleans, we need some system for encoding in our memory *what* is in our environment and *where* it is located. *What* information registers in memory that an object exists and describes or depicts its important features for future reference . *Where* information provides a description or depiction of the object's location in space. The words description and depiction are both appropriate because environmental information can be processed both as words with verbal (language) systems and as images with visual systems to store it in memory. The rest of this chapter is used to define some basic terms that will be used throughout this book and to demonstrate that both verbal and visual systems are used to process spatial information.

1.2. Cognitive Processing

Although isolated researchers had provided discussions of the perception of orientation (Gulliver 1908), imaginary maps (Trowbridge 1913), imaginary worlds (Kirk 1951; Wright 1947) and the perception of hazardous environments (White 1945), geographic researchers had traditionally focused on questions related to the objective structure of the geographic environment. Geographic information related to both human and physical environments was observed, recorded, described, classified, and, with methods provided by the quantitative revolution, analyzed, but relatively little concern was traditionally given to processes that directly connected humans with their environments. Behavioral geographers asked new questions about *spatial behavior* (the rules of decision making) rather than *behavior in space* (patterns on maps) (Rushton 1969). This required new assumptions to be made about both geographic environments and the humans living in them. The behavioral geography movement focused attention away from a quest to understand and explain objective reality and toward the goal of understanding subjective worlds and the decision-making processes of the less-than-perfect people who live in them.

Words such as perception, cognition, and image tended to be used by researchers almost as if they all meant the same thing when cognitive studies started to appear in the geographic literature. Gold's (1980) discussion of this problem suggested other disciplines were experiencing the same initial problems coping with cognitive science terminology. Downs and Stea (1973), Downs (1981) and Golledge and Stimson (1987) have made a particular effort to provide specific definitions and meaningful distinctions among various cognitive science terms. A similar effort is made here to identify important terms and discuss their meaning.

The purpose of this book is not to provide a discussion of the development of behavioral geography or to make a formal comparison of behavioral research done by geographers and psychologists. These goals have already been accomplished in an excellent fashion by other authors (Golledge and Stimson 1987; Gärling and Golledge 1993). The purpose of this book is to discuss cognitive theories that are relevant for geographic researchers interested in understanding and explaining spatial behavior. The simple thesis is that we can explain the spatial behavior of humans living in geographic environments through an understanding of processes these humans use to acquire, represent, and use spatial information.

1.2.1. COGNITION

Martindale (1991, p. 1) began his book, *Cognitive Psychology: A Neural Network Approach*, by simply stating "Cognitive psychology deals with questions about how people learn, store, and use information." Psychologists consider cognition a general concept that is concerned with acquiring information about the world, representing and transforming this information as knowledge, and using this knowledge to direct our attention and behavior (Solso, 1979). Note the three components of cognition, (1) Acquisition, (2) Representation, and (3) Use, reflect the theoretical interests shared by geography and cognitive psychology through spatial cognition. The general interests of cognitive psychology are more focused for geographers. Geographers are interested in asking specific questions about the acquisition, representation and use of information for specific geographic environments (Figure 1.1).

1.2.2. PERCEPTION

Perception is a cognitive process that is directly involved with the detection and interpretation of sensory information. Perception requires a direct connection between a person and the object being perceived so that sensory signals can be directly processed. Viewing a map and acquiring information about the sizes of cities is perception. Hearing an animal in a forest environment running ahead of you and to your right is a direct sensory experience and, therefore, perception. Thinking about how long it would take you to drive to the mall is not perception. Acquiring information about Boston while reading a *Spenser* novel (Parker 1973) is also not perception because you are not in direct sensory contact with Boston. Except for map reading studies, geographers rarely do what could truly be called perception studies. For example, a study that would have subjects rank their preferences for urban neighborhoods could not be called a perception study because subjects could not perceive (have direct sensory contact with) all the neighborhoods simultaneously.

1.2.3. MOIST MAPS

The advancement of geography as a discipline has been linked by some influential authors to the ability of geographic researchers to create significant spatial theory (Downs and Stea 1973; Morrill 1987). If this challenge is to be met, geographers must develop a perspective that allows them to (1) observe processes that directly connect humans with their environments, (2) form hypotheses that explain these processes, and (3) develop theories that connect these processes and ultimately provide an explanation of various types of human spatial behavior. Cognitive theories have traditionally been developed and tested using experimental methods that make it possible to control stimuli that have been specifically designed for the experiment. A typical experiment might monitor some simple behavior when a person is exposed to various stimuli that represent the effect being studied. Suppose a cartographer wanted to test the notion that large map symbols will be remembered more efficiently (faster and more accurately) than small map symbols. A person could view both large (4 square inches) and small (1 square inch) map symbols for a fixed amount of study time and instructed to remember the symbols. The person then could be shown a series of symbols and asked to determine as quickly and accurately as possible if each trial symbol was part of the memorized set. Only half of the trial symbols would be from the memorized set. One could then analyze the data to determine if large symbols produced significantly faster reaction times and more accurate responses than small symbols.

Now suppose, as a second example, that a person is given a complex reference map and asked to search for a particular symbol. The target symbol has a size, but it also has a shape, color, and orientation. Many other symbols are also on the map and they also vary in theses characteristics so that many different sizes, shapes, colors, and orientations are represented on the map. The target symbol shares some characteristics with these other symbols on some dimensions. The location, size, shape, color, and orientation of each object provide meaningful information to the map reader that is related to a particular symbol on the map. In addition, the pattern formed in space by the set of objects may also provide significant information to a map reader. The second example is closer than the first to real-world problems that might be studied by geographers, but it is also much harder to study using experimental methods that control environmental effects that produce variation in behavior.

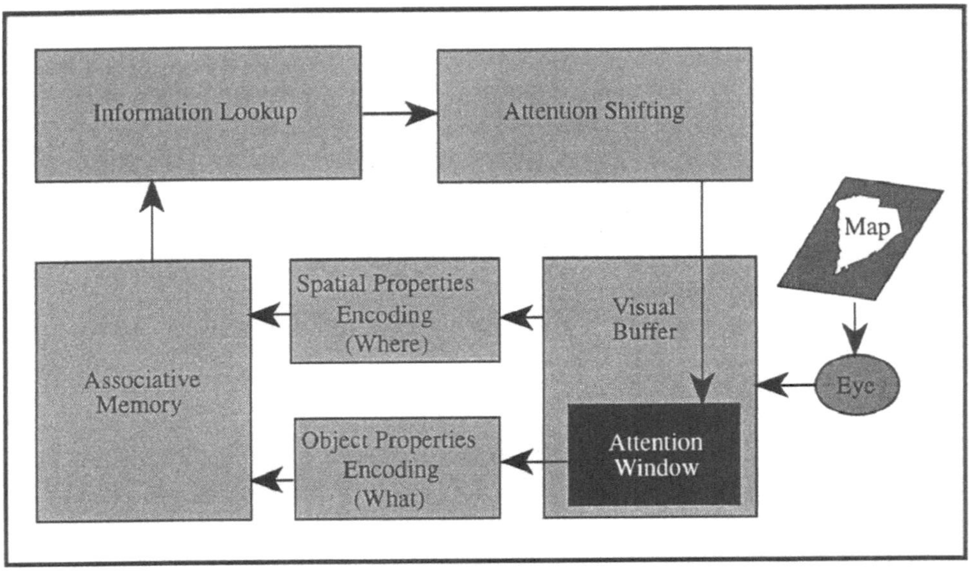

Figure 1.3: Subsystems used in high-level vision as defined by Kosslyn, Flynn, Amsterdam, and Wang. After Kosslyn, Flynn, Amsterdam, and Wang (1990).

Researchers trying to test the validity of cognitive theories in more complex environments will generally find it difficult to completely control the factors that affect behavior. One way to enhance the probability that a theory will survive transplantation from the laboratory to the real world is to base it on biologically correct principles. Kosslyn and Koenig (1992) have recommended a "Wet Mind" approach to cognitive neuroscience. The "Wet Mind" approach argues that theories that explain how sensory information about stimuli is translated into behavior should be consistent with what is known about how information is processed by the sensory systems and the human brain. Acquiring information from an environment or map usually relies on the visual system for the initial input of information. Kosslyn, Flynn, Amsterdam and Wang (1990) have identified groups of neural subsystems related to high-level visual object identification. If the activations of neurons in a map reader's brain were monitored while he or she was searching a map, networks of neurons related to these hypothesized subsystems would, theoretically, input, process, and output information during various stages of the search. A simplified representation of these subsystems and their connections, hypothesized by Kosslyn and Koenig (1992), is illustrated in Figure 1.3. Moist maps are maps as they exist in wet minds. The subsystems in the following discussion should be used when searching a map for a target.

Visual Buffer

Information on the map is sensed through the eyes and produces a pattern of activity in a visual area of the brain located in the occipital lobes. This area is referred to in Figure 1.3 as the *Visual Buffer*. The spatial organization of the information coming from the eyes is thought to be maintained on the surface of the brain in the *Visual Buffer* (Kosslyn, Segar, Pani, and Hillger 1990). If more specific details are needed, then we focus our attention on part of the visual buffer with the *Attention Window* (Posner 1988). Whether or not focused attention is needed for decision-making during a search task is a critical issue for search theories (Duncan and Humphreys 1989; Treisman and Galade 1980; Wolfe, Cave, and Franzel 1989). More specific information about the content of the *Attention Window* is obtained through two separate *what* and *where* systems (Müller and Rabbitt 1989; Sagi and Julesz 1985; Ungerleider and Mishkin 1982).

Spatial Properties

The *dorsal system*, areas in the brain running from the occipital lobe up to the parietal lobes, process spatial properties (Kosslyn and Koenig 1992). Since the information in the visual buffer is specified relative to the retina rather than space, it may not be directly useful for some tasks. "A spatiotopic representation of the location of objects is necessary for coordinating separate objects or parts in a single frame of reference, for navigation, and so on. Thus, there must be a subsystem that takes as input a retinotopic position, distance (computed using stereo and via other bottom-up processes), eye position, head position, and body position, and uses such information to establish where an object or part thereof is located in space" (Kosslyn, Flynn, Amsterdam, and Wang 1990, p. 218). Besides determining the locations of objects in space, this subsystem is thought to determine the sizes of objects (Maunsell and Newsome 1987) and their orientations (Holmes and Gross 1984).

Other parts of the *dorsal system* are thought to encode spatial relationships between objects or parts of objects. One subsystem is thought to do categorical encoding with propositions such as "connected to," "below," or "to the right of." The proposition "inside of" might be used in map reading to encode that a city was located in a state or a state was located in a region. This type of categorical spatial information can be encoded in long-term associative memory for later use (Kosslyn, Flynn, Amsterdam, and Wang 1990). Distances can also be encoded categorically during map reading as "near" or "far" from a reference location. Since more precise spatial knowledge is sometimes required, e.g., during navigation, another subsystem is thought to encode coordinate relations between objects (Kosslyn, Koenig, Barrett, Cave, Tang, and Gabrieli 1989).

Object Properties

The *ventral system*, areas in the brain running from the occipital lobe down the inferior temporal lobe, processes object properties (Kosslyn and Koenig 1992). One important subsystem preprocesses the incoming information to consider "nonaccidental properties" defining the shapes of objects (Biederman 1987; Lowe 1985, 1987a, 1987b). The encoding of "nonaccidental properties" allows us to recognize familiar objects in any orientation. This is an important concept that may explain observed differences between spatial knowledge acquired from navigation within an environment and spatial knowledge acquired from maps with fixed north-at-the-top orientations (Carpenter and Just 1986; Evans and Pezdek 1980; Lloyd 1993; Lloyd, Cammack, and Holliday 1995; Presson and Hazelrigg 1984; Presson, DeLange, and Hazelrigg 1989; Sholl 1987; Throndyke and Hayes-Roth 1982). Kosslyn, Flynn, Amsterdam, and Wang (1990) call nonaccidental properties of images "trigger features" because they are features of an image that are likely to remain constant under different viewing conditions and, therefore, can be used to trigger the identification of objects. Since maps are usually only experienced in a north-at-the-top orientation, the typical map reader has not had sufficient opportunities to encode nonaccidental properties and, therefore, has difficulty recognizing maps at other orientations (Conerway 1991; Holmes 1984; Lloyd and Steinke 1984; Steinke and Lloyd 1983a).

Another subsystem is needed for the pattern activation of previously stored visual information. Patterns of stored information must be compared to current sensory input to access the probability of a match. Recognition occurs when a sufficiently close match is made. In some cases, e.g., determining the gender of an approaching person, we may want to activate a very generalized pattern that would match a range of possible inputs. In other cases, e.g., finding a friend in a large crowd, the pattern activated must be narrowly constrained. Suppose a polygon representing the outline of a state were presented to a person. Previously stored information could be used to activate patterns representing outlines of states for potential matches. One might first try generating patterns of states in the East when trying to match a small state and states in the West when trying to match a large state. If recognition does not occur on the whole state outline, parts of the outline, e.g., panhandles or river boundaries, could be processed with the same subsystems with focused attention.

Many other features can be used for identification. These are also visual, e.g., color, texture, but separate from shape. A separate feature detection subsystem that encodes "non-shape features" like color and sends them to associative memory rather than visual memory has been hypothesized (Kosslyn, Flynn, Amsterdam, and Wang 1990). Suppose you are meeting a stranger at the airport. All you know about the person is that his or her hair will be green. You should have little trouble detecting him in a crowd even though you have never seen the him before. You do not need a visual memory of the person's hair. The unusual association of green with hair color should be sufficient for a match.

Associative Memory

The *spatial* and *object* properties subsystems pass on their information to *associative memory* where objects can be represented and stored (Figure 1.3). Object identification will require the information that has been associated with the object. This information might include the object's name, its function, places where it might be found, other objects associated with it, categories it belongs in, its parts, and so on. If objects are stored as propositional structural descriptions, the stored representations later can be compared with objects we are viewing. "To the extent that the input properties and their spatial relations match those of a stored object, one can be confident that one is seeing the object; and to the extent that properties and spatial relations are distinctive for one object, one can reject hypotheses favoring other objects" (Kosslyn, Flynn, Amsterdam, and Wang 1990, p. 234).

Information Lookup

Most of the time we are not actively searching for a particular object. We encounter many objects as we navigate through the environment or on a casual inspection of a map. We need to identify what our attention has focused on. In this case we generate and test hypotheses that the object of our attention is something we have previously encoded in associative memory. Testing a specific hypothesis, e.g., this part of the map is Texas, requires us to look up in associative memory properties the hypothesized object should have. These properties were encoded during previous map reading experiences. Some properties are more useful for identification than others. Its panhandle or the shape of the Rio Grande River boundary might be an important property for Texas. We need to know where to look for these properties to use such distinctive properties to test a hypothesis. Subsystems are needed to generate either categorical or coordinate locations for properties. "A panhandle extends from the top of Texas" might describe a meaningful categorical location. Texas could be rejected if no such property were visible. Knowing the coordinate location for Dallas would enable you to look for a place name. Finding such a place name on the map at that location would support the "Texas hypothesis." Since the hypothesis testing takes us to various parts of the map, another subsystem must exist that controls our shift of attention. Since the location of the properties have been encoded as spatiotopic coordinates, Kosslyn, Flynn, Amsterdam, and Wang (1990) argued the attention-shifting subsystem is capable of converting spatiotopic coordinates into retinotopic coordinates. This is because the shift of the attention window is in the visual buffer and it represents locations not as they exist in space, but as they exist on the retina.

A search process with a known target (top-down search) is simpler because we know what we are looking for and must just determine if what we are looking at is what we are looking for. We are, therefore, testing the hypothesis that we are currently looking at the target. A candidate in the visual buffer, that might be the target, must be inspected and, if rejected, new candidates must be selected for consideration. The attention window is shifted to a new candidate and the process is repeated. A search process ends when the target is found or when the search is abandoned.

1.2.4. IMAGES

Mental images can involve any of the senses, but geographers have usually studied visual imagery for environments or maps. Imagery involves reactivating information that has already been coded and stored in long-term memory. Kosslyn and Swartz (1978, p. 223) argued "images are spatial representations that occur in active (short-term) memory." They also claimed "the image we experience is not simply 'retrieved,' but is generated from a more abstract representation in long-term memory." It is further contended that images are "constructed using both perceptual and conceptual information" that previously have been encoded and stored (Kosslyn and Swartz 1978, p 226). A person can experience an image of an object that is not physically present, e.g., a map of California in a bare room with the lights out and your eyes closed. A person can generate a visual image of a place that he or she has never directly experienced or even one that does not actually exist because information about the place has been acquired through verbal coding, e.g., reading a text that describes the place (Lilliput) or hearing some one describing the place on the radio (Lake Wobegone).

1.2.5 SCHEMATA

Representing spatial information as images has an intuitive appeal because images can function like cartographic maps, which are a familiar way to express spatial information. Another type of internal structure that represents spatial information has also been discussed. Schemata are "internal representations of common relationships among familiar objects in the environment. Such schemata help to organize and give meaning to real-world settings....Schemata for familiar settings appear to be well learned at a fairly young age" (West and Morris 1985, p. 23). Spatial schemata are seen as distinctly different from images (Tuan 1975; Gold 1980) and as a framework for organizing past and present experiences within an environment (Lee 1976). Spatial schemata are more directly associated with behavior taking place in the everyday environment and are able to easily accommodate environmental information currently being experienced. It requires concentrated attention to construct images in the visual cortex from information in long-term memory stored in a more abstract format (Fink and Schmidt 1977; Fink and Kosslyn 1980; Pinker 1979). Forming images requires such concentration that the person forming the image may become temporarily oblivious to current sensory information while spatial schemata may be used without a conscious effort (Gold 1980). For example, people navigating a familiar route in a well-learned environment while concentrating on non-environmental information frequently report arriving at a destination without being able to recall the trip. Lindberg and Gärling (1981) have also demonstrated new environmental information can be encoded during locomotion even when a person has been concentrating on a distracting task. This suggests that we can acquire and use some spatial information without making a conscious effort.

Neisser (1976) argued that we pick up information from the environment in a perceptual cycle. He argued that *anticipatory schemata* are important elements of the perceptual cycle that guide the information acquisition process (Figure 1.4). "A schema

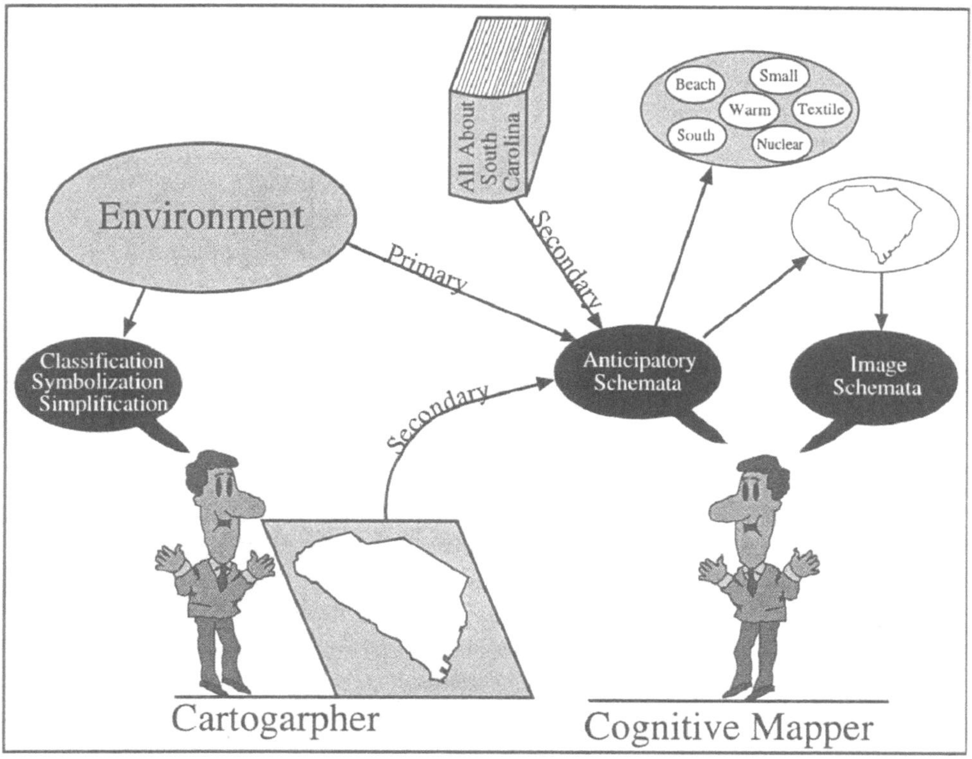

Figure 1.4: Cartographers and cognitive mappers generalize and simplify the information when they encode information about the environment from primary and secondary sources.

is that portion of the entire perceptual cycle that is internal to the perceiver, modifiable by experience, and somehow specific to what is being perceived. The schema accepts information as it becomes available at sensory surfaces and is changed by that information; it directs movements and exploratory activities that make more information available, by which it is further modified" (Neisser 1976, p. 54). Differences in acquiring primary and secondary spatial information can be related to different types of schemata. The person in Figure 1.4 would use schemata developed by experiences with the environment to acquire primary information directly from the environment. Schemata developed by reading experiences would be different from those produced by environmental experiences and would direct the acquisition of secondary information from a book. A third type of schemata would be developed by map reading experiences and would direct someone acquiring secondary information from a cartographic map. Neisser (1976) sees a schema as a *format*, like a format in a computer language, that can be used to acquire information and as a *plan* for obtaining more information. A person would use schemata to accept information from the environment or objects and to guide future behaviors.

Neisser's notion of *anticipatory schemata* as internal representations of spatial information is limited to primary information acquired while interacting with the immediate environment. Primary navigation activities produce internal structures that are orienting schemata because one's ego is involved. As Neisser (1976, p. 116-117) explains:

> *Information about oneself, like all other information, can only be picked up by an appropriately tuned schema. Conversely, all information that is picked up, including proprioceptive information, modifies a schema. In the case of movement through the environment, this is an orienting schema or cognitive map. This means that the cognitive map always includes the perceiver as well as the environment. Ego and world are perceptually inseparable.*

Neisser did not discuss acquiring spatial information from cartographic maps, but one could argue that the schemata used to acquire information from a cartographic map would not involve the ego (Lloyd 1993; Lloyd and Cammack 1996; Lloyd, Cammack, and Holliday 1995). A map is an object that represents the environment constructed by a cartographer using abstraction processes. Object schemata are used to accept information from maps and to direct future map-reading activities. It is easy to believe that people have both orienting schemata, which are based on primary experiences, and object schemata, which are based on map-reading experiences, for environments at the urban scale and larger.

The appropriate schemata are automatically initiated to accept information when a person must consider a particular situation. The orienting schema with the embedded ego is used when we are considering the immediate environment, e.g., turning left or right at the next intersection or giving detailed route directions on the telephone. Neisser (1976, p. 130) argued "the experience of having an image is just the inner aspect of a readiness to perceive the imagined object, and that differences in the nature and quality of people's images reflect differences in the kind of information they are prepared to pick up." This suggests images are not pictures in the head, but plans for obtaining information from objects, environments, or cartographic maps. The imagery experience is a state of visual readiness to perceive something and the image is just what we expect to see.

1.2.6 CONCEPTUAL-PROPOSITIONS

Some researchers have argued that all verbal and visual knowledge is stored as abstract conceptual propositions (Anderson and Bower 1973, Rumelhart, Lindsay, and Norman 1972; Pylyshyn 1973, 1984). The raw information we acquire is abstracted, summarized, and interpreted before it is stored. Conceptual propositions are assertions that make a statement about the nature of the world. Simple examples are *the building is tall* or *Wyoming is square*. Semantic examples like these two are usually used to illustrated propositions, but visual information can also be depicted in memory using conceptual propositions. It is argued that visual information is just an alternative form of input that can be processed and stored as conceptual propositions. It has been suggested that abstract verbal inputs are rather simple when compared with visual information (Raaijamakers and Shiffrin 1981). The basic difference between an internal representation that encode a verbal description of a map and one that encoded a visual depiction of the same map would be the amount of detail. It is argued that imaginable material is better remembered because it has a rich and detailed network of related propositions. To illustrate that a map can be encoded as a visual depiction and as a description of conceptual propositions consider the following example (Figure 1.5 and Figure 1.6).

The map is for the hypothetical state of Dyland. Dyland's two counties, Stone and Hardrain, are shown as are its two major cities, Tambourine and Heaven's Door, and transportation arteries, 4th Street and Highway 61. On the map (Figure 1.5) one can see the absolute and relative locations of these various elements. A quick reference to the graphic scale and north arrow and one knows that Tambourine is about 35 miles north of Heaven's Door on 4th Street. One could also encode most of the basic spatial information for the objects on the map as a network of conceptual propositions (Figure 1.6). The conceptual-proposition network can tell us that Tambourine is north of Heaven's Door on 4th Street, but cannot provide the distance between the two cities.

1.2.7. DUAL CODING

The nature of internal representations of space and whether an imagery or conceptual- proposition theory best accounted for empirical findings was hotly debated for some time (Kosslyn 1980; Pylyshyn 1984). The radical imagery position argued that perception processes act on acquired information to make it simpler and to organize it (Kosslyn and Pomerantz 1977). Larger units that are objects and their properties are created by this organizational process and stored in memory. These units can then be used to assemble images that are picture-like spatial entities that resemble entities experienced during perception. A key argument is that similar internal representations are the basis of both perceptual and imaginal visual experiences. The same processes may be activated when one is using perception to look at a map or long-term memory to imagine a map. The obvious implication of this is that similar behavioral effects could occur with either the map or map image (Shepard and Podgorny 1978; Finke 1979). One could argue that map images are functionally equivalent to cartographic maps in form, if not in specific content. Kosslyn and Pomerantz (1977, pp. 53-57) have distilled the main arguments of the radical imagery theory into five points:

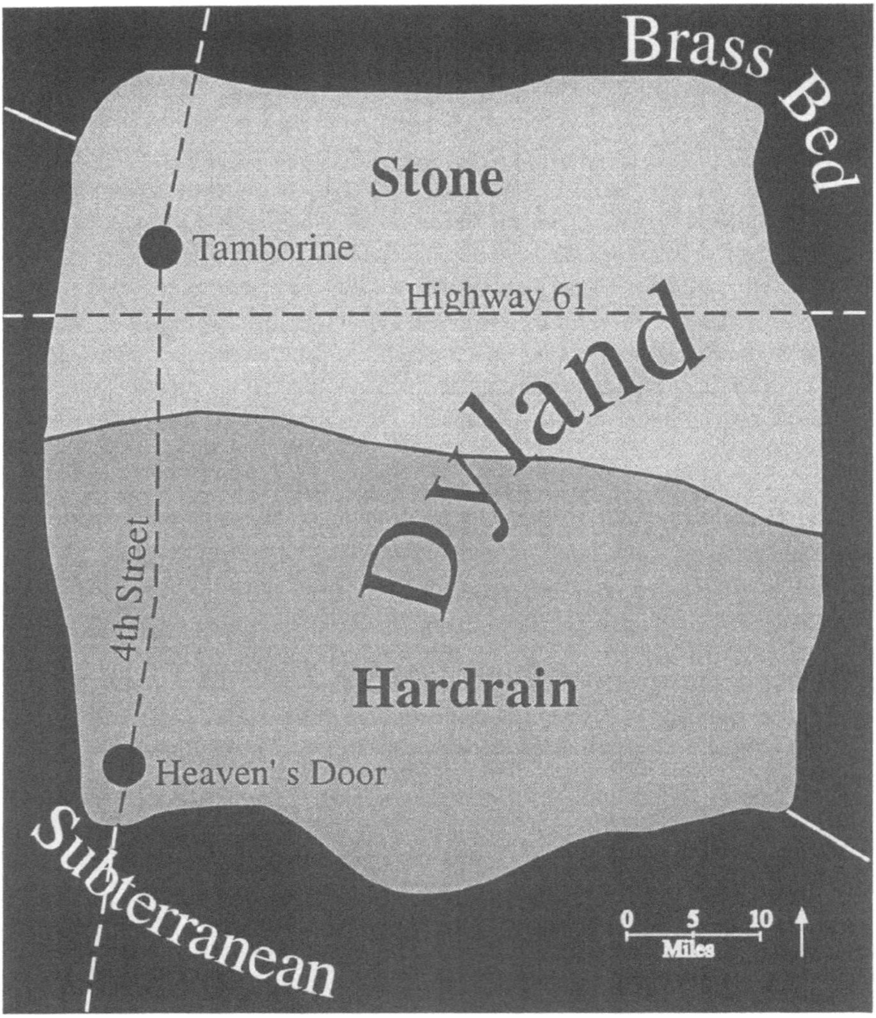

Figure 1.5: Map that depicts a hypothetical state (Dyland), counties (Stone and Hardrain),cities (Tambourine and Heaven's Door), and transportation arteries (4th Street and Highway 61

(1) Images are spatial representations. Visual images are spatial representations that provide an experience similar to seeing an object during visual perception. Images are generated from underlying abstract representations, but the contents of the underlying representations are accessible only when a surface is generated to experience the image.

(2) Images have a capacity limit. Only a fixed amount of memory resources is available for processing when images are constructed and represented. This limits the details that can be activated for an image at any point in time.

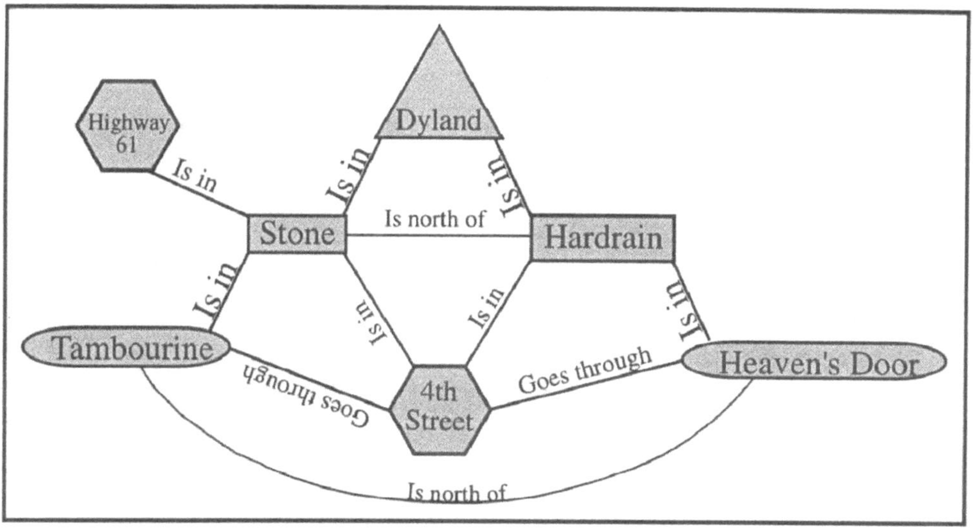

Figure 1.6: Network of conceptual propositions that describe the map depicted in Figure 1.5. Nodes represent hypothetical state (Dyland), counties (Stone and Hardrain), cities (Tambourine and Heaven's Door), and transportation arteries (4th Street and Highway 61). Nodes are connected by a statement of their relationships (Is in, Is north of, and Goes through).

(3) Images are whole entities. Once images are formed they are experienced as wholes that may be compared to percepts in a template-like manner.

(4) The visual system supports imagery. The same structures in the brain that represent spatial information being extracted during vision also are available to support images.

(5) Images are functionally equivalent to the objects they represent. Many of the same cognitive processes that are used to analyzing percepts also can be used to investigate images.

Kosslyn and Pomerantz (1977, p. 66) also have distilled the main arguments of the conceptual proposition theory into five points:

(1) There are no pictures (images) stored in memory. Any information stored in memory is only represented by networks of conceptual propositions.

(2) Propositions have a capacity limit. Active memory can only process a finite number of conceptual propositions at one time. Both time and effort are needed to make propositional structures active.

(3) Both serial and parallel processes can access conceptual proposition networks. Networks may be serially searched for information by accessing nodes in a specified

ordered. Pairs of networks may be compared in parallel to assess their similarity by accessing all their nodes simultaneously.

(4) Visual information can be represented as conceptual propositions. Networks can be constructed so that any desired spatial relationship can be represented.

(5) Propositional networks can be compared to determine the relative similarity of the entities they represent. Networks or parts of networks are similar or different to the extent that their elements and relations are similar and different.

Good arguments can be provided for the existence of both images and conceptual propositions. Anderson (1978) suggested that the question of which position is correct and which is incorrect cannot be answered. A simple resolution of the quarrel regarding which perspective is correct is that both are correct. Work on recalling paired associated words done by Pavio(1969; 1971) first suggested a dual-coding theory. He did not elaborate on the exact nature of images, but did postulate two separate and interconnected memory systems, verbal and imagined, that operated in parallel. Experiments had subjects learn pairs of words that had imagery values. Abstract words, like context or virtue, had low imagery values because it is difficult to generate a visual representation of the word. Concrete words, like elephant or church, had high imagery values because they easily can be visualized. Subjects were given one of the paired words as a stimulus and they responded with its paired associate. The greatest success was achieved when both words were easily visualized. When one of the pair had a high imagery value and the other a low, the greatest success occurred when the stimulus word was easily visualized. This suggested that visual and verbal systems worked together and that recalling the word pair was easier if the stimulus word could be imagined.

A later study used reaction times to investigate differences in visual and verbal processing (Kosslyn 1976). Adult and child subjects were asked to imagine an animal, and asked if it had a particular characteristic. For example, imagine a cat and determine if it has claws. Adults were only slightly faster than children when instructed to use imagery to determine their answers. Younger children's reaction times were about the same when no instructions were given to use imagery, but adults responded much faster. This suggested adults found it more efficient to use verbal coding to answer these simple questions when they could select their own processing method. Younger children used imagery even when not instructed to do so. This could suggest that verbal encoding abilities develop as people mature and that adults have more complex verbal coding structures because they have had more experiences. Obviously we can only use verbal processing to recall geographic information if we have encoded the information with such a system. To illustrate this point consider two questions about the 48 contiguous states of the United States that could be answered by looking at a map. Without looking at a map, read each question and note how quickly you came up with an answer and what mental processes you used.

Question 1: How many states touch the Atlantic Ocean?

Question 2: What is the smallest state?

If you are like most people asked these questions, you had to imagine a map of the United States to answer the first question because you had no verbal coding available to recall the answer. You also probably could have used either an image of a United States map or verbal coding to answered the second question, but answered quickly using verbal coding you learned in the fourth grade. According to a *race model* proposed by Kosslyn, Murphy, Bemesderfer, and Feinstein (1977) we can retrieve and process information as visual imagery and conceptual propositions at the same time. The process that runs to completion first provides the needed information. This competition between processing systems happens frequently and automatically.

Reports of clinical cases, where either visual or verbal memory systems have been damaged, provide the most convincing support for the dual-coding theory. Patients with lesions of the left hemisphere were not successful at recognizing a series of words, but could easily recognize a series of abstract pictures (Milner 1971). Sasanuma (1974) reported Japanese patients with damage to the left hemisphere lost the ability to read Kana (phonetic characters), but could still read Kanji (ideographic characters). O'Keefe and Nadel (1979, p. 493) have postulated that the left hippocampus in the human brain receives and processes verbal information about space into a "left hemisphere cognitive map [that] acts as a semantic map, providing something like a deep structure for connected discourse." The anthropologist Wallace (1989) has suggested this ability developed between 2.5 and 2.0 million years ago when a global reduction in temperature and an alteration in rainfall produced marked environmental changes in eastern and southern Africa. Hominids adapted to an expansion of open grasslands at the expense of forests and woodlands by consuming leftover meat from animals killed by carnivores. He suggested "an early-hominid brain mechanism for communication-while-mapping is the probable direct result of diurnal scavenging" (Wallace 1989, p 522). Since carcasses were more widely distributed in space than previous food supplies, there was a new need to communicate food locations to others. "Thus, there was constant selection pressure for verbal communication. The conversion of the mapping system [left hippocampus] to this task would not have required drastic neurological changes" (Wallace 1989, p 523).

1.2.8. BEHAVIOR

There are many different types of behavior that need to be explained. Geographers frequently consider *overt behavior* the acts that caused some change in the order of the universe (Golledge and Stimson 1987). This change can be detected by noting the locations of objects (people or other interesting things) at time t and time $t + 1$. The movement of people in space provides many examples, e.g., commuting (going to work and returning), shopping (going to stores and returning), recreation (going to an interesting place and returning), or migration (going to new locations and staying). This overt behavior of many individuals can be aggregated and the patterns it makes displayed on maps. These maps of behavior can be compared to maps of other things for clues that might explain the patterns. The numbers that were used to make the maps can be analyzed and models can be computed that predict how many people will go from place *A* to place *B* between time *t* and *t +1*. If we all had tails that dragged behind us and marked our paths through space, this type of overt behavior would be clearly marked in the environment. This "behavior in space" is only a secondary manifestation of the important behavior that preceded it (Rushton 1969). The process that produced a decision that initiated an act of overt behavior was done by a human brain using information that had been encoded from the environment and stored as an internal structure to be decoded and used at the time it was needed. Morrill (1987), in his plea to geography for a theoretical imperative, reported the dominant research theme shared across the social sciences was an interest in decision making. Geographers need to focus on understanding the decision-making processes that produce overt spatial behavior to respond to Morrill's theoretical imperative. This behavior, the decision, is not overt. It takes place inside heads of people who frequently do not know how the decision is made. Geographers, as cognitive scientists, must learn to use experimental research designs and consider dependent variables like reaction times to understand spatial decision-making processes.

1.2.9 COGNITIVE MAPS

Geographers were naturally drawn to discussions of environmental learning and spatial behavior that related to maps stored in the brain and the cognitive map became a central issue in spatial cognition research (Tolman 1948). The frequently quoted definition originally presented by Downs and Stea (1973, p. 9-10) focused on the processes that produced the cognitive map. "*Cognitive mapping is a process composed of a series of psychological transformations by which an individual acquires, codes, stores, recalls, and decodes information about the relative locations and attributes of phenomena in his everyday spatial environment.* " A cognitive map is "the product of this process at any point in time." This influential definition did not state any specific characteristics of cognitive maps and appeared to be limited to the everyday spatial environment. Other researches soon focused attention on specific issues to explore the nature of cognitive maps. For example, differences between procedural and survey knowledge suggested that spatial information may be acquired in different forms and suggested that cognitive maps generated from navigation and map reading experiences may be significantly different (Thorndyke and Hayes-Roth 1982; Lloyd 1989a). It was also recognized that cognitive maps are not exact copies of environments or cartographic maps but are internal representations whose form may reflect a hierarchical network structure (Stevens and Coupe 1978) and whose accuracy may reflect systematic distortions (Tversky 1981; Lloyd and Heivly 1987). Some have suggested that cognitive maps are mental models that integrate both verbal and visual information into a single structure (Lloyd, Cammack, and Holliday 1995; McNamara, Halpin and Hardy 1992; Taylor and Tversky 1992a, 1992b). Some of these notions are incorporated in the recent definition and discussion of cognitive maps provided by Golledge and Stimson (1987, p. 70).

> ***Cognitive mapping** is the process of acquiring, coding, using, and storing information from the multitude of environments external to the mind. A **cognitive map** is, therefore, a person's model of objective reality. This model of reality is a complex one and should not be interpreted as a simple one-to-one mapping of discrete things that exist in the environment into the mind. Thus, it is not assumed that a cognitive map is an equivalent of a cartographic map, nor is it assumed that there is any one-to-one mapping between a piece of objective reality and a person's cognitive map of that reality. It is perhaps more appropriate, therefore, to regard the cognitive map as a set of stored propositions about the environment, each having been assigned a truth value by a given individual. In this sense, the cognitive mapping process is one designed to receive and code environmental information, store it in an accessible manner, and decode it in such a way as to allow spatial behaviour to take place.*

Expressed in the simplest terms possible, cognitive mapping processes encode in memory the existence of objects, their characteristics, and known locations in space. A cognitive map is, in these simplest terms, the encoded structure in our long-term memory of *what* is *where* (Kosslyn and Koenig, 1992; Sagi and Julesz, 1985)

1.2.10. NEURAL NETWORKS

Some of the following chapters will develop ideas related to neural networks in some detail. In this introductory chapter some fundamental definitions and ideas related to neural networks will be provided that will be useful for understanding these later discussions. An artificial neural network processes information similar to the way a brain processes information. This is quit different from the way a standard desk-top computer processes information. Desk-top computers have a central processing unit that acts on instructions usually provided by a software program. It may search for needed information stored in its memory on a disk drive or compact disk and move the information to process it. The various instructions in the program are executed very quickly but one at a time in a serial process. "The mind is *not* like a general purpose computer that runs all kinds of different software; it more closely resembles a huge number of dedicated computers that are wired up to perform specific tasks" (Martindale 1991, p. 13). There is no need for software since each computer (neuron) is wired to do one specific task. Artificial neural networks are refereed to as parallel-distributed processing models because the information is distributed among many individual processors that all run at the same time (McClelland and Rumelhart 1986; Rumelhart and McClelland 1986). Individual processors (neurons) are connected to other processors (neurons) and generally act simultaneously to produce a parallel process. There many be some order to the processing caused by the fact that the output from one neuron becomes the input for another neuron.

Neural network models may become complex at the global level as the number of neurons and connections increase, but they remain relatively simple at the local level. Rumelhart, Hinton, and McClelland's (1986) discussion of the various components of a parallel distributed processing model provides a basic understanding of key elements and processes associated with artificial neural networks.

Processing Units

The basic structure of any neural network is provided by the processing units. These fundamental units have been called processing units, artificial neurons, nodes, cognitive units, and neurodes by various authors (Caudill and Butler 1992a, 1992b; Hewitson and Crane 1994; Martindale 1991; Rumelhart and McClelland 1986; Lawrence 1994). Units in a network can represent some conceptual objects like places, words, or events, but they can also represent abstract elements that can define some meaningful pattern. Although artificial neurons are comparatively simple, these processing units function in a neural network model like neurons function in the brain. The biological neuron receives input signals from other neurons and produces an output signal for other neurons (Figure 1.7)

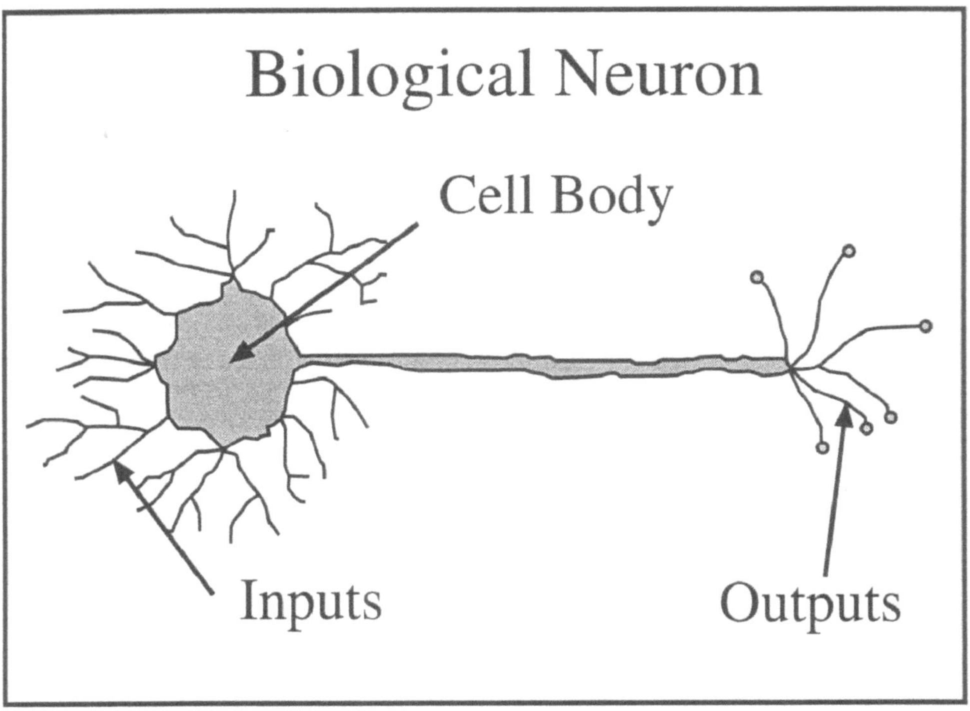

Figure 1.7: A representation of a neuron in a biological neural network. The cell receives inputs from other cells and produces outputs for other cells.

An artificial neuron in a neural network model does the same thing with digital information and mathematical functions (Figure 1.8).

State of Activation

Processing units take on an activation by receiving inputs from outside the network or from other units in the network and, in some cases, pass on an output to influence the activation of other units in the network. Figure 1.9 represents a very simple forward-feeding network that can illustrate three types of units. This forward-feeding network has an input, hidden, and output layer. Information enters the network by activating the input units (**A**, **B**, **C**, and **D**) from outside the network. This initial information is processed through units $\mathbf{H_1}$ and $\mathbf{H_2}$ in the hidden layer to unit **O** in the output layer that takes on some final activation value. Units in the middle layer are called "hidden" because they have no direct connection with the outside world. In this simple network the influences are only in one direction. Input units are influenced from outside the network, input units influence hidden units, the hidden units influence the output unit, and its final activation is presented to the outside world. Processing units

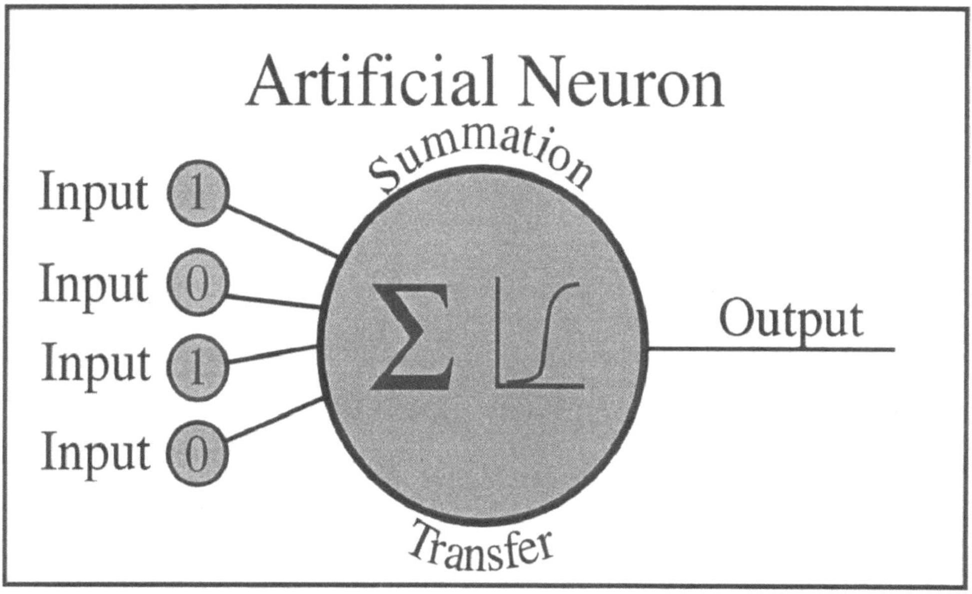

Figure 1.8: A representation of an artificial neuron in a neural network model.

can be activated to varying degrees. If neurons in the brain "are activated beyond some threshold, we are conscious of whatever they code. The set of these activated nodes corresponds to the contents of consciousness. The most activated nodes represent whatever is being attended to at the moment" (Martindale 1991, p. 12).

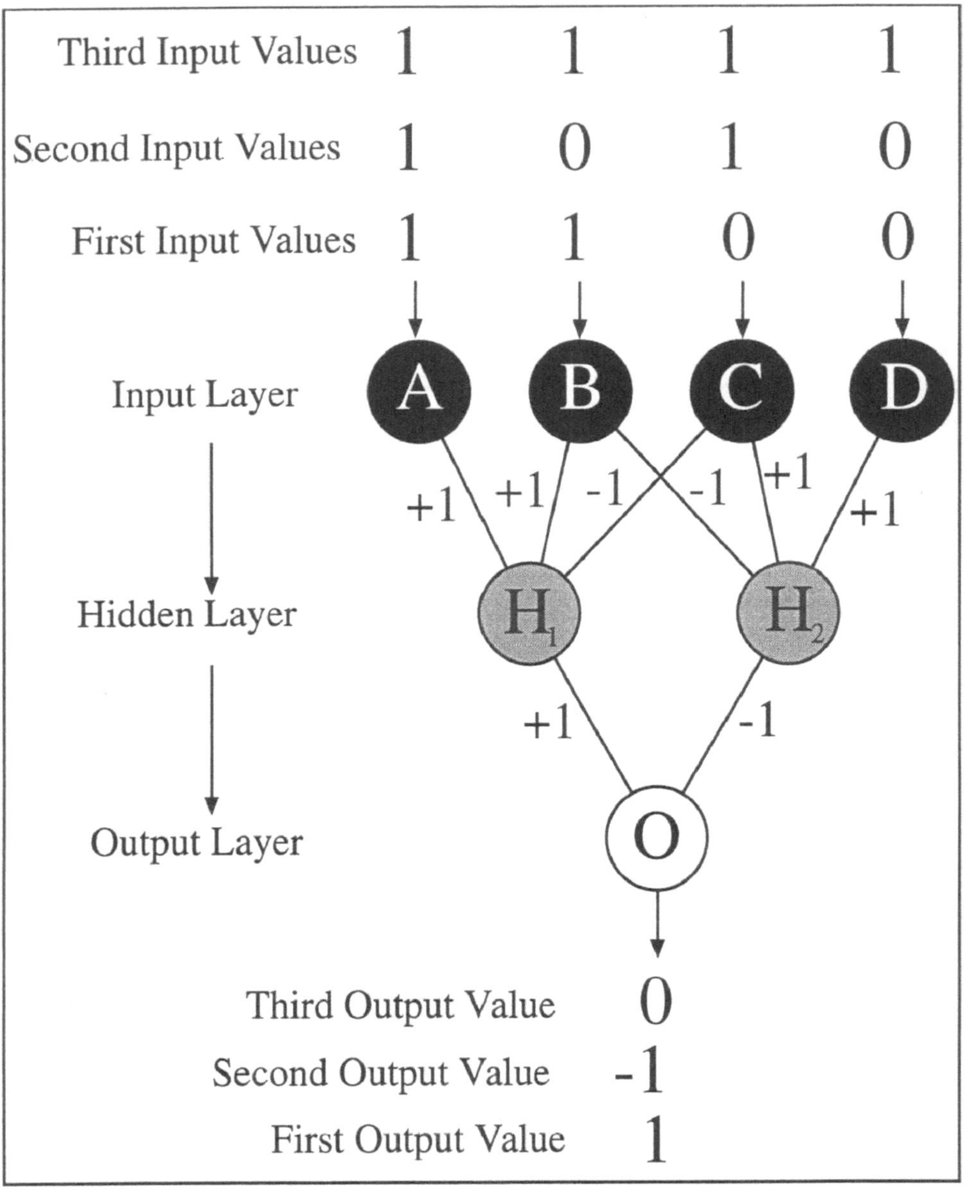

Figure 1.9: A simple three-layer network with four input neurons, two hidden neurons, and one output neuron. The first example turns on input neuron **A** and **B** and turns off input **C** and **D**. Multiplying these activations by the connection weights and summing results in hidden neuron $\mathbf{H_1}$ having an activation of 2 and hidden neuron $\mathbf{H_2}$ having an activation of -1. Multiplying the outputs (1,0) of the hidden neurons by the connection weights and summing results in output neuron **O** having an activation of 1.

Pattern of Connectivity Among Units

Within a network units are connected to each other by either an excitatory or inhibitory connection. These connections are used as weights to compute the influence one unit has on another unit. In the example network (Figure 1.9) hidden unit $\mathbf{H_1}$ has an excitatory connection (+1) to input units **A** and **B** and an inhibitory connection to input unit **C**. $\mathbf{H_1}$ has no connection with input unit **D**. Hidden unit $\mathbf{H_2}$ has an excitatory connection (+1) to input units **C** and **D** and an inhibitory connection to input unit **B**. $\mathbf{H_2}$ has no connection with input unit **A**. Hidden unit $\mathbf{H_1}$ has an excitatory connection to output unit **O** while hidden $\mathbf{H_2}$ has an inhibitory connection. In the brain the strengths of the connections between neurons constitutes long-term memory for associations between whatever the neurons represent (Martindale 1991).

Activation Rules

There must by some method for merging the inputs from various units that are connected to a particular unit and combining this with its current activation to set a new activation level. In the current example it is assumed that all activations are initially zero and that initial values are set for the input units from outside the network. Three sets of initial values are shown along with the resulting final activation for output unit **O**. The first set of initial values (1,1,0,0) provides one activation pattern for the input units. For this simple example, the activation of hidden units $\mathbf{H_1}$ and $\mathbf{H_2}$ is computed by multiply the activation of each input unit by it connection weight and summing over the four input units. The activation of $\mathbf{H_1}$ would be computed as (1 x 1) + (1 x 1) + (0 x -1) + (0 x 0) = 2 and the activation of $\mathbf{H_2}$ would be computed as (1 x 0) + (1 x -1) + (0 x 1) + (0 x 1) = -1.

Output Function

Rules must also be established to define the relationship between a unit's activation and its output. If a unit just output its activation it would not be very useful. In the current example a hidden unit's output is either 1, if the activation is some positive number, or 0, if the activation is not a positive number. The first set of initial values for the input units produced an activation of 2 for $\mathbf{H_1}$ so its output is set to 1. Those same initial input values produced an activation of -1 for $\mathbf{H_2}$ so its output is set to 0. Multiplying these outputs by the corresponding weights connecting the hidden units to the output unit and summing produces an activation of 1 for **O**. Resetting the activations for all units to 0 and processing the second or third set of initial values for the input units produces a different activation for the output units.

Learning Rule

Learning just means changing the weights with experiences. A simple learning rule provided by Hebb (1949) stated that the weight of the connection between two connected units should be strengthened if they are both activated simultaneously.

The basis of this rule is the correlation of the activation of pairs of units. The delta rule suggested by Widrow and Hoff (1960) makes an adjustment in a weight proportional to the difference between the current activation and some expected activation. The minimization of error is the basis of this rule. The expected activation is supplied by an expert who knows what the activation of output unit **O** should be for a given pattern of initial settings for the input units. After experiencing patterns and making adjustments to the weights many times, the "best" weights that minimize differences between achieved and expected activations for the output units should be produced.

1.3. Conclusions

The first chapter has introduced some important ideas that are related to the cognitive processes humans use to relate to geographic environments. Many of these ideas will be discussed in more detail in later chapters. The second chapter considers how cognitive processes are connected and presents some simple examples of sensory, perceptual, and conceptual analyses of spatial information. The notion that artificial neural networks can be used as a tool for modeling cognitive processes was introduced in this chapter. Later chapters will expand on this idea in a number of contexts related to cognitive maps, searching cartographic maps, learning spatial information, and creating categories and prototypes. More details on the use of neural network models for geographic research are provided for specific examples in the next to last chapter.

CHAPTER 2

A CONNECTIONIST APPROACH TO SPATIAL COGNITION

2.1. Introduction

Geographers became interested in environmental cognition and the relevance of behavioral concepts in geography approximately 30 years ago. As ideas related to spatial cognition became known to a greater number of geographers (Downs and Stea 1973; Ittelson 1973), their proper place within geography became the subject of some debate (Tuan 1975; Graham 1976; Bunting and Guelke 1979; Downs 1979, 1981; Rushton 1979; Saarinen 1979). Lloyd's (1982) review of early progress made by behavioral geographers pointed out two general weaknesses. First, much of the early research was not based on explicit cognitive theories on how spatial knowledge is encoded, structured in human memory, or used to make decision. This was not because geographers refused to acknowledge and investigate spatial cognition theory. Research topics such as imagery had only just become a popular and important topic for psychologists (Cohen 1977; Solso 1979) because the behaviorist paradigm had not considered introspection to be an essential part of psychology. Subjects such as consciousness, mental states, and images were not considered to be worthy research topics (Woodworth 1948). Geographers who were trying to investigate cognitive issues at this time were not completely aware of the theoretical progress that had only recently been made in cognitive psychology. This chapter considers a theoretical approach that could provide a fresh perspective for many geographic problems. Connectionist theories and the neural network models that provide the practical context for such theories could provide useful insights for a variety of geographic problems.

Behavioral geographers have always accepted the notion that all adults have an enormous amount of knowledge about the world they occupy. Geographic researchers began to focus on questions such as how is this information acquired (Gould 1975), how is it structured in our memory (Golledge 1978), and how is it used to make decisions (Lloyd and Jennings 1978). The human brain, with its several hundred billion elements, allows humans to interact with and adapt to a wide variety of environments (McNaughton and Nadel 1990). Any understanding of spatial cognition is clearly related to an understanding of how the brain processes information in general and, more specifically, how it processes information relevant to geographic research problems. This chapter will provide an overview of the connectionist approach to information processing and provide some geographic examples to illustrate the relevance of this approach to geography.

2.2. Memory Hardware

In his effort to simplify a human brain, Kosko (1993, p. 206) considered it "three pounds of meat" that can store and process information. In these simple terms the brain can be reduced to about 100 billion neurons (brain cells) and accompanying synapses (connecting wires). Since each neuron can connect to up to 10,000 other neurons, synapses number in the quadrillion and make up about 40 percent of the brain mass. These connections between neurons are of primary importance because they store the information that we learn (Figure 2.1). "It is not correct to think of neurons in the

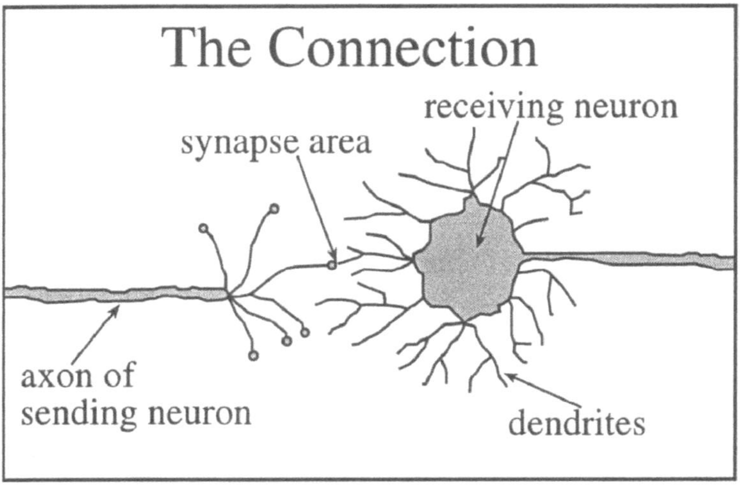

Figure 2.1: The connection between two neurons is the synapse. This is where information transfers from one cell to another.

human brain as computer memory sites. Learning and memory lie in the great tangled webs of synapses. Not in cells, in webs" (Kosko 1993, p. 206). There is no neuron that holds a picture of important environmental objects like your home, the shopping mall, or the public library. The information we have about the environment is distributed in networks of cells and connecting synapses that change every time we have a new environmental experience. Kosko (1993, p. 205-206) discusses learning and physical changes in the brain as follows:

> *To learn is to change. And to change is to learn. You cannot learn without changing or change without learning... Everything you sense changes your brain. Your brain measures things and those things change your brain. It learns new changes and forgets or unlearns old changes. A single photon of light changes your brain.*

At the micro level of neurons and synapses cognitive maps are imbedded within complex networks that represent the spatial information and adjust it with each new experience. The information we experience also has to be analyzed to give it meaning.

2.3. Map Reading

Map reading is done to acquire spatial information. The information on a paper map in its simplest form is just reflected light. The brain processes this raw data (light and dark patches on a page) to turn it into meaningful information by processing it through specialized systems. Martindale (1991) suggested that the brain has a large number of *analyzers* or specialized modules that are all constructed in a similar way. These analyzers can be classified by function into systems that have specialized uses. He identified sensory systems, perceptual systems, the semantic system, the syntactic system, the episodic system, and the action system. An example will help illustrate how the various systems might be used during a map reading experience. Suppose an attractive, but cryptic, cartographer gives you a map (Figure 2.2) with instructions to

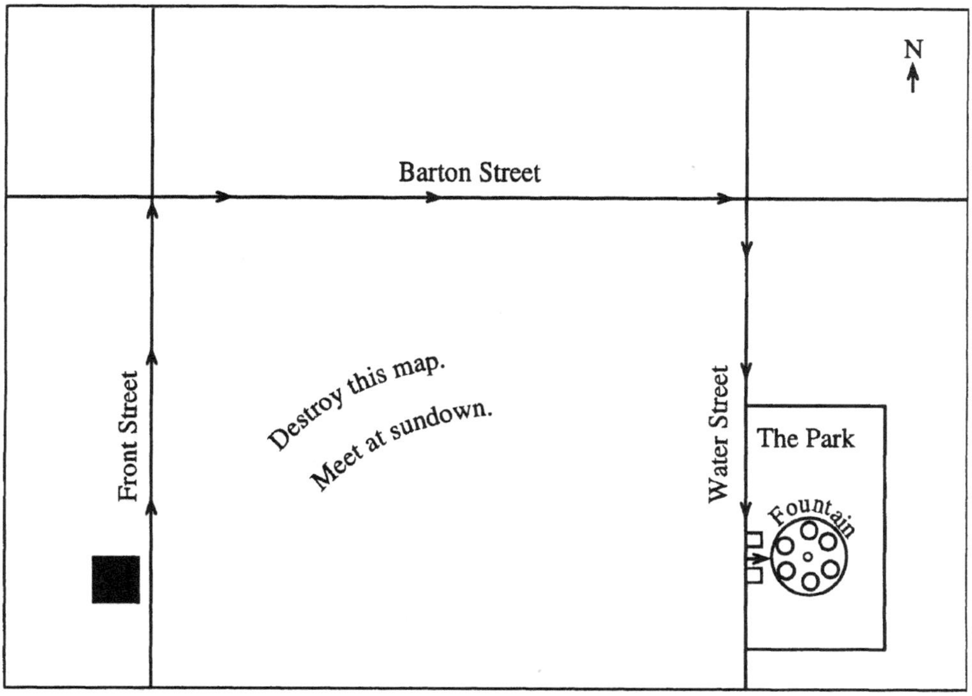

Figure 2.2: Map reading may require a number of systems for processing spatial information. This example illustrates the use of sensory, perceptual, semantic, syntactic, episodic, and action systems.

"Destroy this map. Meet at sundown." Analyses of the map would be required at the sensory level (you would sense dark and bright patches on the paper), the perceptual level (you recognize some words and lines), the semantic level (you understand some lines are streets with names), and the syntactic level (you understand the streets form a route from your current location to a destination). At this point other systems that analyze the possible intentions of the cartographer and your own emotions might be evoked. If you are to follow the instructions, you would have to encode the map as either a map image or as procedural knowledge of the route before destroying it and store the message in *episodic memory* so you would not forget it. You would use the action system that has learned programs for walking or driving a car. The action system also manages the motor system that controls the specific muscle movements used to execute walking or driving behaviors needed by the action system.

Your response to the map would involve sequential activity in a number of different analyzers that would process information in a particular order. Each analyzer is thought to be a hierarchical neural network made up of nodes and connections among nodes in a lattice-like structure. Martindale (1991) argued that most analyzers have four layers. The vertical connections among neurons on different layer usually *excite* or cause the sender neuron to increase the activation of the receiver neuron and lateral connections among neurons on the same level usually *inhibit* or cause the sender neuron to reduce the activation of the receiver neuron. Lateral inhibition is important within a layer where the activation of one neuron must emerge above others to produce a meaningful result. Connections within an *analyzer* are thought typically to be bidirectional but not symmetrical. This means that, if neuron X is connected to neuron Y and can affect its activation, neuron Y is probably also connected to neuron X and can likewise affect its activation. The relative strengths of the connections, however, may not be equal. In other words, neuron X may have more influence on the activation of Y than Y has on the activation of X.

2.4. Analyzers

Each of our senses has at least one analyzer associate with it that processes raw sensations (Martindale 1991). Visual sensory analysis processes information related to such things as color, shape, size, orientation, location, and motion (Kosslyn and Koenig 1992). Evidence of the existence of analyzers in distinct areas of the cortex have been provided for location (Ungerleider and Miskin 1982), spatial frequency (Tootell, Silverman, and DeValois 1981), motion (Zeki 1974), depth (Blakemore 1970), color (Livingston and Hubel 1984), and detection of distinct features such as lines, edges, and angles (Hubel 1988).

2.4.1 SENSORY ANALYSIS

An example of an analysis of sensory information that automatically and instantly identifies a boundary in a visual display can easily be illustrated (Figure 2.3).

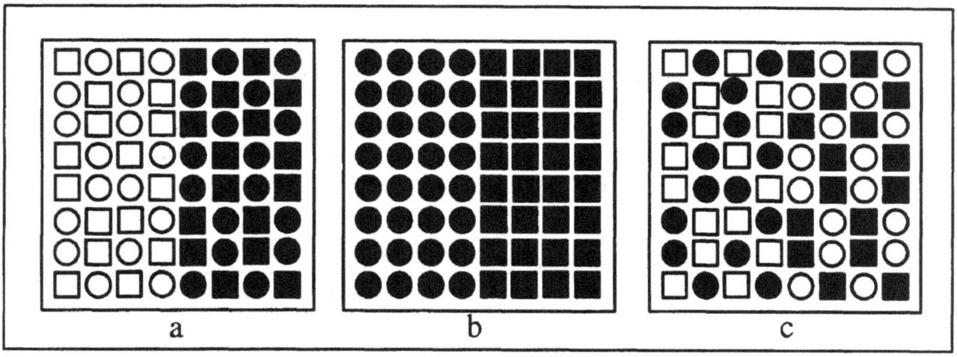

Figure 2.3: The boundary separating the left and right half of a display can be automatically and rapidly identified by segregating unique brightness differences (a) or shape differences (b). The boundary does not emerge from the display automatically and rapidly if differences between the halves are not unique. Display c has white squares and dark circles on the left and dark squares and white circles on the right.

When the left and right sides of a display are uniquely different based on either brightness (a) or shape (b) the boundary is immediately apparent to any viewer. The boundary is isolated during the earliest preattentive stage of processing without conscious effort. The display on the right (c) also is different on the left and right sides with the left side having dark circles and light square and the right side having dark squares and light circles. The boundary does not automatically and instantly emerge from the display, however, because neither the brightness nor the shape analyzer can segregate the two sides of the display. Both sides have dark, light, circles, and squares.

2.4.2. PERCEPTUAL ANALYSIS

Sensory analyzers produce an output that enters perceptual analyzers. Konorski (1967) has suggested that separate visual analyzers exist that process information related to things such as human faces, handwriting, small objects that can be manipulated, and printed words. Perceptual analyzers for objects and words would be particularly useful for map reading. An example can be provided using the map in Figure 2.2 and a hypothetical perceptual analyzer for printed words (Figure 2.4). Output from sensory analyzers serve as input to the bottom layer of perceptual analyzers. Suppose someone was reading the street name for the east-west street in the northern part of the map. Sensory information would enter the bottom layer of the printed word analyzer and activate primitive shapes for horizontal and vertical straight lines and concave upward and downward curved lines. Activation from the bottom layer would be input to the above layer and activate letters. Lateral inhibition among competing letter units would allow the most active letter unit to turn off the competition. Output from the letter level would be input to the syllable units in the above level and used to activate syllable units. Lateral inhibition among the syllable units would allow the most active syllable unit to turn off the competition. Finally, output from the syllable layer is input to the top layer activating word units. In this example lateral inhibition would allow the *barton* unit to turn off its competitors (Figure 2.4).

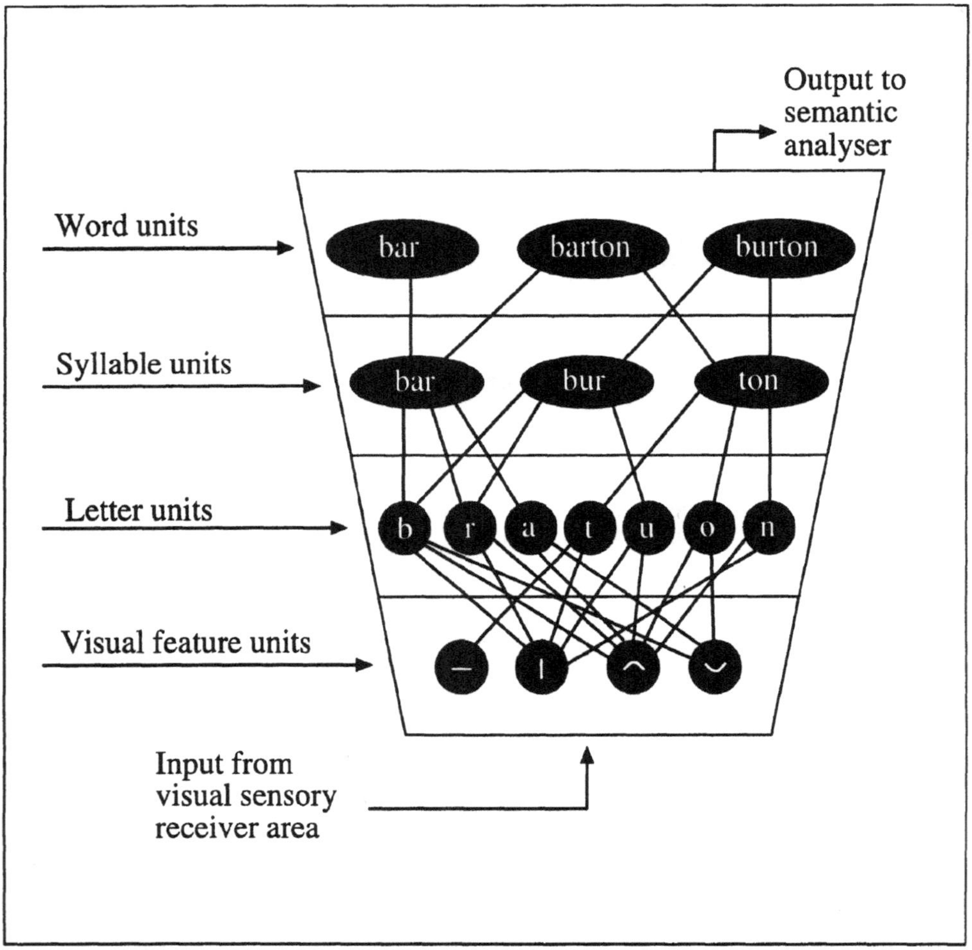

Figure 2.4: A perceptual analyzer that might be used to process the names of streets from the map shown in Figure 2.2. Only a few example nodes are shown at each level that could relate to reading "Barton" street from the map.

2.4.3. CONCEPTUAL ANALYSIS

As can be seen from the above map-reading example, the output from one analyzer becomes the input for another analyzer. The level of processing can easily be illustrated with another map reading example (Craik and Lockhart 1972). Stimuli on a map can be processed to varying depths. Shallow processing would be done by sensory analyzers before focused attention. If processing stopped at this level, as when a map is flashed on a monitor for a fraction of a second, you might be aware that the color red was on the map, but not know its location on the map or what objects had a red color. This would require focused attention and deeper processing. Perceptual analyzers allow us to sense a stimulus and recognize it as an object. If you are reading a map that is labeled in a foreign language that you are not able to understand, you would only be able to do sensory and perceptual processing for the labels on the map. The letters and words might be activated, i.e., you know they exist, but processing stops at that level. If the map were labeled in your native language, deeper processing would be possible that would allow you to understand and interpret the labels. This would be done by conceptual analyzers that deal with mental process that are more abstract than sensation and perception (Martindale 1991).

Semantic Analysis

Semantic analyzers store knowledge that we have learned relating to the meaning of concepts and percepts. Semantic analyzers, like the example illustrated in Figure 2.5, are also multilayered systems. This example uses a few geographic units to illustrate the structure of semantic analyzers, but many more units are possible. The upper layer of the semantic analyzer receives information from a variety of perceptual analyzers. There is a node in the top layer for every concept a person knows. This could be a large number of units even if we only consider the geographic units known by a person. Semantic units receive input from many different perceptual analyzers. The semantic unit encoding information for the concept of Africa could receive input from the printed-word analyzer when someone is reading about Africa, from the speech analyzer when listening to a person talk about Africa, or from the image analyzer when viewing a map of Africa (Figure 2.5). Everything one knows about Africa is coded by connections between the Africa semantic unit and semantic units coding other relevant concepts. The concepts at each lower level in the semantic analyzer are increasingly abstract. Thus, Africa is connected to the continent unit in the next deeper level. This level contains basic-level geographic categories that have a special importance for storing spatial information and communicating about geographic space (Rosch, Mervis, Gray, Johnson, and Boyes-Braem 1976; Lloyd, Patton, and Cammack 1996). Other basic-level geographic categories in addition to continent and ocean at this level would be country, region, state, city, neighborhood, etc. Basic-level theory is discussed in more detail in Chapter 7. The continent unit is connected to the land unit at the next deeper level as the concepts become more general. Finally, the land unit is connected to the place unit, the most general geographic concept, at the lowest level in the semantic analyzer.

Syntactic Analysis

Recall the map given to you by the cryptic cartographer (Figure 2.2). After using the relevant sensory and perceptual analyzers, the meaning of each word and graphic symbol on the map would be looked up in semantic memory. To understand the message, you need to know more than the meaning of individual words and graphic

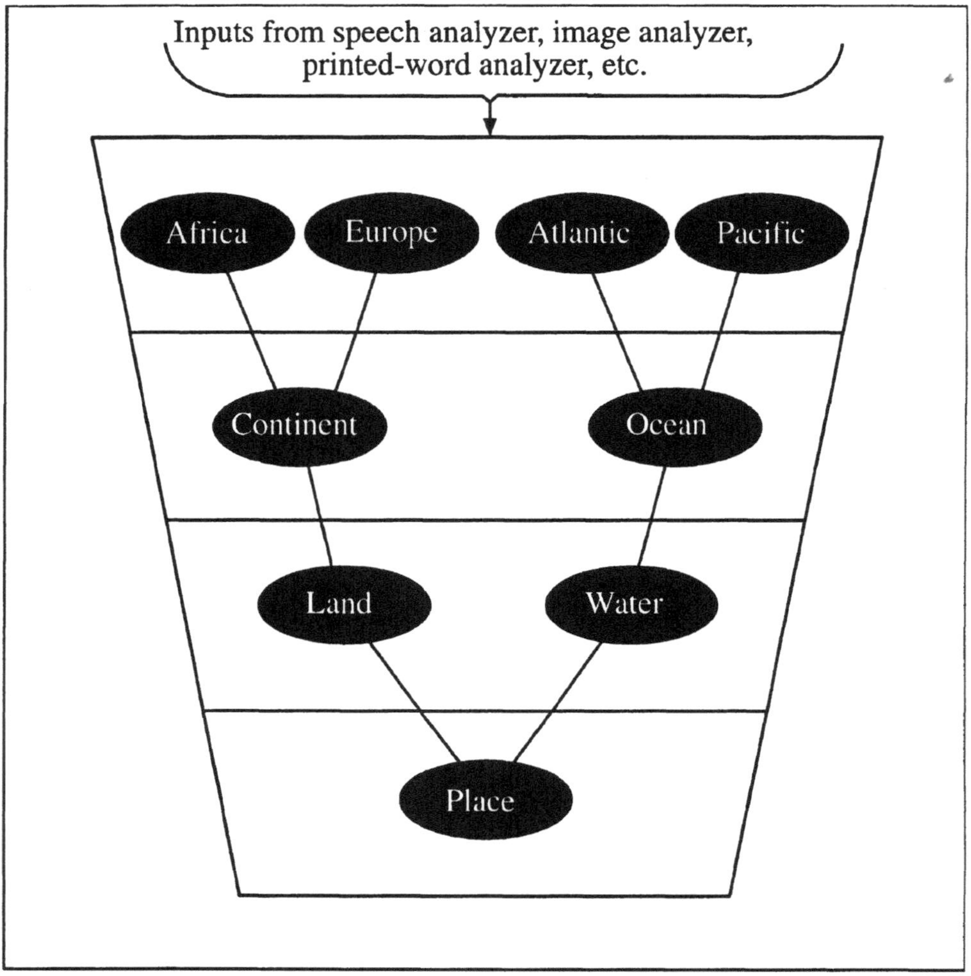

Figure 2.5: A semantic analyzer with geographic examples shown at each level. Only a few sample nodes are shown.

symbols. You need to know what needs to be done, when it needs to be done, where it is to be done, and who is supposed to do it. Understanding comes about by combining the meaning of the individual parts into an abstract conceptual-proposition. Such a proposition describes an event in terms of an action, agent, object, time, location, etc. (Anderson 1985). What action is to take place? The request for a meeting and the route on the map combine to suggest a trip is necessary. This understanding would not be possible if you only had the map or only had the message. The message says to "Destroy the map." Who should do this and when and where should it be done? Although it is not clear why this action is necessary, it is implied that you, the possessor of the map object, are to be the agent of its destruction. Although there is no direct information on when the map should be destroyed, common sense (and many old spy movies) might suggest it should be before making your trip and after you have memorized the route. The route might be described as propositions expressed as a series of left or right turns at street intersections and landmark identifications. "Go north on Front Street and make a right turn on Barton Street. Proceed east on Barton Street to Water Street and then turn right. Go south on Water Street until you see the Park's entrance. Turn left into the Park and stop in front of the water fountain". The message also says to "Meet at sundown." When the meeting is to take place is clear, but who is to meet and where is the meeting to take place? Again common sense suggests that you are to meet the cartographer at the end of the route on the map--the fountain in the Park. This commonsense understanding of the map and its message would be the result of a syntactic analysis. Any one unit in the syntactic analyzer would receive input from units in the printed-word analyzer, image analyzer, and from semantic memory units (Martindale 1991).

2.5. Types of Memory

Psychologists generally agree that there are different types of memory. Martindale's (1991) discussed memory in connectionist terms and argued that consciousness refers to the currently activated cognitive units in sensory, perceptual and conceptual analyzers. The activation in many of the units is directly caused by the sensations and perceptions caused by current stimuli. Other units have been turned on by recent, but currently absent stimuli. Others may be active because they are connected to units that are active or that have recently been active. The primary memory is the contents of our awareness minus sensation and perception. Primary memory can be divided into sensory memory and short-term memory. Sensory memory is a positive after-effect that persists about one second after a stimulus is no longer present. Sensory memory does not require effort and is automatic and preattentive. The other part of primary memory is short-term memory. It is the persistence of activation in perceptual and conceptual analyzers. It lasts about 20 seconds, but can be maintained longer with rehearsal. People often transfer an unfamiliar telephone number to the speech analyzer and repeat it over and over to maintain it in short-term memory long enough to successfully call the number. Telephone, license, and PIN numbers that are needed for more than a short time must be transferred to long-term memory. Connectionists consider long-term memory to be the entire neural network. Long-term memories are stored in the connections among the neurons. Rather than thinking of short-term memory as a place in the brain, connectionist consider short-term memory to be the currently activated neurons. Information is transferred from short-term to long-term memory by the neurons staying activated long enough for the connections among the neurons to change. Repeated experiences can bring about this needed change.

Long-term memory can be divided into three types (Turving 1985). Semantic memory contains general knowledge, such as the knowledge that ice floats in water. There is no temporal sequence to these facts so one usually does not recall when such general facts were first learned. Episodic memory contains knowledge that has a temporal tag with it. Historical events or autobiographical memories of the sequence of things that happen to you. Procedural memory contains knowledge of how to do things such as go from you house to work, ride a bicycle, or compute the square root of numbers.

2.6. Episodic Analyzer

Episodic analyzers are used to encode a proposition input from semantic and syntactic analyzers with a time tag. An example of a multilayered episodic analyzer for urban travel is illustrated in Figure 2.6. The top layer has units that encode events

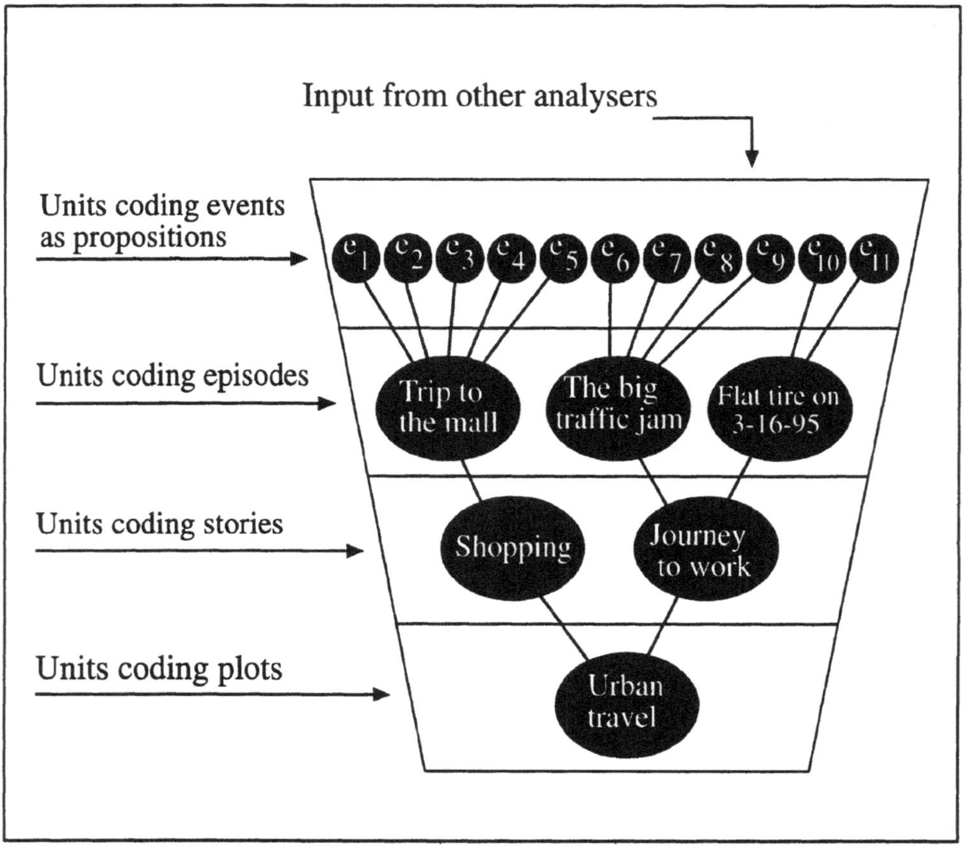

Figure 2.6: An episodic analyzer for urban travel. Specific events are in the top layer and lower levels become more general or abstract.

1 through 11 that happened in a time sequence. Suppose you left home in the morning and drove to the local mall to buy a new coat. Events 1 through 5 occurred during this sequence. Maybe the car did not start the first time, you found a great parking space, you had to wait five minutes for the mall to open, the store was having a sale on coats, and the sales person smiled at you and approved of your purchase. Events 6 through 9 relate to a traffic jam that occurred after you left the mall. Perhaps the traffic light was not working, the driver behind you kept blowing his horn, a bus was stalled in the

intersection, and you made an illegal U-turn to get to the freeway ramp. Events 10 and 11 relate to a flat tire you got just before entering the parking lot at work. Possibly you heard a loud noise and drove the last 50 yards to the parking lot with a flat tire. The events in the top layer are connected to deeper layers that get more generalized or abstract from top to bottom. The layers are given literary labels with the second layer being episodes. The earliest events (1 through 5) connect to the trip to the mall episode. The middle events (6 through 9) connect to the big traffic jam episode, and the late events (10 and 11) connect to the flat tire episode. The trip to the mall episode connects to the more abstract shopping stories unit one layer down while both the traffic jam episode and the flat tire episode connect to the journey to work story in that layer. Both of these stories connect to the urban travel plot in the lowest layer in the analyzer. This multilayered system allows you to generalize information from the events of your life into memory. This allows you to recall specific travel episodes and retain general knowledge about traveling in urban areas.

2.7. Action System

Shallice argued there should be an action unit for any action described by a transitive verb (Shallice 1978). An action system for recreation behavior would have units in its top layer that represented basic acts engaged in during recreation (Figure 2.7). These units code actions like swimming, hitting a ball, walking, or diving a car. Once the appropriate action is determined, the output from action units is sent to the motor system that automatically executes the action (Martindale 1980). Like the previous analyzers, the layers get more generalized or abstract from top to bottom. In this hypothetical analyzer the units represent more generalized actions related to recreation. The second level codes are called scripts (Shank and Abelson 1977). These are a sequence of actions for activities like playing golf or hiking. The two lower layers of the hypothetical analyzer reflect a perspective that sustains the recreational behavior. Note that the golf and auto racing units are connected to the achievement unit in the disposition layer and the going to the beach and hiking units are connected to the affiliation unit. This suggests the former activities are pursued competitively while the later two activities are pursued for amusement. Subselves compete for control in the lowest layer (Martindale 1980). Lateral inhibition within this layer allows the ambitious self and relaxed self to compete for control. Sometimes the wrong unit wins and causes your behavior to seem inappropriate for the situation. Doing a celebration dance when the client's tee shot goes into the woods would not be good for business.

2.8. Interacting Systems

The map reading example (Figure 2.2) can illustrate how the above systems can be linked together so that output from one system becomes input for the next system (Figure 2.8). Looking at the map (physical stimulus) would cause a signal (information) to be input to your senses (vision). At any point in time sensory analyses would be done by one or more of the vision analyzers used to process

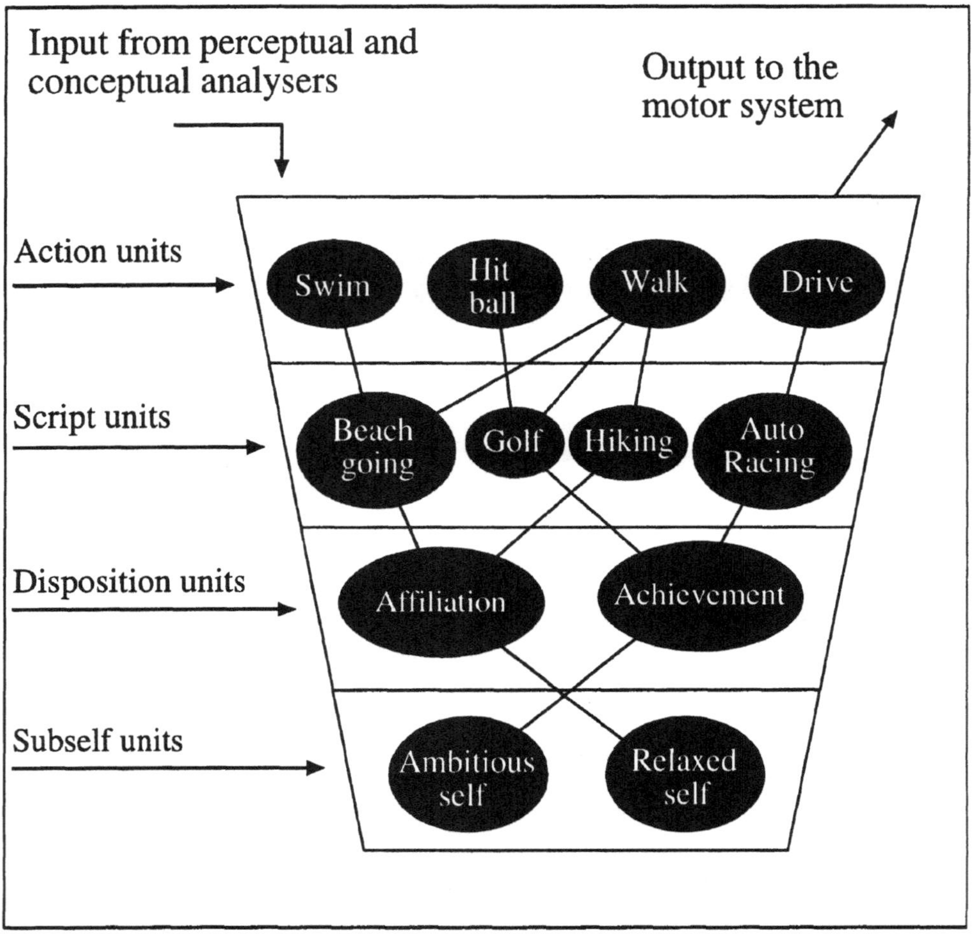

Figure 2.7: An action system for recreation behavior. Specific actions are in the top layer and lower levels become more general or abstract.

brightness, location, etc. The output from one of these processes (sensation) would be input for a perceptual analysis done with objects (lines) or words (names) on the map. The output from one of these processes would be input for conceptual analysis. The semantic analyzer would process the meaning of the object or word and produce an output. The episodic analyzer could code the information in long-term memory with a time tag. For example, the route to the park with the order of streets and landmarks extracted from the map could be encoded in memory. The output from the semantic and episodic analyzers then could be processed by the action system to produce a decision. The signal would interact with your personal disposition (curious self) to produce a number of outcomes including instructing the motor system to walk the route or to drive a car over the route to the park.

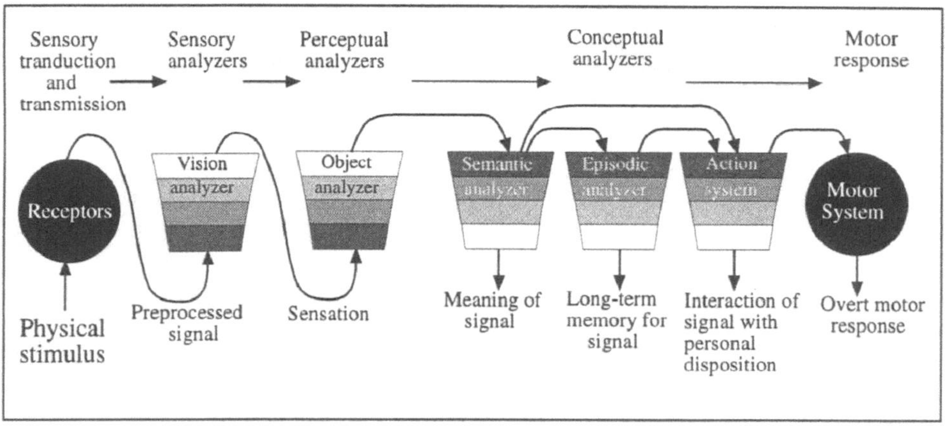

Figure 2.8: Flow chart of the cognitive processes used to read the map represented in Figure 2.2. After Martindale (1991).

2.9. Fuzzy Cognitive Maps

Once knowledge is encoded into memory we can process it in creative ways to think about patterns and relationships. Our ability to predict how complex events might interact gives us an ability to think about the future and make plans. Kosko (1993, p 222) uses the term "fuzzy cognitive map" to describe a type of network that can be used to simulate this type of mental activity. He argues that fuzzy cognitive maps draw a causal picture by tying "facts and things and processes to values and policies and objectives." Kosko's examples of fuzzy cognitive maps are not maps in the usual sense. They define political spaces rather than geographic spaces. One example considered political policies relating to South Africa in 1986 (Figure 2.9.). The model defined nine nodes related to apartheid politics and connected them with a positive linkage if a sender node had a positive influence on a receiver node or with a negative linkage if a sender node had a negative influence on a receiver node. For example, an increase in Black Tribal Unity was thought to decrease Apartheid, but increase White Racist Radicalism. The model was used to simulate what might happen if foreign investments were changed. At the time some political oracles favored "divestment" to pressure the government to abolish apartheid laws while other oracles argued withholding investments only hurt workers you were trying to help by reducing the number of jobs. By turning on the Foreign Investment node and allowing the dynamic system to run until is converges on an equilibrium the "hidden pattern" in the system can be revealed (Kosko 1993, p. 226).

Another fuzzy cognitive map example provided by Kosko (1993, p. 230) presents a causal picture of the "drug war " (Figure 2.10). Nodes represent factors in the cocaine trafficking system for supplying the United States. Connections between nodes indicate if the sender node has a positive or negative influence on the receiver node. For example, *American Police Interdiction* has a negative impact on *Street Gangs* and a positive impact on the *Cocaine Price* on the street. To complete the loop, *Cocaine Price* has a positive influence on *Street Gangs*. Simulation could be done to see the impact of *Cocaine Price* or *Acres Coca* of on the other nodes. How much change occurs in *Drug Usage*? Are *Profits* up or down?

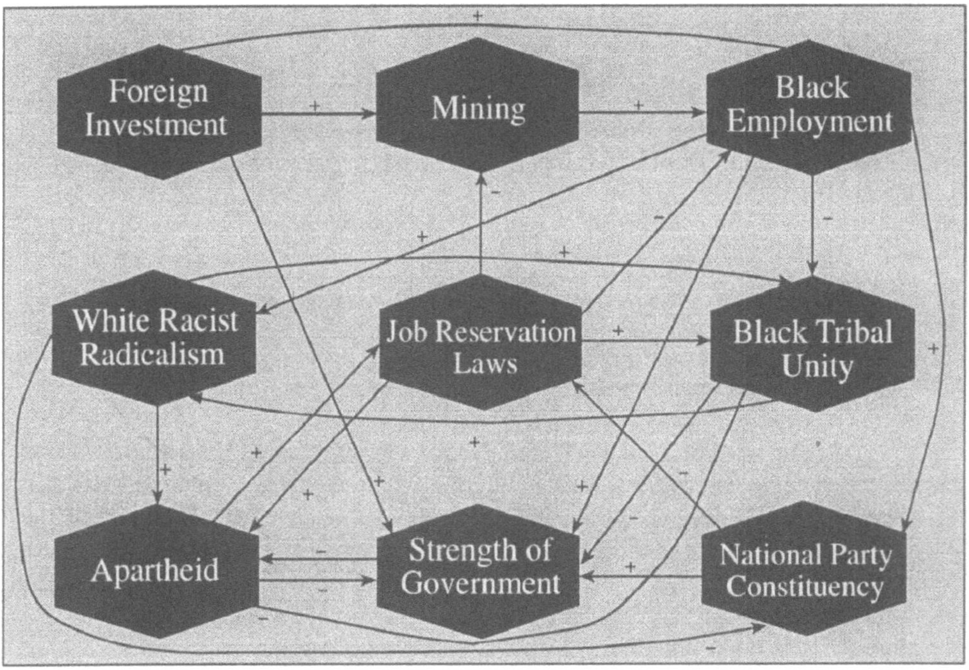

Figure 2.9: A fuzzy cognitive map of South Africa with nodes related to apartheid politics. After Kosko (1992).

These models represent the model builder's beliefs about what factors are important and how these factors relate to one another. This is not unlike an individual who has encoded knowledge into memory and is using it to make an intuitive decision. The decision is only as good as the knowledge on which it is based. Intuitive thinking is also frequently based on hidden patterns in one knowledge structure that are unknown to the decision maker.

2.10. Interactive Activation Competition Models

A type of neural network that simulates associative memory can be used to store information and retrieve it from memory. McClelland and Rumelhart's (1989) interactive activation competition neural network model (IAC model) stores knowledge about objects, e.g., people, places, or things, and can be used to retrieve generalized

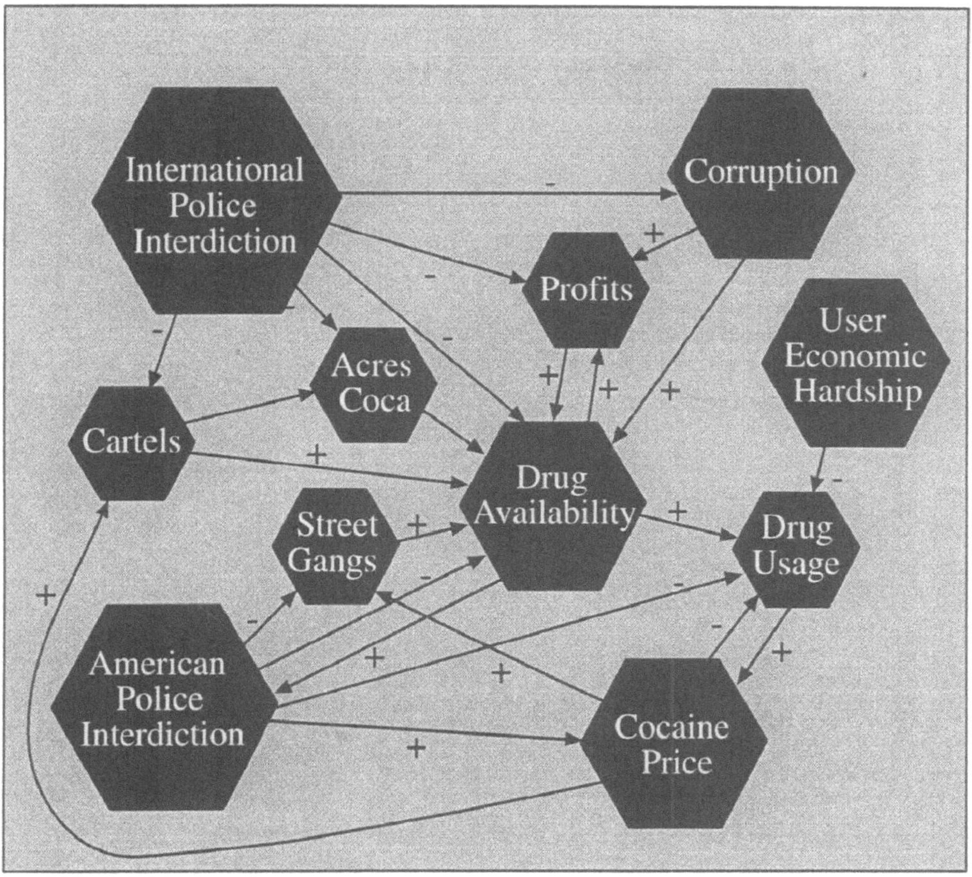

Figure 2.10: A fuzzy cognitive map of the drug war. Nodes represent elements of the cocaine trafficking system. After Taber (1991).

knowledge implicit in the network. Their famous example involved two street gangs, the Jets and the Sharks (McClelland 1981; Rumelhart and McClelland 1986). One obvious and useful feature of human memory is that it is content addressable. We can access a representation of an object in memory based on any of the attributes associated with the object. A word association game illustrates this point. One person presents a word or phrase and the other person responds with the first thing that comes to mind. The response presumably comes to mind because it is activated through its connection with the stimulus. McClelland (1981) discusses an undesirable neighborhood inhabited by a number of hypothetical characters (Figure 2.11.). The IAC model has pools of units to represent the characters (central pool) and their characteristics (peripheral pools).

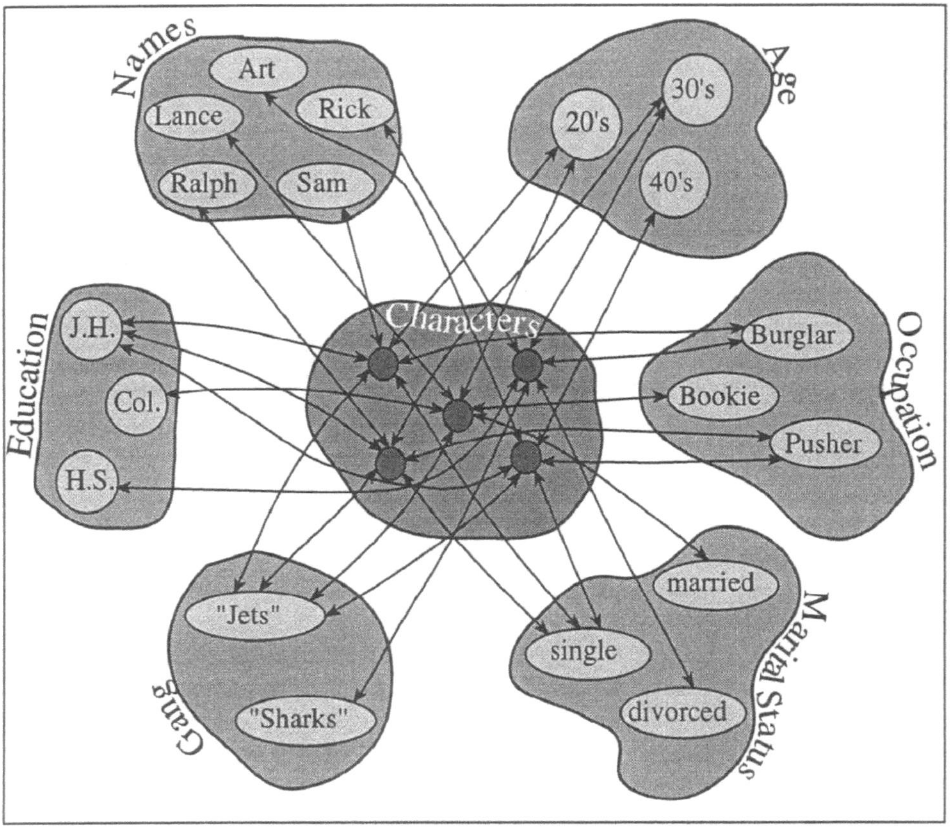

Figure 2.11: McClelland's example of an interaction activation competition neural network model to represent knowledge about street gangs in a neighborhood. After McClelland (1981).

The characteristics represent the name, age, occupation, marital status, gang affiliation, and education of the characters. Each of the individual items in the pools, e.g., Burglar in the occupation pool or single in the marital status pool, represents a neuron in a neural network (Figure 2.11). The character units in the central pool are each connected to one appropriate unit in each characteristic pool. For example, characters that are members of the Jets gang are connected to the Jets unit and characters that went to college are connected to the college unit. Each character unit is also connected to his name unit in the name pool. All the connections between pools are positive and operate in both directions. This means a particular unit in the character pool that is connected to the Shark unit can activate the Shark unit and that the Shark unit can also activate that particular unit in the character pool. There are no connections between units in different characteristic pools. For example, an age unit cannot directly affect an education unit. Within all pools the connections are negative and operate in both directions. This means all units within a pool try to turn each other off. This is because an individual character can only have one age or marital status. It is assumed that a character cannot be a member of both gangs or have more than one occupation. To start a simulation one or more characteristic units can be activated from outside the network. For example, activating the name Rick is like a person thinking of the name. This starts the process and sends information through the connections to other units. This can be repeated for a number of cycles until the activation levels of the network's units stop changing. Character units are considered to be hidden units. This means they are not directly accessible from outside the network. They can only receive information from other units in the network and they can only send information to other units in the network.

The simplest application of the IAC network is to retrieve information about an individual by thinking of his name. First, assume that all units in the network are not activated, i.e., set to zero. By activating the Sam name unit, i.e., setting it to one, and cycling the network the connections will pass energy (information) through the network and activate Sam's characteristic units. Sam is a single burglar in his 20's with a junior high school education who is a member of the Jets gang. The network will also reprocess information as it cycles back through the character units and turn on the names of individuals that have characteristics similar to Sam's. Information can be retrieved in a similar fashion by activating one or more units in any of the characteristic pools. Activating the Shark unit and cycling the network will provide the general characteristics of that gang and the names of its members. Activating the divorced marital status unit, the college education unit, and the 40's age unit and cycling the network will give you the individuals with those characteristics and also activate their typical occupation, and gang affiliation.

2.11. Interaction Activation Competition Models as Cognitive Maps

Cammack and Lloyd (1993) used an IAC model to produce a neural network that coded information about the 48 contiguous states of the United States and operated like a cognitive map. Their model had a pool of hidden units that were places, i.e., the states. Characteristics of the states were represented by separate pools of units. Connections were made for the model by considering objective data available for the states. A simplified model, similar to the one used by Cammack and Lloyd (1993), is presented in Figure 2.12 to reduce the graphical complexity of the connections. In this hypothetical model there are five place units in the hidden unit pool, but it would be possible to have any appropriate number of places represented as units in this pool. This hypothetical neural network might represent a knowledge structure in someone's memory. Since it represents information the person has learned about places, it functions like a cognitive map. The place units are connected to characteristic units in six pools, i.e., name, population, politics, economy, temperature and precipitation, that represent types of information one might acquire about the states. For example, the place unit connected to the Nebraska name unit is also connected to the rural population

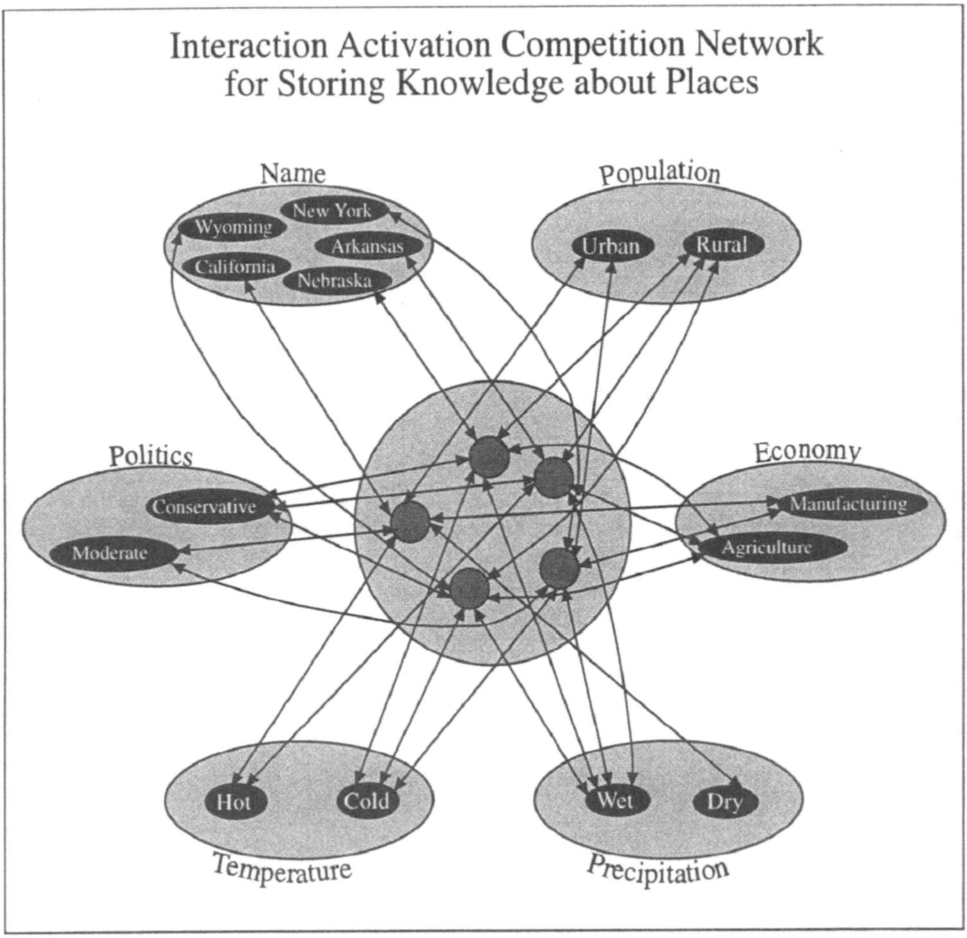

Figure 2.12: An interaction activation competition neural network model for storing knowledge about places.

unit, the conservative politics unit, the agriculture economy unit, the cold temperature unit, and the wet precipitation unit. In a person's cognitive map the connections would be learned from experiences with the places, e.g., personal visits, talking with other people, reading texts, viewing maps, etc. The IAC model for states (Figure 2.12) can be used like the IAC model for the street gangs (Figure 2.11). Information about a particular place can be retrieved by activating its name and running the model through a number of cycles. This will activate units in each characteristic pool associated with that particular state and also activate state name units that have a similar characteristic. If a set of state name units were initially activated, e.g., states in New England or the South, the character of the regions would be revealed through the characteristic units that became activated. This is generalized information implicit in the network connections that is not obvious by examining individual units and their connections. One could also activate units in two or more characteristic pools to produced generalized cognitive maps. For example, what state name units would be activated if the cold temperature unit and the agriculture economy unit are initially activated? Since the coding in a person's cognitive map is based on experiences it may be biased, incomplete, and systematically distorted (Lloyd and Heively 1987; Tversky 1981). One could imagine a GIS system, however, that is based on the type of neural network discussed above and coded with accurate information. The user of the GIS system could input some initial activations for characteristic units to represent some concept and the activations produced for the place units could be mapped on the screen to reveal the result of thinking about that particular concept.

2.12. Conclusions

This chapter has focused on the cognitive processes used to analyze spatial information. Martindale's (1991) notion that separate connected analyzer systems input, process, and output information was discussed in a geographic context. A number of spatial examples were discussed and served to demonstrate the connectionist approach to information processing. Kosko's (1993) conception of a fuzzy cognitive map as a memory structure that connects important information was also discussed. Cognivitve mapping can be thought of as information processing by various systems in a neural network. The next chapter considers more on cognitive maps and the processes that produce distortions in them.

CHAPTER 3

COGNITIVE MAPS

3.1. Introduction

The purpose of this chapter is to consider the nature of cognitive maps. The need to simplify a complex world has caused researchers to produced a number of classification schemes for spatial information. If we consider a cognitive map the knowledge we have learned about an environment that is stored in our memory, a number of immediate questions arise. What does it mean to learn? How is the knowledge stored? Are there different types of memory? Are there different types of spatial knowledge? Given that the human brain has approximately 100 billion neurons that are connected in a network to process information (McNaulton and Nadel 1990), and following a wet mind approach (Kosslyn and Koenig 1992), a simple definition of a cognitive map for a particular chunk of geographic space would be those neurons and connections that represent information about that chunk of geographic space. For example, Figure 3.1 shows part of a hypothetical neural network for landmarks in a city and some of their characteristics. One pool of neurons represents urban landmarks and others represent characteristics or a list of names (McClelland and Rumelhart 1989). One could access knowledge related to a particular landmark by thinking of its name. For example, turning on *Capitol* from the name list connects through the object pool to the color *brown*, the size *Small*, the category *Building*, and the location *North*. Thinking of a category of landmarks allows you to access specific examples. For example, turning on *Parks* from the category pool connects through the object pool to *Finlay Park*. It would also connect to all other parks that were encoded as objects in the person's neural network. Since the neurons and connections in the network were created through the person's experiences with the urban environment, they reflect those experiences. Different parts of this cognitive map as neural network might have been encoded at different times using different encoding processes and much of the information has been abstracted. Particular views of cognitive maps can be illustrated by considering differences in encoding processes for the various types of spatial information discussed in the literature, internal structures that organize and store the information, and distortions that occur when information is decoded and used to make judgments for decisions.

Like cartographic maps, cognitive maps may express spatial information at many different geographic scales. Cartographer's define a map's scale as the representative fraction made by dividing a number of linear distance units on a map by the number of earth units they represent. A large scale map would be for a room and a small scale map would be for the earth. This same notion can be applied to cognitive maps. The information encoded into a large-scale cognitive map of a room might be for objects like furniture. The spatial information might be expressed as propositions stating *what is where*, e.g., the table is near the window, or as a picture-like image of the room (Figure 3.2). The word *near* at this scale refers to a distance appropriately

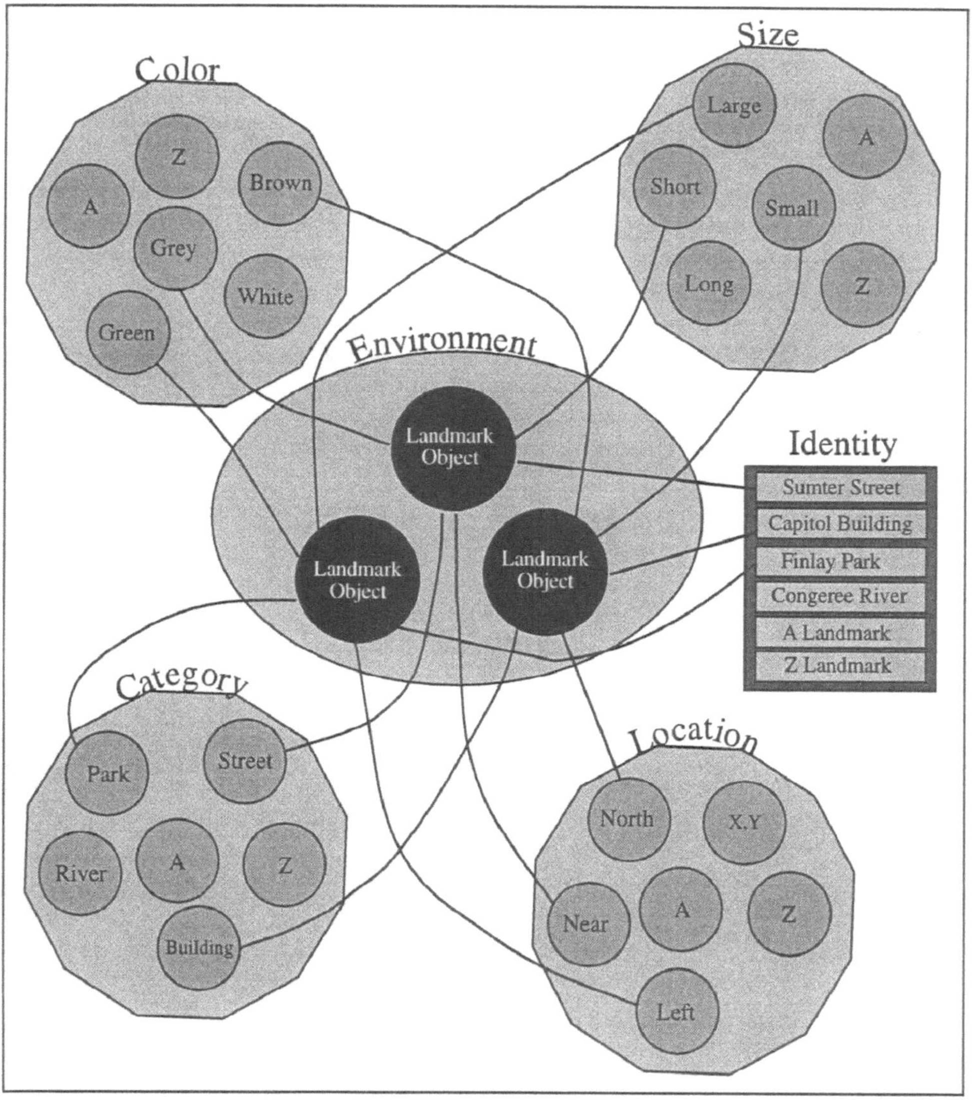

Figure 3.1: A portion of a neural network illustrating landmark objects represented as neurons in a central pool with other neurons representing characteristics of the landmarks.

measured in inches or feet and objects in the room separated by such distances are easily distinguished on the image. The objects of interest in as small-scale cognitive map of the world might be countries or cities. Verbal propositions also could be used to express information at this scale, e.g., the large communist country is in the East. Picture-like world map images could also represent small-scale spatial knowledge and be functionally equivalent to cartographic maps of the world (Figure 3.3). The word *large* at this spatial scale would refer to millions of square miles and only objects, e.g., cities, separated by relatively large distances could be distinguished on the image.

3.2. Encoding Processes

A cognitive map that was produced by one particular type of experience may be different from a cognitive map produced by a different type of experience. This seems reasonable because a different set of cognitive processes would be used to access, analyze, and store the spatial information.

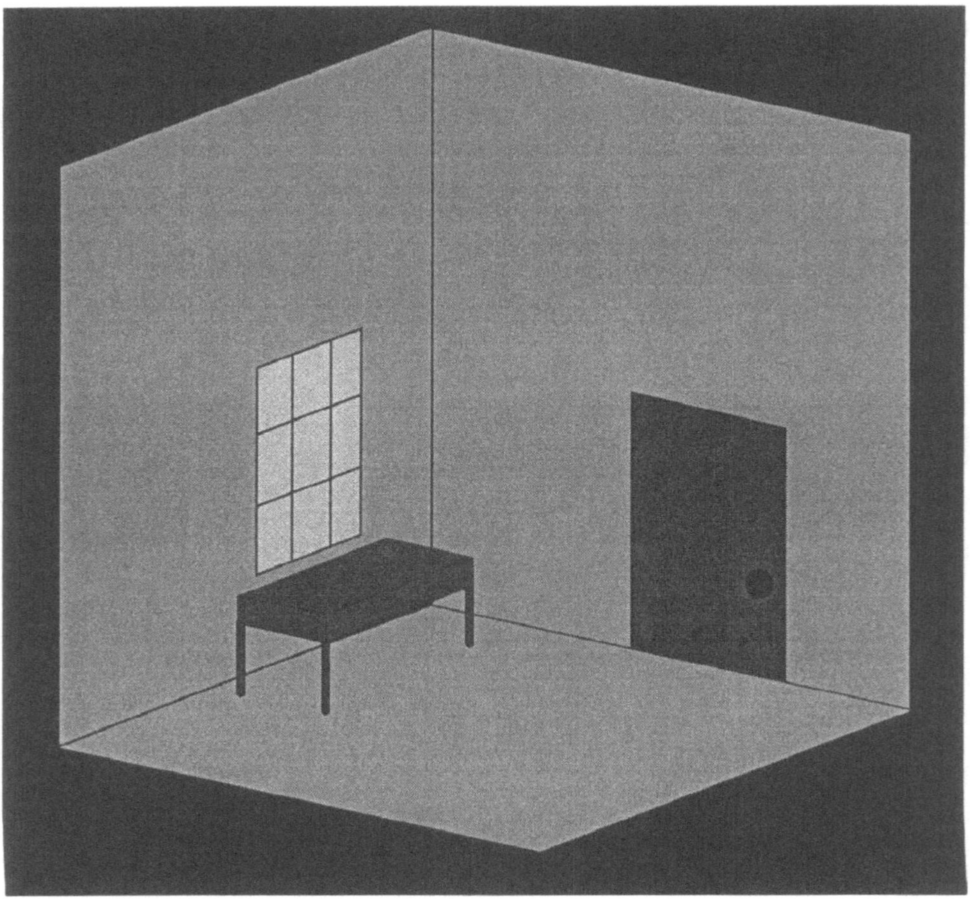

Figure 3.2: A large-scale cognitive map might be for a small space like a room

3.2.1. ORIENTATION

Considerable attention has been focused on the learning situations that produce cognitive maps with fixed orientations and those that produce cognitive maps that are orientation free. A number of authors have offered theoretical interpretations that explain the differences between these apparently different types of cognitive maps (Carpenter and Just, 1986; Evans and Pezdek, 1980; Lloyd 1993; Lloyd and Cammack 1996; MacEachren, 1992a; Lowe, 1987a; Presson and Hazelrigg, 1984; Sholl, 1987).

Viewers and Spaces

Conventional cartographic maps are depicted with structures that link them directly to the earth. Cartographers have traditionally produced their maps with a north-at-the-top orientation and a viewer's perspective 90° above the plane portraying the map. This notion of a north-at-the-top orientation relates a map to vertical and horizontal axes defined as the map might appear hanging on a wall or being viewed on monitor (Figure 3.4). We persistently map spaces onto such vertical and horizontal axes. Shepard and

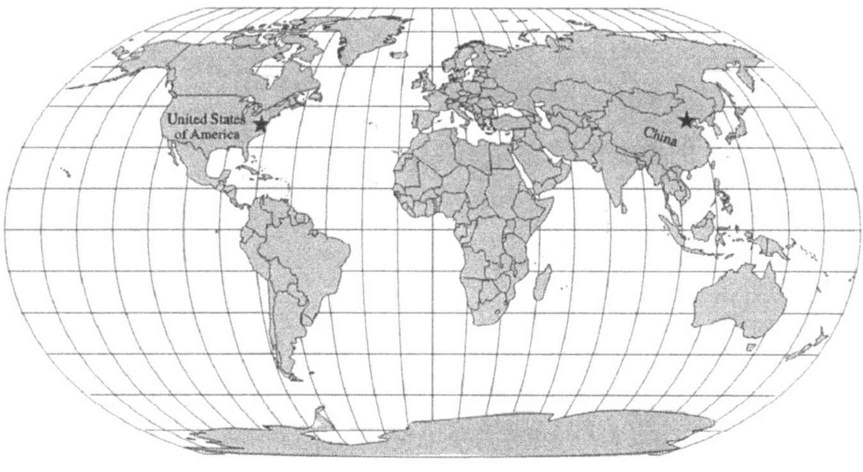

Figure 3.3: A small-scale cognitive map might be for a large space like the earth.

Hurwitz (1984) argued this view is preferred because gravity defines up versus down. Since we always have immediate knowledge of the vertical axis through our sense of gravity, we can easily distinguish top from bottom and left from right. This human preference accounts for the convention of expressing relative location on cartographic maps with north to south along the vertical axis and west to east along the horizontal axis. The convention provides a consistency for map reading, but there may be no good reason for having north at the top. South at the top may seem just as reasonable had the convention originated from cartographers in the southern hemisphere.

The viewer experiences multiple perspectives when navigating through a space. Imagine a person standing on a flat plane. Although many views are possible, a particular view is from some specific point on the plane facing in a certain direction (Figure 3.5). Orientation is expressed within an egocentric spatial structure, i.e., left, right, ahead, or behind, during navigation experiences. Objects naturally have a top and bottom and a front and back when viewed ahead of us. It has also been argued that we map navigation experiences onto vertical and horizontal axes with straight ahead naturally considered as "up" within an egocentric frame (Shepard and Hurwitz 1984).

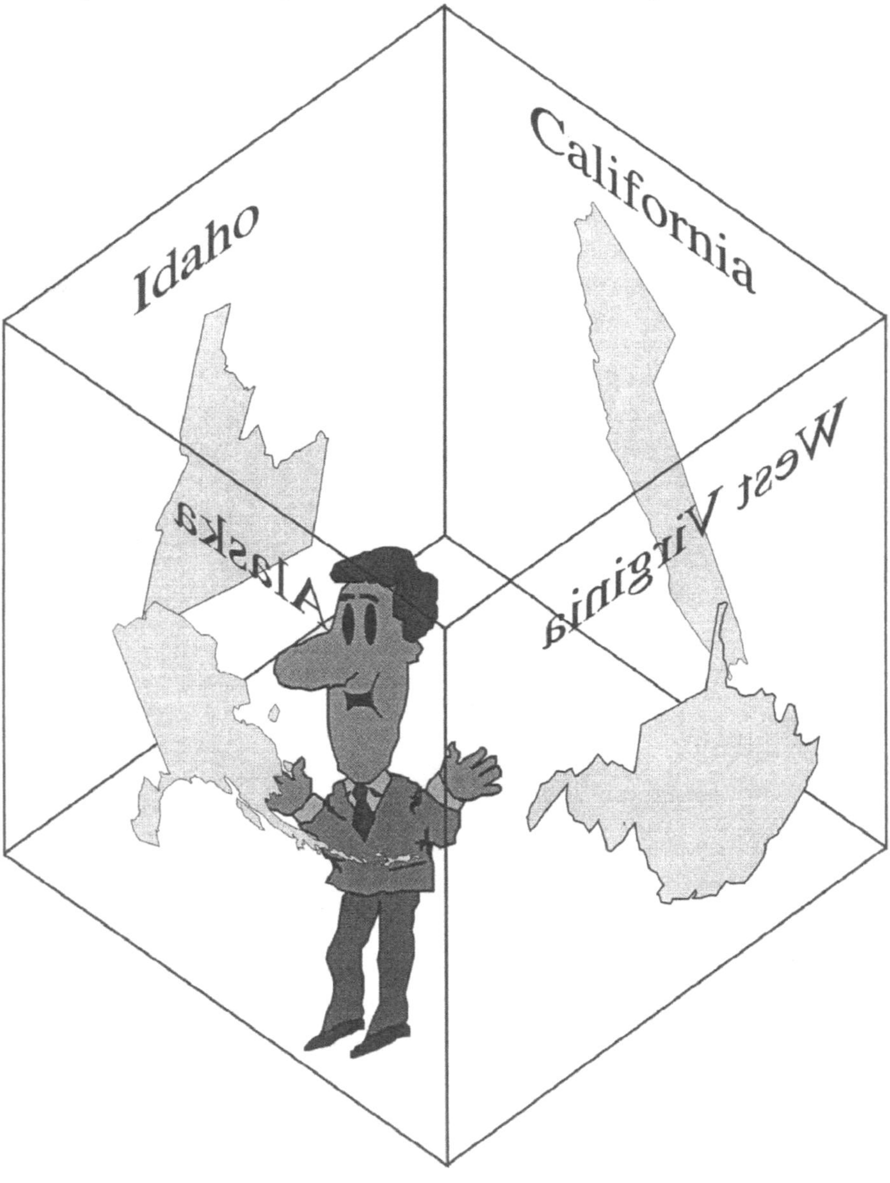

Figure 3.4: Person looking at some north-at-the-top maps on the walls in a room.

Away from the viewer and the backs of viewed objects are associated with the top of a scene through our awareness of gravity along the vertical axis. A relationship between a viewer and a viewed object, however, is dependent on the viewer's perspective within the space. The relationships of environmental objects to vertical and horizontal axes, therefore, are not consistent during a navigation experience. Imagine you are in the center of a space looking at an object at location *A* in the space (Figure 3.5). The object *B* will be behind you if you are facing *A*, but on your left if you turn to face an object at location *L* or on your right if you turn to face an object at location *R*. Walking straight ahead past object *A* and turning 180^{0} to face it again changes all the relationships (Figure 3.5).

Figure 3.5: Person in a space with objects ahead, behind, left, and right of his location.

Types of Spatial Knowledge

Spatial knowledge has been categorized by some researches by the processes used to encode and decode the information. Thorndyke and Hayes-Roth (1982) argued that two types of information can be used to encode cognitive maps. They argued that navigation experiences provide us with *procedural knowledge* to encode in our cognitive maps. We encode into memory the procedures used to go from one location in the environment to another. An episodic analyzer similar to the one discussed in the previous chapter for urban travel (Figure 2.5) might be used to encode the order of events along a route. Thorndyke and Hayes-Roth (1982) argued that we store *procedural knowledge* as verbal information and decoded it using a serial process. They argued that a distinctly different type of knowledge is encoded during a map-reading experience. Reading a map provides us with *survey knowledge* that is encoded into our memories as mental images. Survey knowledge provides us with holistic impression of the environment and is decoded using a parallel process. Although it has been argued that *procedural knowledge* can be reprocessed into *survey knowledge* over time (Thorndyke and Hayes-Roth 1982), some evidence has suggested this transformation does not automatically occur at the urban scale.

Lloyd (1989a) had one group of subjects, who had lived in Columbia, South Carolina for a long time, locate 15 well-known city landmarks relative to three equally well-known reference points (Figure 3.6). These subjects did not look at any map during

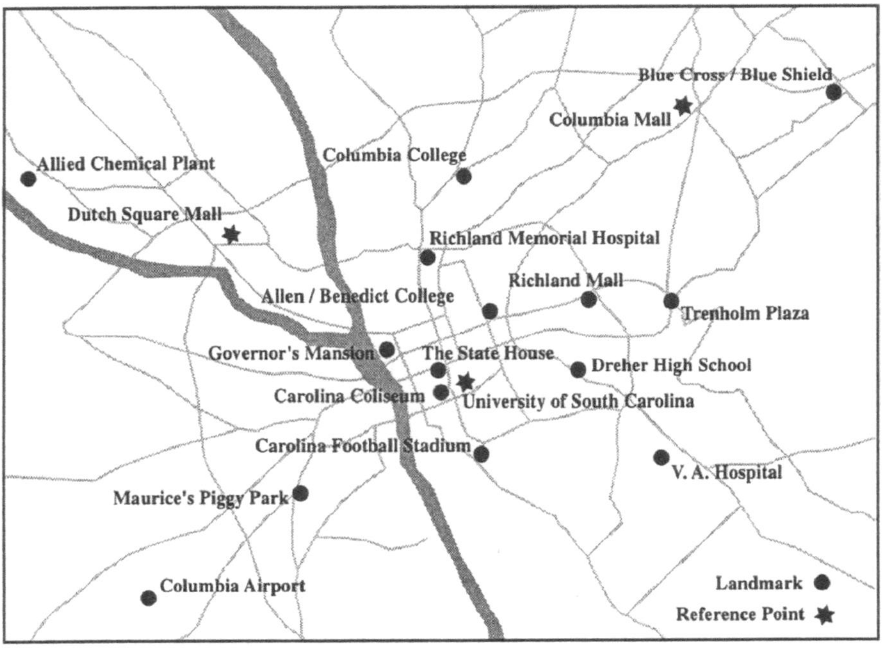

Figure 3.6: Map of Columbia, South Carolina with 15 landmarks and 3 reference points. After Lloyd (1989a).

the experiment and had presumably encoded the landmarks into their cognitive maps of Columbia by navigating through the city. Another group of subjects was shown a map that was identified as Fargo, North Dakota (Figure 3.7). This map was actually a map of the Columbia landmarks reflected around the horizontal axis and labeled with fictitious landmark names. These subjects learned the landmark locations on the map by studying it a few minutes and were tested to verify they were able to match each location with its appropriate name. Like the other group of subjects, this group also located the 15 landmarks relative to the 3 reference points. The experiences that allowed the two groups to encode their cognitive maps of the city were distinctly different. The Columbia group learned from years of navigational experiences and the Fargo group learned from minutes of map-reading experiences. Analyses that considered the relative accuracy and consistency of the two groups indicated that the Fargo group was significantly more accurate and consistent than the Columbia group. Map reading had provide a common experience for individuals that took relatively little time and produced superior results for the particular task of encoding survey knowledge of landmark locations.

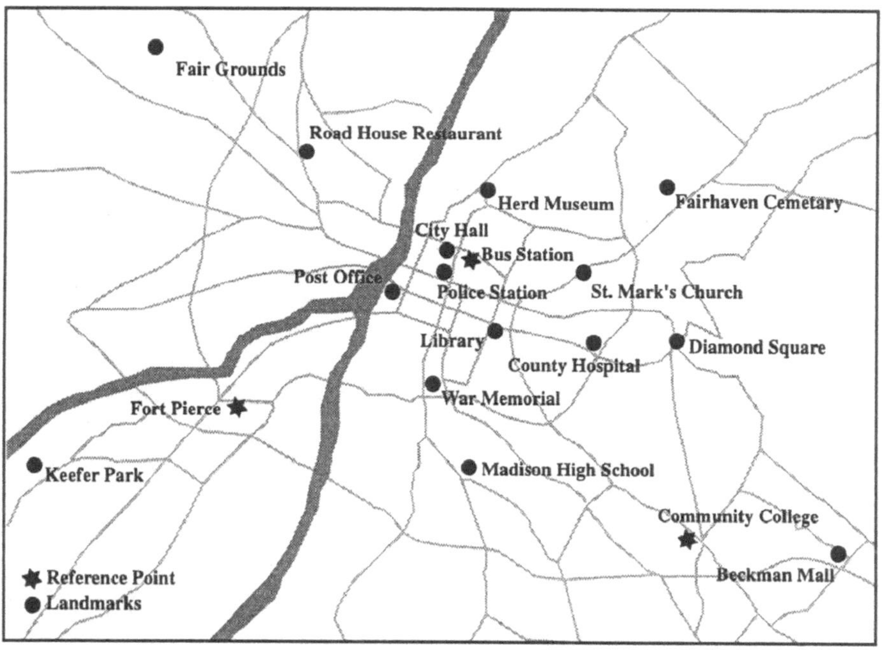

Figure 3.7: Fictitious map of Fargo, South Dakota with 15 landmarks and 3 reference points. After Lloyd (1989a).

Another categorization scheme for spatial knowledge has been suggested by Presson and Hazelrigg (1984) and Presson, DeLang and Hazelrigg (1989). In their studies they define *primary knowledge* as any spatial knowledge a viewer acquired directly from the environment. Primary knowledge may be encoded from a single perspective, e.g., looking from the top of a tall tree, or from multiple perspectives as when navigating through the environment. *Secondary knowledge* is defined as spatial knowledge indirectly acquired through an experience like map reading. Since the map only represents the environment, the contact between viewer and environment is not direct. The distinction between direct and indirect contact between the viewer and the environment may be important, but these initial studies have only considered very simple cartographic maps defined in traditional ways, i.e., line maps with a fixed orientation and vertical perspective. Computer generated map displays can easily provide viewers with *secondary knowledge* in a variety of forms. Maps with multiple orientations and viewer perspectives are easily generated. It is also possible to simulate indirect navigation experiences through map animations. MacEachren (1992b) considered whether the north-at-the-top orientation bias could be eliminated if map readers encoded a cognitive by viewing multiple versions of the same map produced with different orientations. His results indicted that viewing a map displayed with multiple orientations did eliminate the orientation bias. He also reported an interesting side effect. Subjects viewing the maps shown with multiple orientations performed map-reading tasks with decreased accuracy and slower reaction times compared to subjects viewing only a north-at-the-top version of the map.

Orientation Biases

Some studies have suggested the process used to encoded a cognitive map and the nature of the cognitive structure represented in memory is determined by the type of information available (Lloyd 1982, 1989a). People could also selectively choose to encode certain types of information based on the intended use of the cognitive representation at a later time (Taylor and Tversky 1992a). For one example, suppose you have a cartographic map of a particular country that has the usual objects represented on it, e.g., cities, rivers, and highways. You study the map with the general goal of learning about the cities in the country. You encode their names, locations, and how they are connected by the rivers and highways. You make a conscious decision to encode the map into your memory as an image that depicts the map you have studied (Kosslyn, 1980). If you later needed to answer a question related to particular cities, e.g., city A is north of city B, you could generate the image of the map and "look" at it like you had looked at the original map. If the cartographic map had been created with the conventional north-at-the-top orientation, your map image would be encoded in your memory in the same orientation. There would be a functional equivalence between the cartographic map and the map image you encoded into your memory. Your use of the cognitive map would be biased by its fixed orientation (Kosslyn, 1980; Lloyd, 1982).

Evans and Pezdek (1980) assumed that the maps previously used by college student subjects to learn about cities in the United States had been oriented with north at the top. Displays showing three cities were presented to the subjects one at a time. Each display represented the three cities as either a correct presentation or a mirror presentation reflected around the north-south axis. Both correct and mirror displays were also rotated a specific number of degrees away from north at the top (Figure 3.8a). Rotating the display counterclockwise 90° would present it with east at the top (Figure 3.8b) and rotating 180° would present it with south at the top (Figure 3.8c). The experimental task had subjects decide whether each presentation was a correct or mirror display. The results indicated map images have properties similar to images of other

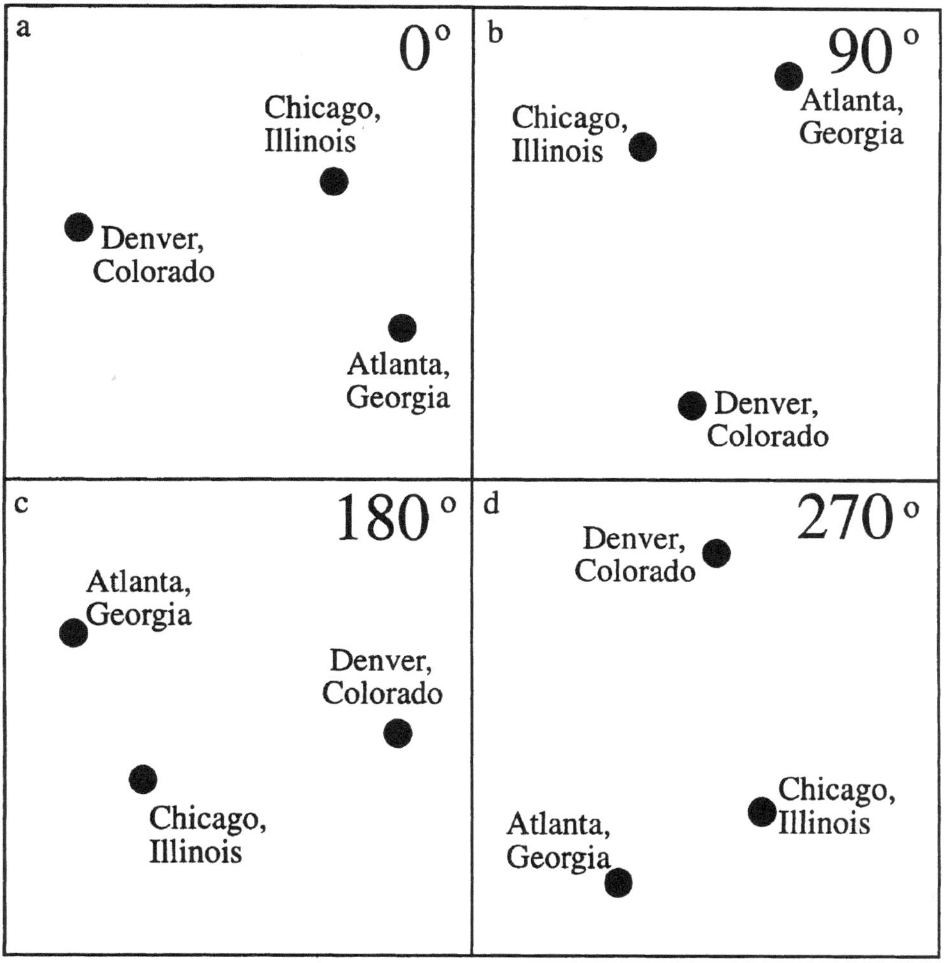

Figure 3.8: The cities of Denver, Colorado, Chicago, Illinois, and Atlanta, Georgia shown with north at the top (a), east at the top (b), south at the top (c), and west at the top (d).

two- and three-dimensional objects (Shepard 1978). The mean reaction times for the subjects and the number of degrees of rotation for maps were linearly related. North-at-the-top map displays produced the fastest times and south-at-the-top displays the slowest times. This result strongly suggested that experiences with maps in a fixed orientation create functionally equivalent map images with the same fixed orientation. When a cognitive map with an orientation bias is used to make identification decisions the bias is demonstrated by the additional time needed to identify maps presented in other orientations. For example, identifying a map presented in a west-at-the-top orientation (Figure 3.8d) would require the map to be rotated 90° so that it is congruent with the cognitive representation before it can be identified. This, of course, takes additional processing time (Figure 3.9b).

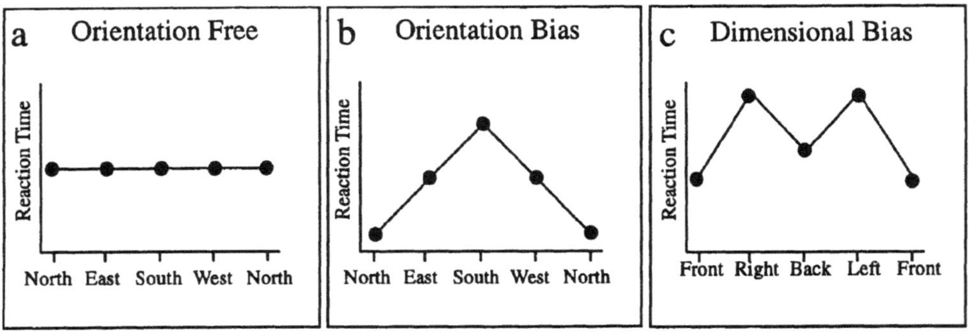

Figure 3.9: Patterns of reaction times for identifying maps at various orientations when the process has no orientation bias (a), an orientation bias for north at the top (b), and a dimensional bias for front and back (c).

Suppose, for another example, you have recently moved to a new home and are learning about your new environment. One might normally expect to have a set of initial landmarks in the environment that require your immediate attention. You would probably begin learning the locations of the landmarks by navigating within the environment from one landmark to another. With enough experience with the landmarks you will eventually have encoded enough information into your cognitive map to comfortably go from any origin landmark to any destination landmark.

Evans and Pezdek (1980) assumed that college students interacted with school landmarks to encode a cognitive map of the campus. Subjects were shown sets of three campus landmarks and asked to determined if a display was correct or a mirror image. Both correct and mirror image displays were rotated away from north at the top. Reaction time results showed no indication of an orientation bias similar to the one discussed above for cities in the United States. Subjects responded as quickly for south-at-the-top displays as they did for north-at-the-top displays (Figure 3.9a). A fundamental difference between the two example situations is suggested by the two results. Map reading experiences produced cognitive maps with a fixed north-at-the-top orientation and navigation experiences produced orientation free cognitive maps. Evans and Pezdek (1980) confirmed this difference with a third experiment that had subjects not familiar with the college campus study a map to learn the campus landmarks. These new subjects then identified the same correct and mirror displays of campus landmarks. Results indicated that learning from the map produced cognitive maps with a fixed orientation bias. The new subjects' reaction times were fastest for north-at-the-top displays and slowest for south-at-the-top displays (Figure 3.9b). The authors concluded from their results that the multiple perspectives provided by navigation experiences is an important factor in producing orientation free cognitive maps (Figure 3.9a).

The research paradigm used with the above examples considered if a relationship exists between the time needed to identify a stimulus as a correct or mirror version of a memorized object and the degree of rotation of the test object from its initial orientation (Shepard and Cooper, 1983). Geographers have also used it successfully to consider if imagery is used to identify outline maps (Steinke and Lloyd 1983a). It has also been shown that reaction times are affected by the type of information displayed on maps. Results indicated that more complex distributions requiring more processing time at all orientations (Lloyd and Steinke 1984).

Conerway (1991) performed a correct or mirror experiment to determine if the complexity of boundaries for states affected a map reader's ability to identify the state. The results indicated an orientation bias for all states and that boundary complexity significantly affected reaction times. States with simple boundaries took more time to identify than states with complex boundaries (Figure 3.10). A comparison of regression

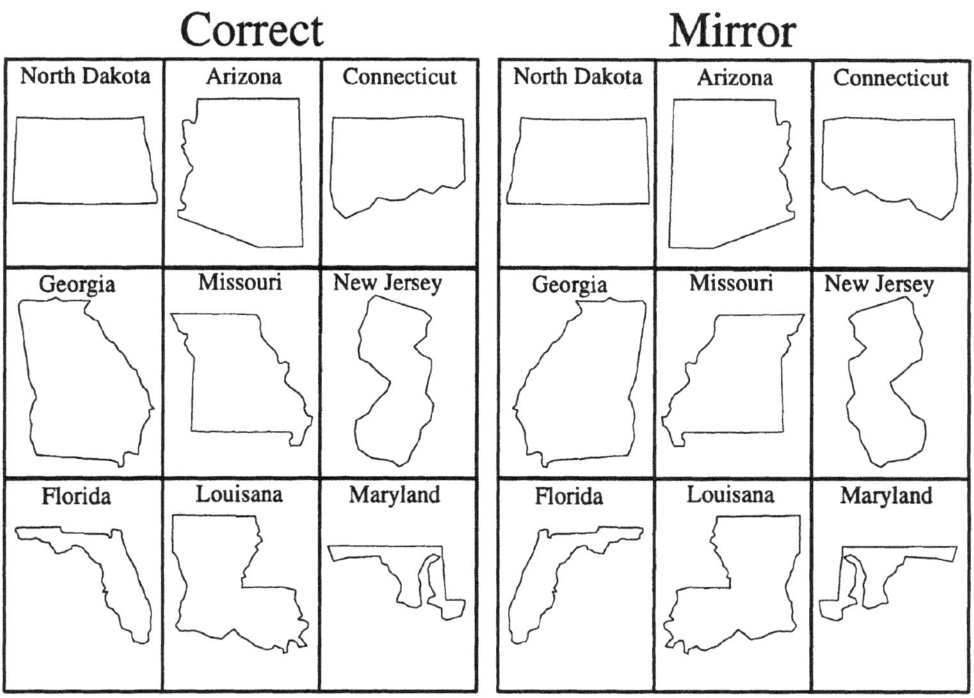

Figure 3.10: Correct and mirror maps for states that vary in the complexity of their boundaries. After Conerway (1991).

coefficients relating mean reaction time and degrees of rotation from north at the top for individual states, however, indicated no significant difference among states. It was concluded that map readers rotated images of all states to north at the top at the same rate, but that states with complex boundaries were easier to identify once they were rotated. Clues provided by the characteristics of the boundaries made the identification easier. Rice (1990) used this method to study the identification of three-dimensional prism maps. His results also indicated map readers used prominent boundary features as part of the identification process.

The Equiavailability Principle

Decision-making processes, such as the identification of polygon shapes discussed above, appear to be biased when cognitive maps are represented as images. This is because the spatial information is encoded and stored in a fixed orientation. There is, however, another important characteristic of map images that make cognitive maps encoded this way preferable. Map images can be parallel processed. This means information from all locations on a map image is equally available (Levine, Jankovic, and Palij, 1982).

Some researchers have argued that navigation experiences encode cognitive maps that are *orienting schemata* (Neisser, 1976; Shepard, 1978). Cognitive maps encode during navigation do not have the information at all locations equally available. Sholl (1987) has demonstrated that cognitive maps that have campus landmarks encoded through a navigation process are biased in a way consistent with the *orienting schemata* argument. Her research design had subjects assume they were at a particular location on campus facing a particular direction. Results indicated that information in front of the viewer could be accessed faster than information behind the viewer. A similar experiment done with cognitive maps of cities learned from cartographic maps did not have this type of bias. Information ahead and behind a viewer was equally accessible for these cognitive maps, suggesting they were encoded as map images (Sholl, 1987).

Hintzman, O'Dell, and Arndt (1981) performed experiments that had subjects imagine themselves in a particular spot in a learned environment facing in a particular direction. The experimental task was to point to targets that surrounded them in the environment. Results indicated subjects were faster answering along the vertical axis (front and back) than the horizontal axis (left and right) for a number of research designs, e.g., visual maps, cognitive maps, and tactile maps. Subjects could point to objects in front of them faster than behind them, but pointing behind was faster than pointing in other directions. A plot of mean reaction times for quadrants around a circle (front, left, back, right, front) looked like the letter "M" (Figure 3.9c). The researchers made a distinction between a visual (perception) task, i.e., pointing to an object that is physically present and can be seen and a cognitive (memory) task, i.e., pointing to an object that is not physically present using information previously encoded into memory. They argued that mental rotation was used with perception tasks, but not with memory tasks. The authors suggested "cognitive maps are not strictly holistic, but consist of orientation-specific representations, and-at least in part-of relational propositions specific to objects" (Hintzman, O'Dell, and Arndt, 1981, p. 149).

Two- and Three-dimensional Depictions

The conventional cartographic map presents the environment it represents in a two-dimensional space. The conventional map gives the viewer a parallel perspective planimetric view of the plane depicting the map. Because the viewer of the conventional map has a vantage point orthogonal to all points on the plane simultaneously it does not reflect actual human experiences (Muehrche 1986). Only the tops of objects are rendered as visible on conventional maps, making all objects on the plane appear as two-dimensional. The third (vertical) dimension of objects on the plane can be rendered as visible to the viewer by changing the view to a single perspective at some angle less than 90^{0} above the plane. Experimental studies that have used the correct or mirror task with three-dimensional map displays have reported a non-rotational strategy was used by some subjects to identify objects (Holmes, 1984; Goldberg, MacEachren, and Kotval, 1992). One research design had subjects judge if pairs of three-dimensional maps were the *same* or *different*. Particular displays could have either two *correct* maps, two *mirror* maps or one of each. The usual rotational strategy was used by some subjects, but other subjects used a verbal strategy. They decided if the two maps were the same by determining if objects on the maps were organized in a clockwise progression or a counterclockwise progression (Holmes, 1984). Cartographic maps with unconventional oblique views may present information in a manner that makes it possible for viewers to encode a more complex cognitive map. Map readers who are viewing cartographic maps from an oblique perspective might easily be able to encode map images for rotational solutions and also encode verbal information about relationships among objects in the space.

Rice (1990) argued that his subjects rotated three-dimensional prism maps of Wisconsin to identify them in a correct or mirror task. He also indicated, however, that the longest reaction times were not always produced by south-at-the-top maps. A closer inspection of his results indicated that foreground objects blocked critical information from the viewer for some views of the three-dimensional space. Longer reaction times for these cases suggested that verbal coding was being used as part of the identification process.

3.2.2. THE ACCURACY OF COGNITIVE MAPS

Cognitive maps are internal structures in our memories that express the spatial information we have learned. Like cartographic maps, cognitive maps can be constructed using many different sources of information and encoding processes (Lloyd 1993). Some cognitive maps may be stored as semi-permanent structures in long-term memory, e.g., a cognitive map of a familiar city, while others may be extremely temporary structures of the current state of a dynamic environment, e.g. a parent keeping track of the locations of children as they play in a park. In either case the characteristics of objects are thought to be stored along with their spatial locations (Kahneman, Treisman, and Gibbs, 1992; Lloyd and Hooper, 1991). Cognitive maps frequently do not reflect the environment they represent with complete accuracy. This is true for both *what* and *where* information. A cognitive map may be incomplete, if important information has not been learned, or it may be inaccurate, if the learning process was biased in some way. For example, Orleans (1973) had residents from different socio-economic and ethnic neighborhoods in Los Angeles draw sketch maps of the city. The maps of neighborhood residents that were less mobile, as reflected by there economic and ethnic status, were very incomplete when compared with highly mobile residents. Their urban cognitive maps were incomplete simply because they had never experienced much of the city outside of their neighborhoods. Leiser and Zilbershaltz (1989) argued that even the most sophisticated cognitive maps are strongly biased by the skeleton of basic and secondary routes in the transportation network.

Lloyd and Hooper (1991) performed an experiment that had subjects respond to a distribution of locations within a familiar city that corresponded to the centroids of census tracts. Subjects, who were long-time residents of the city, were randomly presented the locations one at a time on a map of the city on a monitor. The subjects determined if each location was above average (high) or below average (low) for variables related to race, income, and age. Results indicated that subjects were consistent and accurate for race and income variables, but inconsistent and inaccurate for age variables. It was concluded that race and income characteristics might be easier to generalize in a cognitive map than age. These characteristics are more visually apparent because neighborhoods tend to be more homogeneous with respect to race and income than they are for age.

Lloyd and Heivly (1987) reviewed studies that explained errors in cognitive maps and found three basic arguments. Some studies had focused on the characteristics of the environment and how particular features of the environment caused people to encoded spatial information incorrectly. A number of cognitive distance studies have suggested that the existence of barriers in the environment, e.g., rivers, produce exaggerated distance estimates (Cohen, Baldwin, and Sherman 1978; Kosslyn, Pick, and Fariello 1974; Thorndyke 1981). One study that considered cognitive distances in a number of urban environments concluded that "road and rail systems, general topography, and major geographical features" caused distances to be estimated incorrectly (Canter and Tagg 1975, p. 78). Other studies had traced the sources of errors in cognitive maps to the nature of objects in the environment. The cognitive distance between landmarks was thought to be influenced by the status of the environmental objects (Sadalla, Burroughs, and Staplin 1980). The function and familiarity of landmarks (Briggs 1973; Canter and Tagg 1975), the number of turns on routes (Sadalla and Magel 1980), and the number of intersections on routes (Sadalla and Staplin 1980) have been reported as causing errors in estimating cognitive distances.

Other studies have focused on the characteristics of cognitive mappers. It has been suggested that differences between individual cognitive maps encoded by residents living in the same environment are related to the varying experiences of the individuals, e.g., mobility and length of residence (Downs and Stea 1977; Golledge 1978; Gould 1975). If errors in cognitive maps can be eliminated by experience, then one might expect residents of an environment that are very mobile and have many years of experience to have cognitive maps that are relatively free of distortions (Thorndyke and Hayes-Roth 1982). Lloyd's (1989) study disucssed above suggested this may not always be true. The ability of long-time residents of an urban area to locate familiar landmarks was found to be poor compared to individuals who had recently learned the locations from a cartographic map.

Although the unique characteristics of environments and people may contribute to errors in specific cognitive maps, another approach to explaining more general distortions had focused on the cognitive processes used to encode, store, and decode spatial information (Moar and Bower 1983; Tversky 1981). These studies had identified systematic errors in cognitive maps not related to variations in the environment or the experiences of the cognitive mapper. Tversky (1981, p 407) argued that systematic errors in cognitive maps should be viewed "not as signs of failure of processing, but as consequences of normal processing."

3.2.3. SYSTEMATIC DISTORTIONS

Tversky (1992) has suggested some distortions in cognitive maps are the natural product of the way the human brain processes information. A distinction must be made between errors in cognitive maps that are based on incorrect information and errors that are systematic distortions (Lloyd and Heivly 1987). The first type of error could occur if false information is encoded or if no information is encoded into a cognitive map. A person could be told that the new microbrewery is located in the Northeast Mall when it is actually located in the Northwest Mall. A person drawing a sketch map of a city might not include a large subdivision south of the river because she was unaware that it existed. Both the false information encoded into cognitive maps and the missing information are likely to vary from person to person and be based on the person's unique experiences. These unique errors in a person's cognitive map can be corrected easily with new and accurate information.

The second type of error is likely to be shared by people living in the same environment. Systematic distortions are found in aggregate cognitive maps when a group of individuals has used normal cognitive processes and complete information to consistently produce the same errors (Lloyd and Heivly 1987). Systematic distortions in cognitive maps are shared by individuals because the individuals use similar cognitive processes to encode spatial information. Systematic distortions are not easily corrected because they are the product of correct information and normal cognitive processing

Categorization

One source of systematic distortion can be linked to categorization processes. Information is not just acquired through the sensory systems and stored. The initial information is processed and related to more abstract categories. This learning process stores information initially learned about individual locations in these more abstract categories. To illustrate the abstraction process consider Figure 3.11 as part of a neural network that has learned information by experiencing geographic objects. This portion of the network shows neurons related to some individual cities (C_1 through C_6) and states (S_1 through S_6), more abstract categories (City and State) and most abstract category (Place). The three geographic categories are shown connected to other neurons

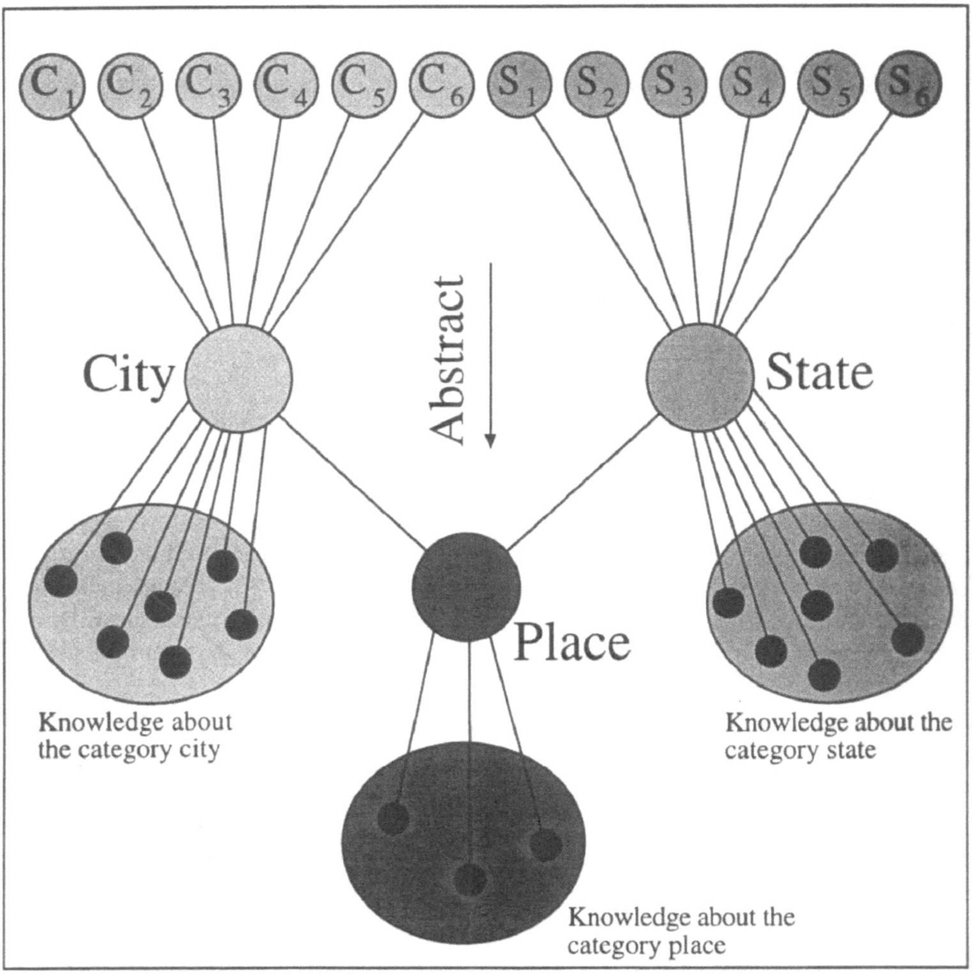

Figure 3.11: A portion of a neural network storing generalized knowledge about the geographic categories city, state, and place. Information is acquired though experiences with specific cities or states at the top. The knowledge about a geographic category is stored in connection to that category. The place category at the bottom of the hierarchy is the most abstract category. Knowledge about the place category is based on information generalized from information learned about cities, states and other geographic categories.

that represent knowledge that has been generalized about that category from experiences with the individual cases. The connections between the geographic categories and other neurons provide generalize knowledge that is true for all cities or all states. Fewer connections are shown for the place category because it represents the most abstract geographic category and its connections are to neurons representing information true for all places. The only connections shown for the individual cities or states are to their geographic categories. This is based on the notion that much of what we know about any particular city or state is true for all members of that category (Lloyd, Patton, and Cammack 1996; Rosch, Mervis, Gray, Johnson, and Boyes-Braem 1976). Every city or state, of course, has some information that is unique for that location, e.g., the

Gateway Arch for St. Louis. In a more complete representation of a network this unique information would be represented as connections directly to the individual cities or states.

Categorization has been offered as an explanation to why cognitive maps cannot always provide precise information. Research on urban cognitive maps has reported that people know that streets exist and intersect, but frequently do not know the precise angle of the intersection (Byrne 1979; Moar and Bower 1983; Tversky 1981). Procedural knowledge encoded during navigation experiences as conceptual propositions selectively encodes only the important information needed to repeat the navigation experience at a later time. Specific angles or precise distances are not encoded into memory because they are not needed for navigation. Remembering the order of landmarks and left or right turns is enough information to navigate most routes in a street network. The two routes (black and grey) illustrated for a hypothetical city illustrates this point (Figures 3.12). Landmarks along the black (A through F) and grey (Z through V) routes represent decision points along the route. The landmarks could be any visible objects in the environment such as a building or an intersection sign. The arrows represent the overt movement along the street once the landmark has been identified and the choice of left or right turn has been made. The two routes take the navigator to the front door (black route) and back door (grey door) of the same house. All the intersections along the black route intersect at some angle other than 90° and all the intersections along the grey route intersect at exactly 90°. Note also that the black route is somewhat longer than the grey route. Encoding each of the route as procedural knowledge might result in structures similar to those shown in Figure 3.13. The code for the black route (top) and grey route (bottom) both simplify the navigation procedure to executing a left or right turn when encountering a specific landmark. The black route has one more decision point than the grey route but differences related to angles or distances between landmarks are not encoded in the structures. It is more efficient to encode a prototype and use it to represent a category. Since the actual angles of the street intersections is not available from memory, we use the 90° category prototype to fill in the missing information. We have generalized this prototypical intersection from experiences with many streets in cities that do cross at right angles. It is easier to remember this than to remember all the details of the large amount of spatial information we encounter every day (Huntenlocker, Hedges, and Duncan 1991; Lloyd 1994). This notion is frequently demonstrated in sketch maps that have all the streets drawn at right angles.

Spatial Hierarchies

A number of studies have argued that spatial information is stored in a hierarchical structure based on a container principle (Eastman 1985; Sevens and Coupe 1978; Wilton and File 1975). This hierarchical structure is formalized on most reference maps, e.g., countries are shown as containing states and states are shown as containing cities. Eastman had subjects learn the locations of cities from cartographic maps with

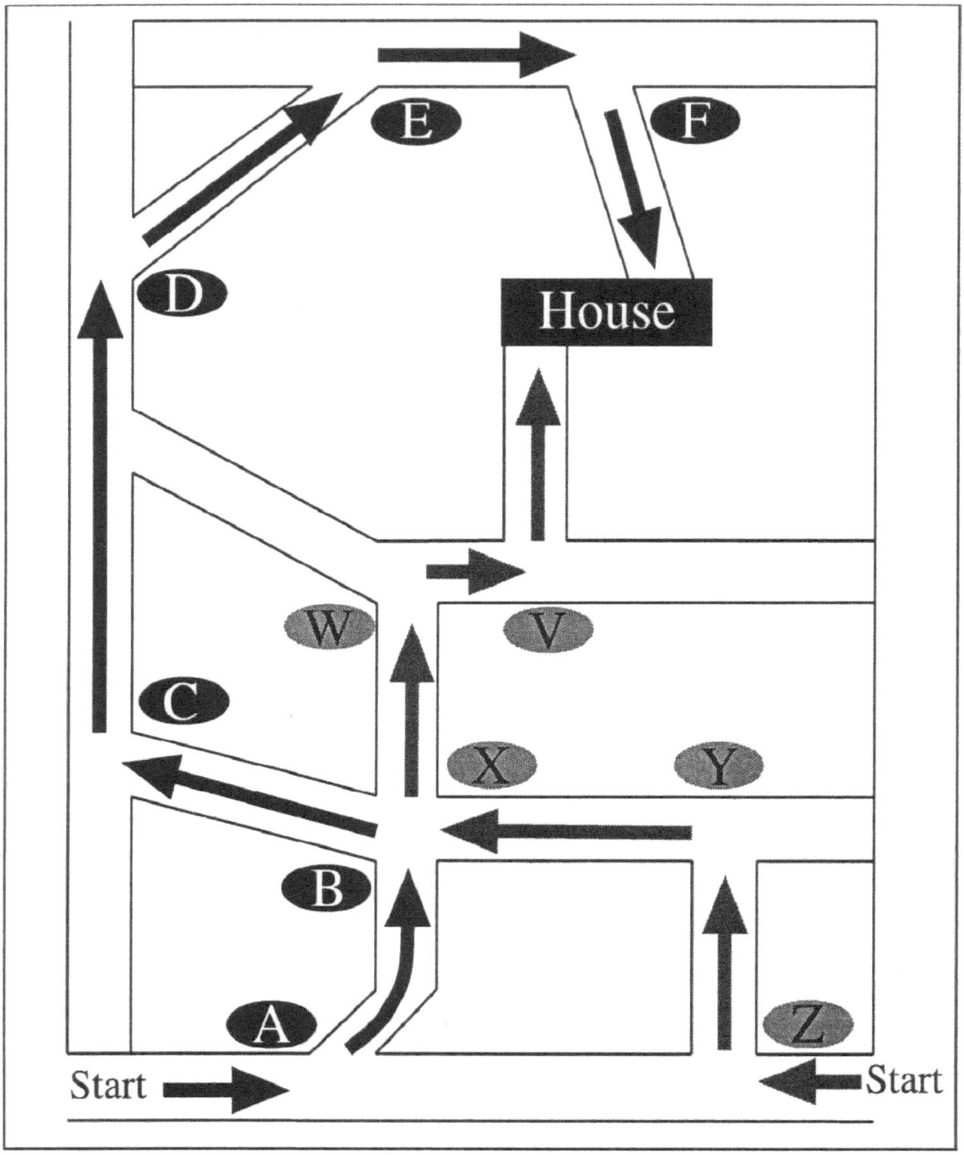

Figure 3.12: Two routes in a hypothetical urban area that could be encoded as procedural knowledge. The Black route is along streets that intersect as angles deviating from 90 degrees. The grey route is along streets that intersect at 90 degrees.

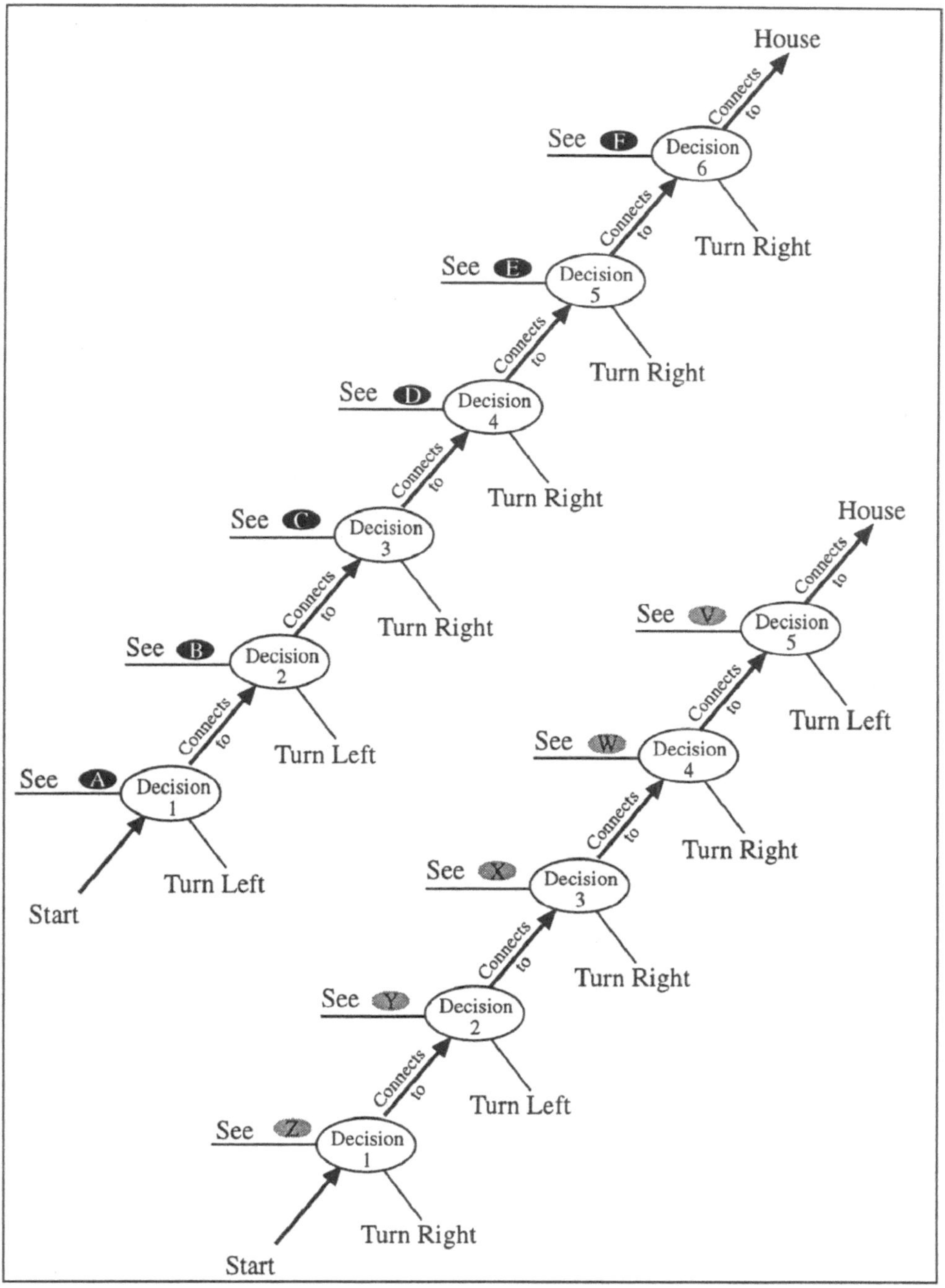

Figure 3.13: The black route, on the top, and grey route, on the bottom encoded as a sequence of decisions. Note it is not possible to extract the precise angle of the intersections or the distances between landmarks from this simplified structure.

different graphic organizations to determine the effect of the graphic organizations on the structure of the information in memory. He concluded, "memory structures were distinctly hierarchical, regardless of the graphic organization encountered" (Eastman 1985, p. 13).

Although hierarchical memory structures allow for efficient encoding and storage of spatial information, it has been convincingly argued that the reorganization of spatial information into hierarchical memory structures can result in errors when the information is used to make judgments (Stevens and Coupe 1978; Tversky 1981, 1992). For an example, consider the fictional island of Shamus (Figure 3.14). It contains two countries named Spenser and McGee. Spenser is north of McGee and contains the states of Hawk and Silverman. Hawk is west of Silverman. Hawk contains two cities named Quirk and Belson. Quirk is south of Belson. Silverman contains two cities named Fiori and Loring. Fiori is south of Loring. McGee's structure could be described in a similar manner. Assume someone has studied the Shamus map and encoded the information in a hierarchical structure like the one illustrated in Figure 3.15. This tree-like structure shows the cities along the bottom of the structure. Following the branch for any city links it to its container at the next level up in the hierarchy, e.g., Belson is contained in Hawk. Relationships between geographic units are thought to be explicitly encoded only within specific containers. Spatial relationships between geographic units in different containers are not explicitly encoded, but implicitly encoded by the relationship between their containers. Using direction as an example, the hierarchical structure (Figure 3.15) encodes that Quirk is south of Belton because both cities are in Hawk. One could recall this information directly from the memory structure. If, however, one wanted to know the relationship between Belton and Fiori, there is no direct encoding of this information. We can obtain, however, that Hawk (which contains Belson) is west of Silverman (which contains Fiori). Using this implicit information for the relationship, we would conclude that Belton is also west of Fiori. This example is similar to the real world example reported by Stevens and Coupe (1978). They asked subjects, who lived in San Diego, California, to recalled the direction from San Diego to Reno, Nevada. Most remembered Reno as east of San Diego when it is actually to the west. The subjects' memories for the spatial relationship between the two cities are consistent with the relationship between the two containers California and Nevada.

Relative and Absolute Error

O'Keefe and Nadel (1978) argued that we encode spatial information for two types of spaces. An *absolute space* is independent of the observer and acts as a container or framework for the objects in the space. An absolute space exists independent from the objects in it and the movement of an observer within the space to a new perspective does not change the relationships among the objects or alter the framework. A *relative space* does not exist independent of the objects in it. A relative space emanates from the observer and changes when the observer changes perspectives. Lloyd (1989a) argued that

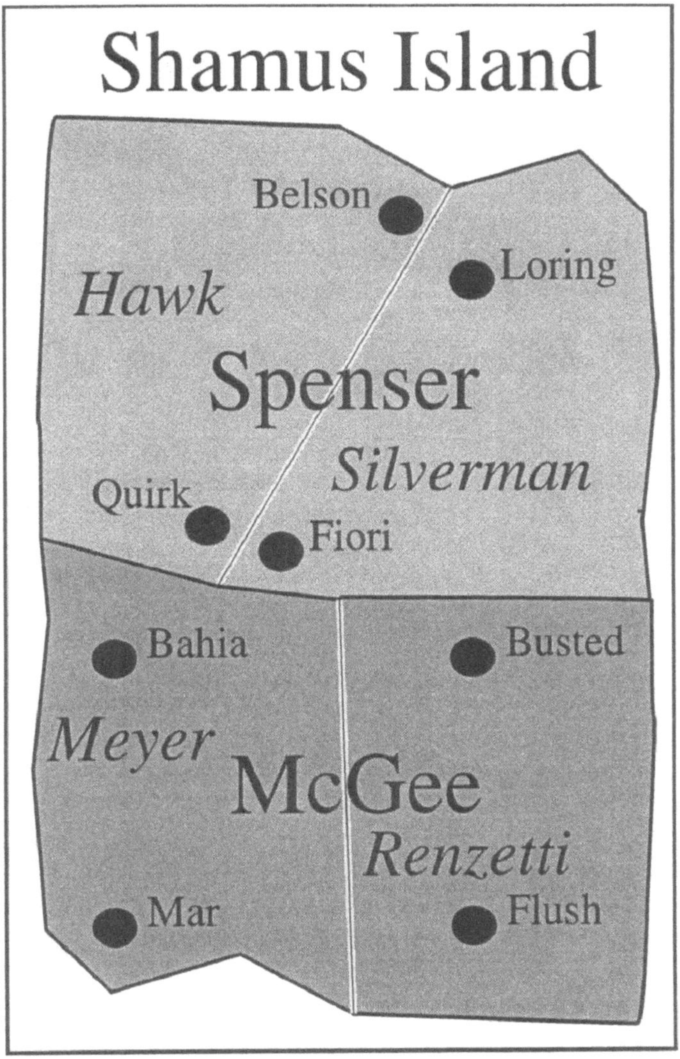

Figure 3.14: Shamus Island has countries, states, and cities that can be encoded into a hierarchical structure.

distortions in cognitive maps can also be defined as absolute and relative errors. He argued that some distortions are caused when the frame of reference of a cognitive map is rotated, translated, or scaled during the encoding process. Rotations can be clockwise or counterclockwise and would cause a systematic movement of objects in the space away from their correct locations. Tversky (1981) had subjects place an outline of the South American continent within a framework defined by correctly drawn parallels and

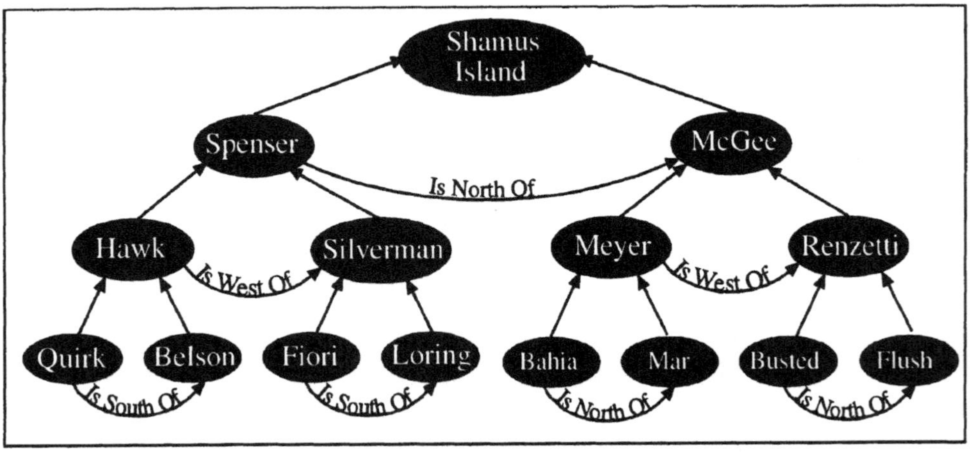

Figure 3.15: Shamus Island (Figure 3.14) encoded into a hierarchical tree structure based on a container principle.

meridians. She reported that most subjects rotated the major axis of the continent to align it with the meridians. This, and similar results (Muller 1985), offer support for her argument that people use alignment and rotation heuristics to simplify their cognitive maps. Lloyd and Heivly (1987) have also shown this to be true for residents of neighborhoods located in different parts of a city. The rotations of the residents' cognitive maps appeared to be a function of where their neighborhood was located in the city and result in an incorrect alignment of major transportation arteries with cardinal directions. Residents who lived in a neighborhood on the eastside of the city appeared to rotated their cognitive maps clockwise (Figure 3.16). This aligned Trenholm Road that connected their home and the downtown with the east-west axis. Residents that lived in a neighborhood on the westside of the city appeared to rotate their cognitive maps counterclockwise (Figure 3.17). This also aligned Interstate Highway 126 that connected their neighborhood to the downtown with the east-west axis.

Translations of cognitive maps occur when the frame of reference for a cognitive map is displaced along a vertical or horizontal axis. This movement also systematically displaces all the objects in the space away from their correct locations. Another experiment conducted by Tversky (1981) suggested that Europe and Africa are shifted to the south and South America is shifted to the west on most people's cognitive maps. These translations simplify our world cognitive maps by aligning the continents. This alignment of the continents systematically moves the other information associated with the continents, e.g., countries and cities, and causes absolute error. Figure 3.18 illustrates the correct and translated continents along with selected cities. Note that Pairs is north of Montreal on the correct map, but south of Montreal on the aligned map.

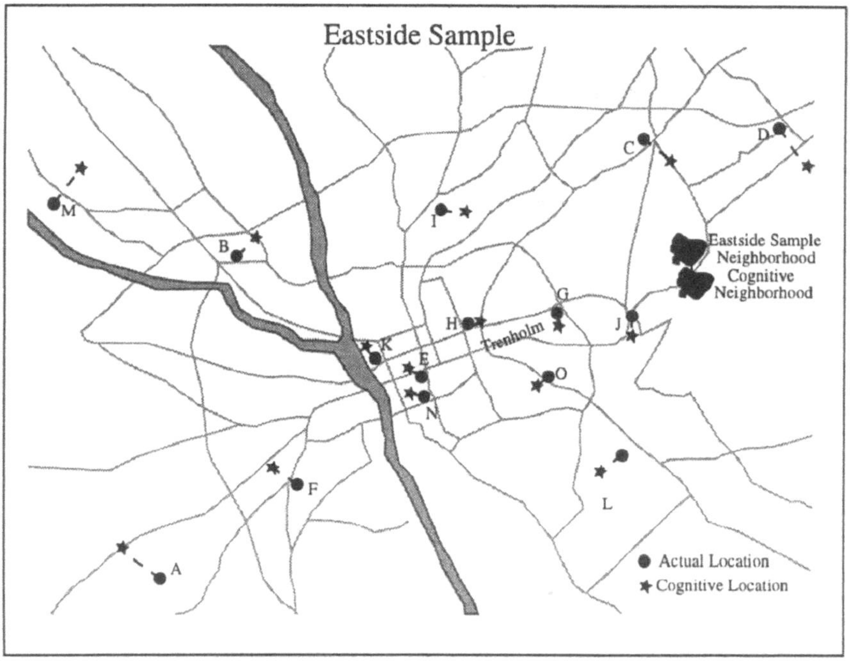

Figure 3.16: Cognitive locations of landmarks for subjects living in an eastside neighborhood show a clockwise rotation. After Lloyd and Heivly (1987).

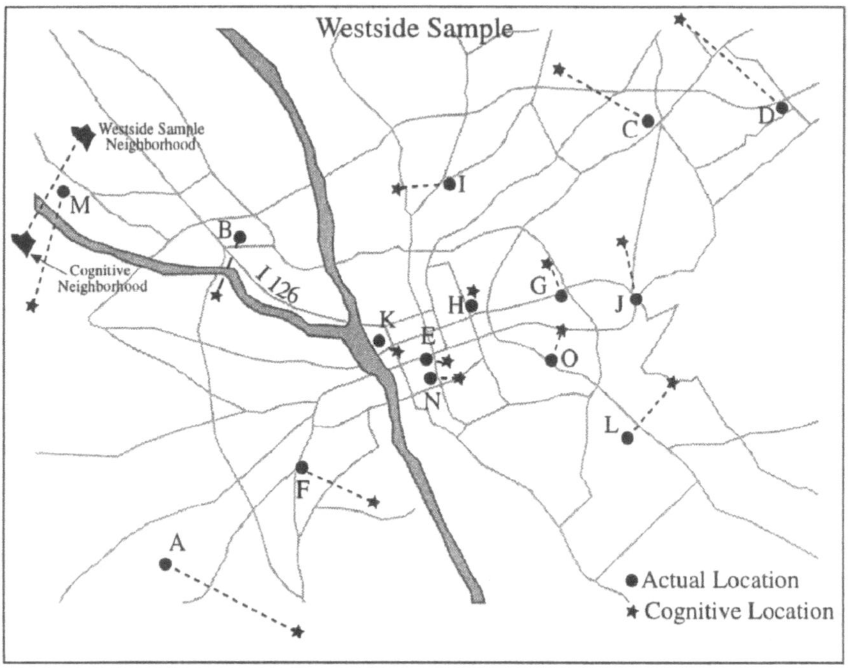

Figure 3.17: Cognitive locations of landmarks for subjects living in a westside neighborhood show a counterclockwise rotation. After (Lloyd and Heivly 1987).

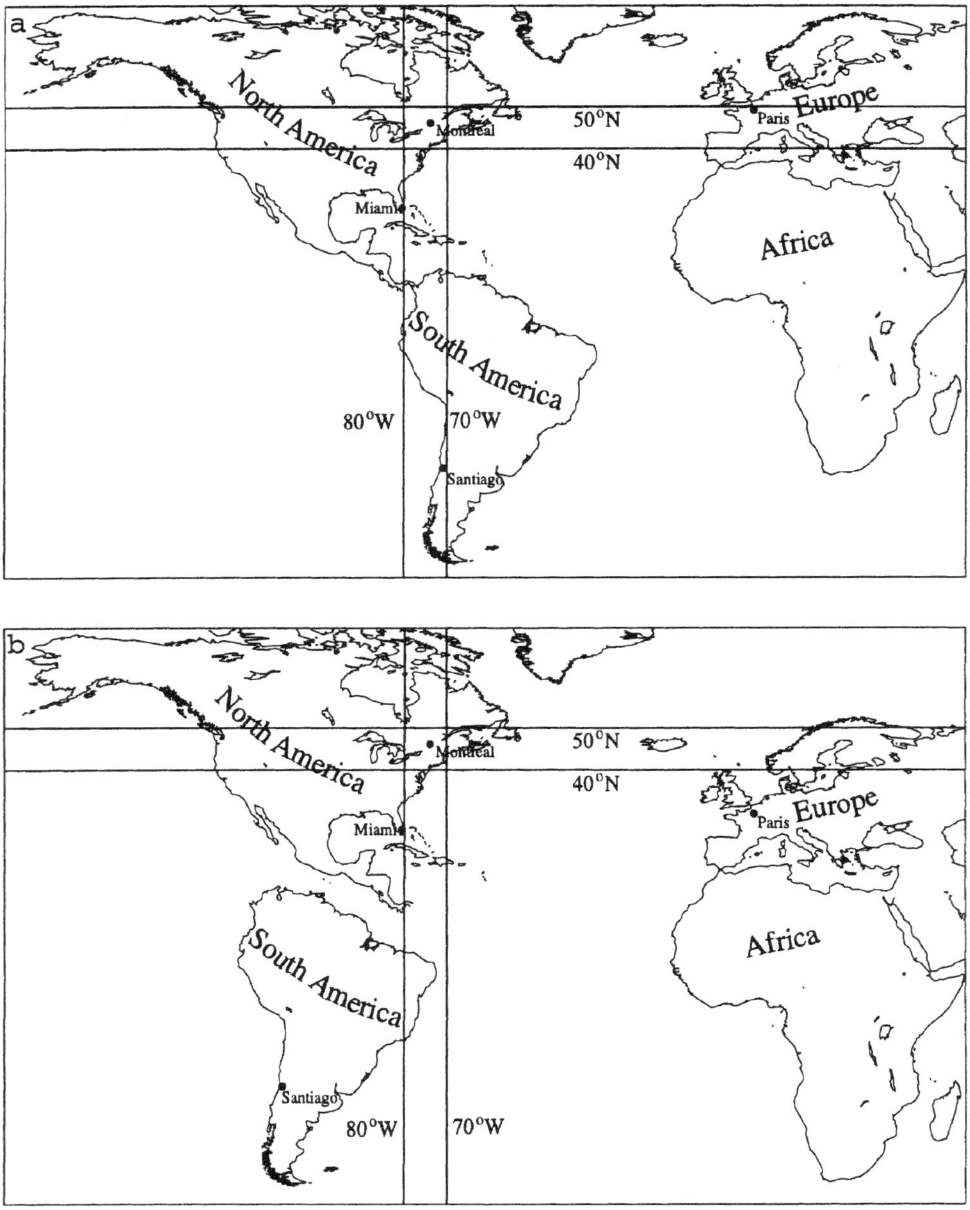

Figure 3.18 The details of the actual locations of North America, South America, Europe, and Africa (a) are encoded as aligned (b) by most people. Alignment errors cause Paris to be remembered as south of Montreal and Santiago west of Miami.

Scaling causes the size of the frame of reference to change by either contracting or expanding the space. Lloyd (1989a reported that subjects' cognitive maps of a city generally were scaled to be larger than their correct size. This exaggeration of the size of the city occurred both when subjects learned the city from navigation experiences and from map reading experiences. A similar result was reported for world cognitive maps by Lloyd and Gilmartin (1991).

Reference Points

Reference points on cognitive maps apparently have a special significance. Some have argued that special landmarks in environments serve as anchor points for encoding other information (Ferguson and Hegarty 1994; Hirtle and Jonides 1985; Couclelis, Golledge, Gale, and Tobler 1987; Golledge 1978). Errors in cognitive maps caused by reference points have generally not been reported as systematic distortions. A number of studies have reported an expansion of space near a reference point and a contraction of space far from the reference. Lloyd and Gilmartin (1991) found this pattern for world cities using Victoria, British Columbia as the reference point. It has also been reported that the distribution of places on cognitive maps changes when considered from different reference points. Holyoak and Mah (1982) asked some subjects to image themselves on the east coast and other to imagine themselves on the west coast Both groups judged distances between cities (San Francisco, Salt Lake City, Denver, Kansas City, Indianapolis, Pittsburgh, and New York City) that formed a linear pattern across the United States. Their results indicated that distances were expanded near a reference point and contracted away from a reference point for both groups. Subjects using a west coast perspective expanded the distances between cities near that reference point (western cities) and contracted distances between cities far from that reference point (eastern cities). The subjects with an east coast perspective did the opposite using their perspective. Since all the subjects were performing the experiments in Ann Arbor, Michigan, the perspective effect was not just a function of the home location of the subjects. A similar result was reported by Lloyd (1989a) for urban cognitive maps. Subject were assigned to use either of two suburban shopping malls located on opposite sides of the city as reference points when estimating the locations of other familiar urban landmarks. Data for both groups showed an expansion of space near the reference point and a contraction of space far from the reference point. Since reference point effects are not consistent throughout the space and change as a viewers position with the space changes they are better described as relative errors than absolute errors.

3.3. Conclusions

This chapter has explored the nature of cognitive maps. Considered were the types of processes used to encode spatial information into memory, the type of information encoded, and systematic distortions caused by encoding (categorization), storage (spatial hierarchies), and decoding (reference points) processes. The next chapter takes a closer look at two theories that consider how spatial information is stored in memory.

CHAPTER 4

STORING SPATIAL INFORMATION IN MEMORY

4.1. Introduction

The complex world we encounter every day contains many objects. We selectively focus attention on some of the objects and may wish to store important information associated with these objects. This information could include *where* information, recording the object's location, and *what* information, recording other characteristics such as the object's shape, color, or size. Encountered objects vary in their relative importance to the person experiencing the objects (Figure 4.1). Many objects that exist in the environment are completely ignored because we never focus attention on them. Other objects may attract our attention long enough to make us briefly aware of their existence just long enough to decide we have no further interest in them. At times we may be forced to abandoned objects we have focused on as we shift attention to new objects because our working memory has a limited capacity (Baddeley 1986). At other times we may have sufficient interest in an object we are perceiving to track it over time and update any changes to the *where* and *what* information associated with it. The location of the object may change while we are experiencing it as well as its other characteristics. Because we have attended to the object and updated information related to it, we know it is the same object. If we wish to store information about the object after perception processes are completed, it would be necessary to encode the information related to the object into long-term memory (Figure 4.1). This could be verbal information expressed as a conceptual-proposition such as *the car is blue* or *the car is in the garage*. It could also be a visual image of the blue car in the garage.

Sometimes the focus is not on a specific object, but on a space that contains a number of objects, e.g., a neighborhood or a city. We need a cognitive structure that can be used to store relationships among specific objects. The structure should allow us to relate to a set of objects embedded in a common space. In this case the cognitive structure in memory is for the space rather than for individual objects. The objects have locations in the space and characteristics that give them an identity.

Some spatial information may be extremely important to an individual while other spatial information may have only a passing interest. On one end of the importance continuum might be where your home is located within the environment. Not being able to recall this information could be a serious problem. How to travel between your home and other important locations such as the workplace is the first important spatial process one learns when moving into a new environment. Such spatial information might be classified under *things we never want to forget* because the information has long-term significance. On the other end of the continuum is spatial information that only has temporary significance. Where your golf ball landed in the rough is only important until the errant ball has been located. Once the next shot is

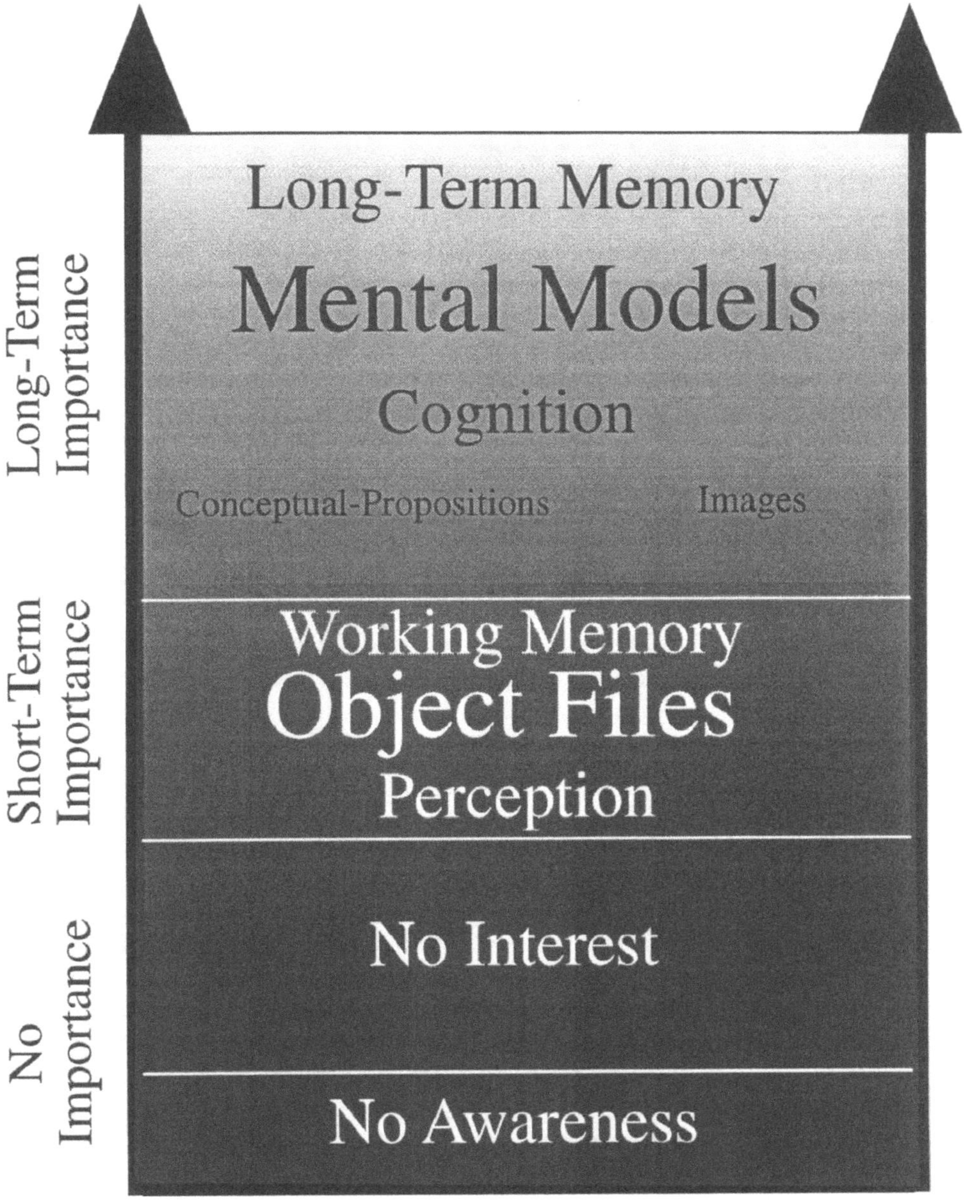

Figure 4.1: Object files are created by perceptual process as temporary structures for objects in working memory and mental models are created to encode spaces in long-term memory.

played the ball's specific temporary location loses significance. You may, however, have been able to encode some generalized knowledge from the experience, e.g., the rough is a bad location. This chapter discusses two theoretical cognitive structures that have been proposed for encoding spatial information. The first, an object file, is used to encode and update information about an object in our immediate environment that has temporarily attracted our attention, but will be forgotten when our attention shifts to other interesting objects. The other, a mental model, is used to encode and update information about spaces that have objects. Mental models are stored in long-term memory so that they might be used at a later time (Figure 4.1).

4.2. Object Files

Object files were first discussed as temporary episodic representations of real world objects by Kahneman and Treisman (1984). They argued that object files are separate and distinct from representations stored in long-term memory that can be used for identifying and classifying objects. They wished to make a distinction between the more specific and temporary object files, which they called *tokens,* and the more generalized and permanent structures used to label the object's identity, which they called *types*. For example, suppose you are attempting to find your car in the airport's parking lot after returning from a week-long trip. There are a number of objects (cars) in the space (parking lot). As you scan the space you open an object file for each car in view and store information about the object in the file. Included in a specific object file might be information on the relative location, color, shape, and size of that object. These are *tokens* in the immediate environment and one of the *tokens* is your car. You have viewed your car as an object many times and have encoded generalized information about that object (car) in your memory. This is a *type* because it represents all the views of your car as an object not any particular one. A *type* is a generalized representation in long-term memory and a *token* is a specific instance currently being viewed (Figure 4.2).

Kahneman, Treisman, and Gibbs (1992) argued that focusing attention on a particular object not only enhances the salience of its current properties, but also reactivates the recent history of the object. They called the process, which causes a current object to evoke an item previewed in a previous visual field, *reviewing.* The reviewing process maintains the perceived continuity of an object as it moves, changes characteristics, or momentarily disappears from sight by relating its current state to its previous states. The reviewing process facilitates recognition when the current and previous states of an object match, but hampers it when they do not match. If someone moved the traveler's car to a different location, or painted it a different color while he was away (Figure 4.2), he might find the car more difficult to identify. The car would still be the same object, but it would have changed some of its characteristics.

An object file is opened for use by the perceptual system once we have focused attention on the object. Information is put into the file to record the changes occurring to the object over time. This is necessary to maintain the continuity of a changing object. Suppose you are to meet a friend in the park at 11:00 AM and you have arrived a

Figure 4.2: A *type* refers to a representation of an object based on many views of the object stored in long-term memory and a *token* refers to a particular view of an object at a particular time.

little late. As you approach the prearranged location you see a person sitting with her back to you on a park bench. You open up an object file for the person sitting on the bench and assume it is your friend. Since you are still far away the retinal size of the object is rather small and other features are not distinct. As you approach the bench, the person turns around and you realize it is not your friend, but a stranger. Kahneman, Treisman, and Gibbs (1992) made a distinction between identifying and seeing an object. You opened up a single object file when you first saw the person sitting on the bench. You identified the object as your friend because you expected it to be your friend. As you approached the park bench you may have encoded other available information into the file such as the color of the person's coat or the presence of a hat. Instead of identifying the object by name let's say you saw an object and opened up object file X. You encoded the available information into the file and labeled it *my friend*. Throughout the episode, as you approached the park bench, you maintained a single object file X. Realizing you made an incorrect assumption, you now need to update the label for the file from *my friend* to *a stranger*. You did not need a second object file you just needed a new label.

The need to determine if there are two objects in the environment, or one object that has changed it characteristics, is a frequent choice we must make. Normally the perceptual system can maintain the continuity of a single file and keep it open or know to close the first file and create a second file with relatively little difficulty. We have learned the expected motions and behaviors of objects and use these expectations to make commonsense decisions about objects we encounter. If a friend wearing blue jeans and a yellow T-shirt went into an empty bedroom briefly and returned in a red dress, we would not think of the person as a new object. We would just update some of the information in the current object file. Suppose you are at a costume party and observe a person dressed like the Wolfman walk behind a wall and a person without a costume immediately emerge on the other side of the wall. Given the context, you probably would conclude the person without a costume was behind the wall initially and Wolfman stayed behind the wall. You would need two object files. Now suppose you are watching a horror film and a person goes behind a bush and emerges as the Wolfman. Assuming you know the premise of the plot, you would know it was intended that you think of him as the same object. The director of the film wanted you to create a single object file and to change the label and hair characteristics. "The identity of a changing object is carried by the assignment of information about its successive states to the same temporary file, rather than by its name or by its properties " (Kahneman, Treisman, and Gibbs 1992, p. 178). In other words, an object could change all of its characteristic and its label over time and still be considered the same object under some circumstances. For example, you watch a caterpillar disappear inside a cocoon and emerge as a butterfly.

4.2.1. APPARENT MOTION

Whenever an apparent change is detected in a viewed object, we have to determine if the current information related to the change, e.g., a new location or new shape, needs to be assigned to the existing object file or a new object file needs to be created. Kahneman, Treisman, and Gibbs (1992) used this notion to design perceptual experiments that focused on the predictions of object file theory. One experiment was based on the phenomenon of apparent motion. If a single object is presented in a display on a computer screen at location l_0 at time t_0, then removed, and replaced by a single object at location l_1 at time t_1, the object will appear to have moved under some circumstances. When l_0 and l_1 are sufficiently near each other in space and t_0 and t_1 are sufficiently close in time, a single object is seen to move smoothly from the original location to the new location. The object in the second stationary display must be available before the apparent motion can be experienced. This suggests the presentation of the object in the second display must initiate a *review* of the information for the object in the first display. If a plausible match occurs, then a single object is perceived and the new location of the object updated in its object file. If the object in the second display is sufficiently different, e.g., a large distance between l_0 and l_1 or a long time between t_0 and t_1, then a new object makes a sudden appearance and requires a second object file to be created.

For example, a hotair balloon with spiral markings could be presented at time t_0 (Figure 4.3a), then remove, and a new object, a smaller balloon with the same markings, presented at time t_1 (Figure 4.3b). A viewer should experience the apparent motion of a single object if the time interval was 300 milliseconds. The apparent motion would be to the right and away from the viewer because the object got smaller. If the initial display was the same (Figure 4.3c) and the second display quickly followed showing an object with checkered rather than spiral markings (Figure 4.3d) in virtually the same location, one might conclude their was only one object, which had perhaps rotated to reveal different markings on the back side. If the initial display showed the same spiral marked balloon (Figure 4.3e) and the second display followed after a 5 minutes interval with a small checkered balloon far from the initial location of the first balloon (Figure 4.3f), one probably would conclude a new balloon had appeared and open a second object file to encode it.

Kahneman, Treisman, and Gibbs (1992) proposed three distinct operations that are needed to provide perceptual continuity through change. First, when multiple objects are in a display, *correspondence* must be solved before apparent motion is experienced. The correspondence operation determines if each object in the second display is a new object or an old object in a new location. It should be interesting to geographers that

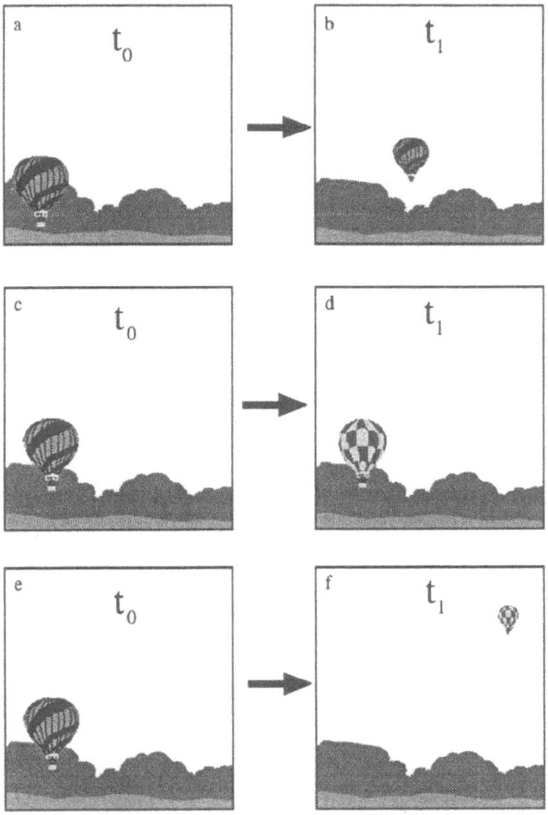

Figure 4.3: When an object is presented in an initial display followed by another object in a second display apparent motion occurs for a single object if the times and locations are in close proximity.

spatial and temporal constraints primarily control this operation rather than other characteristics, e.g., color, size, or labels. Spatial and temporal proximity must be maintained for apparent motion to be perceived. Changes in other properties of the object, however, can be accepted without a new object file being created. Second, reviewing must be completed to retrieve the characteristics for the object in the initial display that is no longer visible. There can be no apparent motion without this earlier information to provide the context for the change. Third, an *impletion* process uses the reviewed information from the first display and the current information to produce the perception of motion that links the first and second displays.

A number of experiments were performed with human subjects to test the object file theory (Kahneman, Treisman, and Gibbs 1992). A typical experiment had subjects view two successive displays. The preview display showed a number of small square objects inside a larger square frame. The small squares had letters in them to make them unique. The letter labels were then removed and the spaces were shown blank for an interval of time between displays. A second display was then presented with a single target letter in one of the small squares. The subject simply named the letter as quickly as possible. The target letter could take on one of three conditions. It could be the same letter in the target location or the same object (SO) (Figure 4.4), it

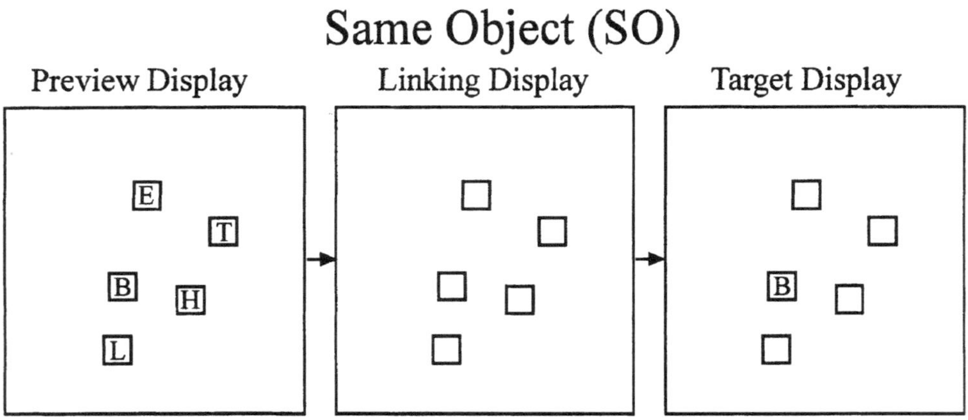

Figure 4.4: A sequence of displays that could be used to produce a Same Object (SO) trial. The location with the *B* in the preview display has the letter disappear and then reappear in the target display.

could be a different letter from the initial display in the target location or a different object (DO) (Figure 4.5), or it could be a new letter that was not in the first display in the target location or no match (NM) (Figure 4.6). In this simplest example the target object is connected to one of the objects in the preview field because it is in the same location. Other experiments connected the same object in preview and target displays through apparent motion and the tracking of a moving blank square. In all these cases it was predicted that there would be preview effects that would affect the time it would take for subjects to name the object in the target display.

The difference in the response times for SO and DO conditions indicated if the target could be named faster if it was connected to the same object in the preview display in some way such as having the same location or apparent motion. Kahneman, Treisman, and Gibbs (1992) called this an *object-specific advantage*. The difference in

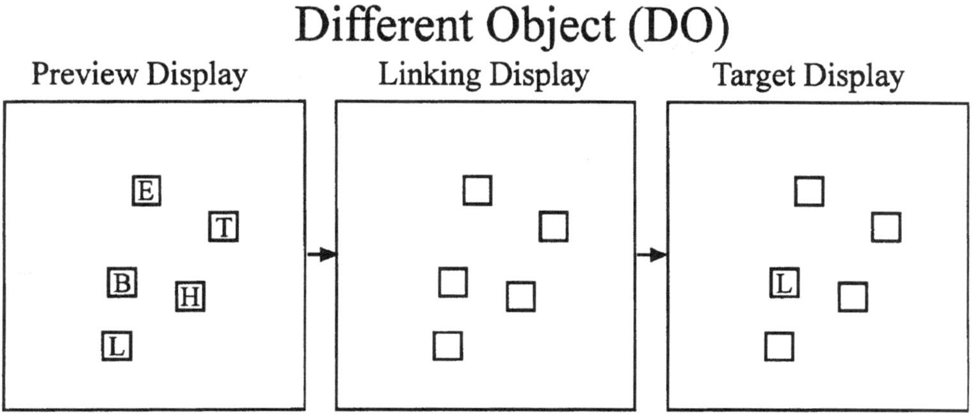

Figure 4.5: A sequence of displays that could be used to produce a Different Object (DO) trial. The location with the *B* in the preview display has the letter disappear and then is replaced by the previously viewed *L* in the in the target display.

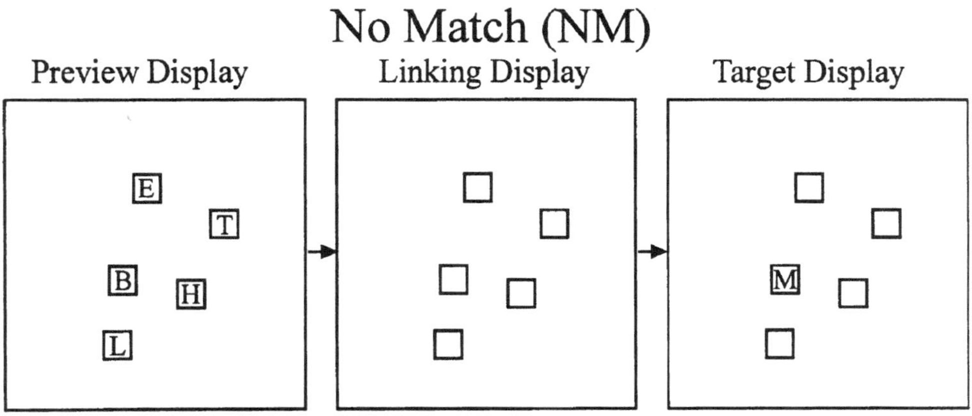

Figure 4.6: A sequence of displays that could be used to produce a No Match (NM) trial. The location with the *B* in the preview display has the letter disappear and then is replaced by the new letter *M* in the target display.

times for DO and NM conditions indicated if the target could be named faster if it had been viewed somewhere in the preview display, but not directly connected by having the same location or apparent motion. This was called a *nonspecific preview effect.* Results indicated an *object-specific advantage* by a significant difference in response times for SO and DO condition over all the experiments. The *object-specific advantage* was consistently produced by objects in displays that shared a common location, shared a relative position in a moving pattern, or appeared in the same moving frame. No consistent results were reported for a *nonspecific preview effect.* A second important result reported by the study was that the *object-specific advantage* sharply reduced as the number of objects in the display increased. This suggested there is a limited capacity for the working memory used to process object files. It was argued that focusing attention on the target automatically causes a reviewing process that selects one of the current object file to review. When the target and the previewed item matched, this reduced response times, and when they did not match, it increased response times. Three important characteristics were suggested for the reviewing process. First, it operates backwards from the last display to the first display. Second, it selects only a single object file to review. Third, it is guided by features of the object that control the unity and continuity of the object over time. Location was found to be a very important feature, but characteristics such as shape and color were found not to be important. This last result has important implications for cartographic animations that display geographic information. As long as the spatial behavior of the object over time suggests it is a single object moving through space, the object could change its other basic characteristics and still be considered the same object. This could be good and bad news because it also means that the perceptual continuity of an object might not be maintained by preserving other characteristics such as color and shape.

4.2.2. ANIMATED MAPS

Cartographers interested in producing animated maps should be interested in object file theory because it explains how map readers should respond to what they see on animated maps (Cammack 1995, Patton and Cammack 1996). Designers of animated maps should be aware of the cognitive process map readers use to view the animations. Such knowledge can point to the advantages and limitations of an animated presentation.

The basic experimental design originally used by Kahneman, Treisman, and Gibbs (1992) was adapted by Cammack (1995) to investigate map reading. He used normal-sized maps for states such as Vermont and Ohio in his experiments and map symbols designed to vary both shape and color. Two experiments presented preview and target maps with a number of map symbols in contexts similar to those illustrated in Figures 4.4, 4.5, and 4.6. In one experiment subjects responded by naming the shape of the symbol in the target display and in the other they responded by naming the color of the symbol. The results indicated no consistent *object-specific advantage* when the object on the target map was related to an object on the preview map by shape or color rather than by a letter name. This gives some support to the contention that changes in characteristics other than location are not of primary importance in maintaining object files. The size of the visual fields defined by the maps in these experiments was also a much larger area than the visual field in the displays used by Kahneman, Treisman, and Gibbs (1992). Cammack (1995) reasoned that his subjects may have had to use eye movements to cover the areas of the preview maps. If working memory has a capacity constraint, the eye movements may have resulted in a person abandoning object files that were currently being used to free resources for new objects. If this interpretation is correct, then cartographic animations that create many objects in a relatively large area may not be able to communicate a coherent message.

In an apparent motion experiment Cammack (1995) showed framed map symbols with different shapes to subjects on a preview map, removed the symbols and moved the empty frames across the map, and then presented a target map with one map symbol. The subjects named the symbol by its shape. The SO trials for this experiment were significantly faster than the DO trials producing the expected *object-specific advantage.* The results also indicated that movements that were short distances were named significantly faster than movements that were long distances and faster movements were named significantly faster than slower movements. This supports arguments that spatial and temporal proximity must be maintained for the apparent motion of a single object to be perceived.

Patton and Cammack (1996) related object file theory to processing *what* and *where* information from choropleth maps. They reasoned that the number of objects on a map was a measure of the map's complexity and that aiding a map reader's ability to chunk information increased his ability to encode the information into memory. Maps that are complex, i.e., have many objects, should be more of a challenge to remember under time constraints than simple maps because more object files must be opened and encoded. The number of objects was defined in this study as the number of faces of contiguous color present on a map. There should also be a greater affect on *where* information than on *what* information because *where* questions require specific *tokens* to be recalled while *what* questions only require *types* to be recalled. For a choropleth map, a *token* is a specific object that has a particular color and a *type* is the class the color represents. The choropleth map in Figure 4.7 has three classes or types (white, grey, and black), but it has six tokens or objects.

The authors had subjects consider a series of choropleth maps on a monitor for a fixed amount of time using either a regular static map or one of two versions of animated maps (Patton and Cammack 1996). After viewing a map on the screen for a fixed amount of time one of two types of questions was asked about the map. For a *what* question, a small square filled with a color was presented and subjects decided if that color had been on the previous map. For a *where* question, a map with one polygon filled with a color was presented and the subjects decided if that color had been in that polygon on the previous map. As hypothesized, the results indicated that accuracy decreased significantly more for *where* questions when the number of objects increased than it did for *what* questions (Figure 4.8).

4.3. Mental Models

For planning and problem solving we frequently need to make deductive inferences from general knowledge. The need to make valid deductions requires a knowledge structure that can provide the pertinent information. We need to understand the situation so we can process decisions and solutions. Johnson-Laird (1993, p. 15) argued we need "to establish a starting point, a set of observations founded, ultimately, on perception, verbal description, or imagination." To visualize the situation, e.g., a geographic space, we need a three-dimensional model of *what* things are *where* in a visual scene (Marr (1982). Our comprehension of the model allows "us to envisage situations that we have not yet perceived and perhaps never could perceive " (Johnson-Laird 1993, p. 15).

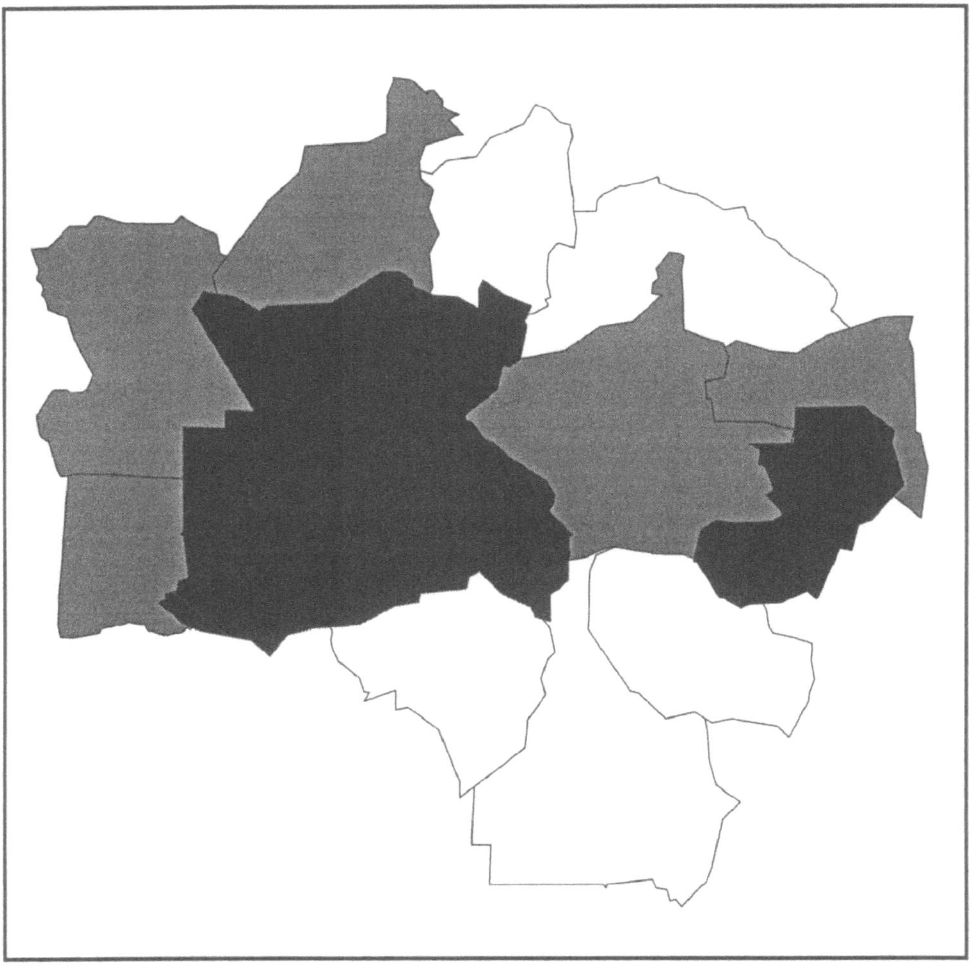

Figure 4.7: A choropleth map with individual polygons filled with three colors to represent classes of data. There are classes of *types* to be remembered, but there are six objects or *tokens* on the map that need to be encoded in object files.

Some mental models contain tokens that are entities in the real world (physical models) while others have tokens that are abstract concepts (conceptual models). Although geographers are often interested in conceptual models, e.g., economic development, discussions here will focus on physical models. Johnson-Laird (1983) identified six major types of physical models. A general *relational model* is a static frame that has a fixed number of tokens representing physical entities. It also has a fixed number of characteristics for the tokens that represent the physical properties of the entities. Also encoded are the relationships among the tokens that represent physical relationships among the entities. A *spatial model* is a *relational model* that only encodes the physical relationships among tokens as spatial relationships. This type of mental model locates the tokens in a two (or three) dimensional space with metric properties.

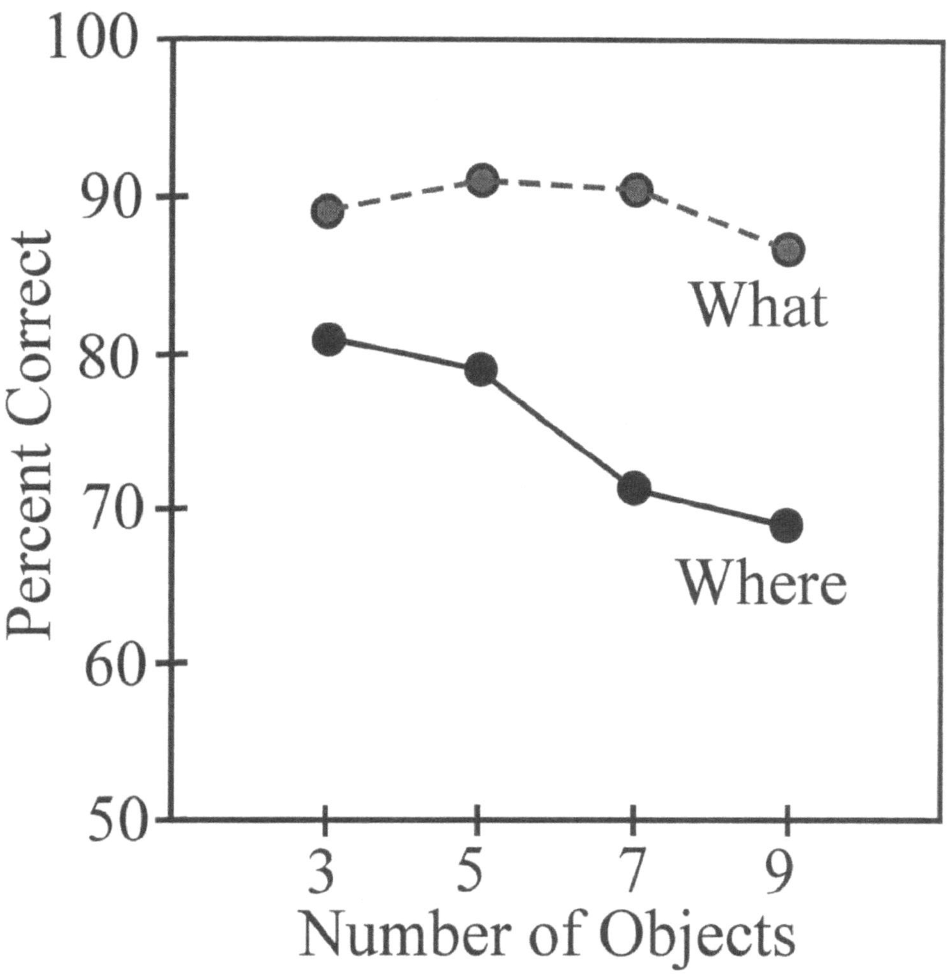

Figure 4.8: The accuracy for answering *what* and *where* questions associated with the number of objects on maps. After Patton and Cammack (1996).

Distances between locations would have to be accurate enough to satisfy the triangular inequality assumption of metric spaces. The distance between any two points must always be less than the sum of the distances between the two points and any third point. A *temporal model* consists of a set of spatial frames that have a fixed sequence that corresponds to the temporal order of events. Although the order of the frames is fixed the temporal order may not correspond to real time. A *kinematic model* is a psychologically continuous *temporal model* that depicts changes and movements of entities in a continuos fashion that can correspond to real time. A *dynamic model* is a *kinematic model* that also encodes relationships between frames representing the causal relations between the depicted events. An *image* is a mental model whose source is visual imagination. It presents a viewer-centered representation of the visible characteristics for a *spatial* or *kinematic model.* It is, therefore, a particular view of the objects represented in a particular model that has been generated. To keep the discussion simple, the general term mental model will be used in the following discussion of cognitive structures that represent geographic environments. In some cases mental models might be encoded using perceptual processes by navigating through an environment or by map reading. Since detailed information can be supplied by navigation and map-reading processes, these mental models could be *spatial model* with metric properties and provide accurate information. In other cases the mental model of a geographic environment could be encoded only by reading or hearing verbal descriptions. The verbal descriptions would have to be coherent and detailed for these mental models to be encoded as *spatial models* of the environment as defined by Johnson-Laird (1983) as providing metric information.

4.3.1. BASIC PRINCIPLES

The nature of mental models that represent geographic environments can be expressed in three basic principles developed by Johnson-Laird (1993). First, each object in the environment is represented by a corresponding token in the mental model. Since geographic spaces can be represented at various scales, it is assumed that a particular model is at an appropriate scale and that only important objects relevant at that scale will be encoded in the model. Second, the characteristics of objects are represented by the properties of their tokens. These properties include any characteristics of the objects that are relevant to modeling the environment. Third, relations among objects are represented by relations among the tokens. The spatial location of tokens in the model corresponds to the spatial locations of objects in geographic space.

Mental models organize and store spatial information as cognitive structures in our memory. Cognitive maps of environments that we have been encoded from both verbal and visual information can integrate this information and store it as mental models that are thought to be perspective-free structures. This means that a person who learned information about locations in an environment that was presented in either a map or route perspective could later process the information equally well in both perspectives. This suggests that the original verbal and/or visual information has not just been memorized. It has instead been transformed into a cognitive structure that efficiently models the environment.

4.3.2. ORIENTATION

We frequently need to solve problems that are related to spatial orientation. One such problem is the identification of objects in an environment. Bryant, Tversky and Franklin (1992) have identified two categories of spatial orientation problems. The first is related to a person who is viewing an object from an *external* perspective. This perspective requires the person to view the object from a fixed vantage point that is detached from the object. For example, a wall map represents an environment, but is viewed as an object from an external perspective (Figure 4.9). The key to solving this

Figure 4.9: A person looking at a map with six landmarks hanging on a wall. The person is viewing the map as an object from an external perspective.

identification problem is the orientation of the object. The viewer may or may not be able to process the information associated with the object efficiently when the object has a particular orientation. Some orientations may allow for an efficient identification and others may be more difficult. For example, identification processes done with cartographic maps presented with north at the top have be shown to be faster than when the maps are presented with south at the top (Conerway 1991; Lloyd and Cammack 1996; Lloyd and Steinke 1984).

A person viewing objects from an *internal* perspective is a second type of spatial orientation problem (Figure 4.10). The viewer and objects are part of the same space for this type of problem. The viewer's orientation within the space relative to the locations of the objects is the key issue for solving this problem. With an *internal* perspective it is the viewer that changes orientations while the absolute positions of objects remain fixed. Given the particular orientation of the viewer, some objects in the space might be easier to identify and others more difficult. An example of this type of problem is a person navigating within an environment like a city and identifying landmarks (Golledge, Dougherty, and Bell 1995; Lloyd and Cammack 1996; Lloyd, Cammack, and Holliday 1995).

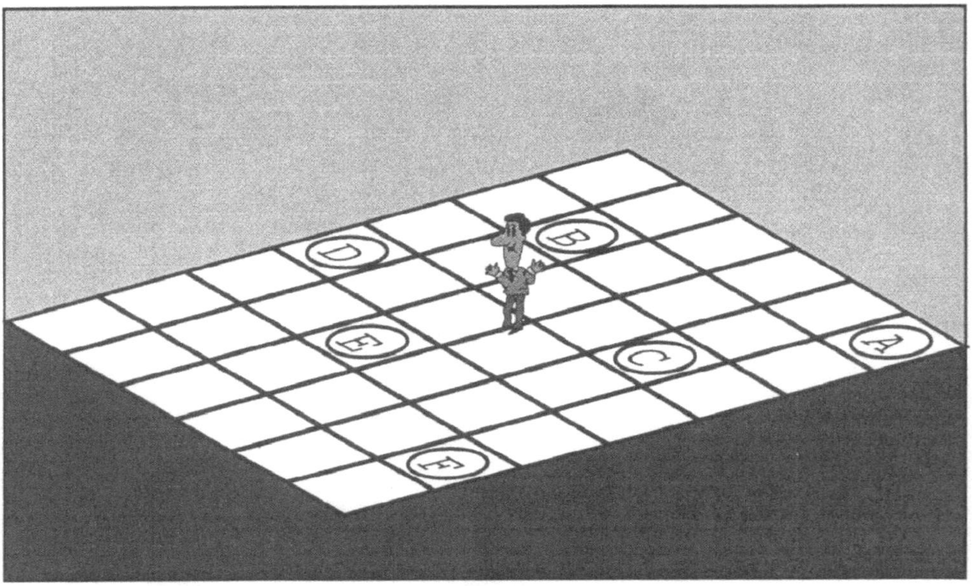

Figure 4.10: A person in a space with six landmarks viewing them from an internal perspective.

The *external* and *internal* orientation problems discussed above are clearly part of larger map reading and wayfinding processes. We do not fine it difficult to identify most familiar objects in our everyday environment no matter what their orientation is in space. These objects are encountered in a variety of orientations and have a familiar look to them from all angles. Objects like your automobile, children, or fishing pole can be identified without hesitation because you have encountered them many times and in many different orientations. Learning about a particular object (token) also gives you information about that category (type) of object (Lloyd, Patton, and Cammack 1995). Having learned the nature of a category of object you can also easily identify an exemplar object as being in that category (automobile or fishing pole) even the first time you view that particular instance. The object you are viewing is compared with a hypothesized representation of the object in your memory to make the identification. The hypothesized object in represented in your memory might be the prototype for a category of objects that has been encoded by experiencing many exemplars of that category or it could be a specific object you have previously encountered (Medin, Altom, and Murphy 1984; Lloyd 1994). A sufficient match between the object being viewed (token) and the representation in memory (type) allows you to associate the name of the representation in memory with the object being viewed. The identification might be for a specific case, e.g., your daughter Beth as opposed to any other female child, or it may be for a category, e.g., a girl as apposed to a boy. Lowe (1987a) has offered a theoretical explanation of our ability to recognize familiar objects based on our knowledge of *nonaccidental features* associated with the object. He defines nonaccidental features as any features that are not a product of a particular orientation being viewed. A similar idea was presented by Gibson's (1950) in his discussion of *invariant properties* that are not changed by a series of transformations. Features like a giraffe's neck or the design of a baseball field, that are consistently distinctive over a range of orientations, make identification easy regardless of the object's orientation (Tversky 1977). Cape Code, the panhandle of Oklahoma, or the "thumb" of Michigan are distinctive shapes found on maps that might be expected to produce similar effects (Lloyd and Steinke 1984; Steinke and Lloyd 1983a).

4.3.3. ENCODING THEORIES

As discussed in the previous chapter, cognitive maps encoded from *external* and *internal* perspectives appear to be vulnerable to different types of orientation biases (Evans and Pezdek 1980; Hintzman, O'Dell, and Arndt 1981; Lloyd and Cammack 1996; Sholl 1987). It has also been suggested by other studies that cognitive maps encoded from cartographic maps provide better information for estimating straight-line distances and cognitive maps encoded through navigation provided better information for estimating route distances (Thorndyke and Hayes-Roth 1982; Lloyd 1989b). The processes we use to encode spatial information would not only appear to have an impact on how information is stored in our memory but also on how efficiently the information can be used to solve specific problems.

The various ways spatial information can be encoded into human memory can be used to explain some of the above results (Lloyd 1982). Theoretical arguments have been presented that propose cognitive representations of spatial environments are stored as images (Kosslyn 1980), as conceptual-propositions (Pylyshyn 1984), and as both (Pavio 1971). Deciding which theory best explains mental representations and processes may be impossible (Anderson 1978). The information processed by studying maps is more likely to be stored as images in our memory because maps are depictions of spatial information. Kosslyn (1980) has argued that images are functionally equivalent to the objects they represent. This implies that processing an internal image of a map in one's memory is very much like processing the actual map with your visual system. Map images, like cartographic maps or pictures, can be parallel processed, but, like cartographic maps or pictures, have a fixed orientation. The navigation of a route can be described in verbal terms as the process one uses to go from an origin location to a destination location. Conceptual propositions that represent routes in our memory are serially processed like written process descriptions, and have no fixed orientation (Pylyshyn 1984). This conventional view of encoding processes considered visual and verbal information to be stored separately and processed by different cognitive systems. A more recent view has considered visual and verbal information to be integrated and stored in a common internal structure.

A mental model is a cognitive structure that has been encoded to represent one's spatial knowledge of an environment. Johnson-Laird (1983, p 165) has proposed a theoretical view with three types of mental representations: "propositional representations which are strings of symbols that correspond to natural language, mental models which are structural analogs of the world, and images which are the perceptual correlates of models from a particular point of view." An important practical characteristic of a mental model of an environment is that it represents a person's learned knowledge of the environment at a point in time. This allows for the mental model to be continuously updated by new information provided by additional experiences with the environment. Another important argument in Johnson-Laird (1983) theory is that either primary experiences, e.g., navigating through the environment, or secondary experiences, e.g., reading descriptions, can provide information for constructing a mental model. Fictional environments encoded from secondary information are just as useful for producing mental models as descriptions of real-world locations, e.g., Krypton. "Human beings can evidently construct mental models by acts of imagination and can relate propositions to such models" (Johnson-Laird 1983, p. 156). The spatial information provided by language describing real or imaginary environments is encoded into mental models of these environments. For example, we might read about city X being east of city Y or that building X will be on your left if you travel from your current location toward building Y. A number of recent studies have demonstrated that mental models of environments can be constructed from verbal descriptions (Bryant, Tversky, and Franklin 1992; Byrne and Johnson-Laird 1989; Ferguson and Hegarty 1994; Franklin, Tversky, and Coon 1992; Taylor and Tversky 1992a, 1992b). That mental models can be encoded by viewing maps or navigating in an environment has also been demonstrated (MaNamara, Halpin, and Hardy 1992; Taylor and Tversky 1992a, 1992b). Johnson-Laird (1983) argued that images we experience of environments correspond to specific views of a mental model. Viewing the same mental model from multiple perspectives results in multiple images of the environment. Mental models may be encoded through perception or imagination, but the specific images we experience represent perceptible features of real-world objects. Taylor and Tversky (1992b) contended that a mental model can not be viewed as a finite entity. They suggested that an architect's three-dimensional model of a city is a good metaphor for what mental models are like. It is possible to create images to provide particular views from specific perspectives, but one cannot visualize all of the space at the same time. Our ability to visualize environments from multiple perspectives supports arguments that images are created from mental models and has been demonstrated for spatial environments by a number of studies (Bryant and Tversky 1992; Franklin, Tversky, and Coon 1992).

It has been convincingly argued that mental models can be created spontaneously when reading a text that describes an environment. This ability is related to the notion that mental models integrate spatial and verbal information into one structure (Glenberg and McDaniel 1992; McNamara, Halpin, and Hardy 1992). Researchers interested in reading skills have also argued that mental models enhance the comprehension of the information in the text (Glenberg and Langston 1992; Waddill and McDaniel 1992; Wilson, Rinck, McNamara, and Bower 1993).

How does one know if someone has used visual and/or verbal information to create a mental model of an environment? This is not a trivial question. The key evidence is related to people's ability to use mental models to make inferences about relationships between objects in the environment.

4.3.4. INFERENTIAL SPATIAL KNOWLEDGE

Johnson-Laird (1983) argued that humans use mental models to make two types of inferences. It takes conscious effort to make *explicit* inferences. They depend on "searching for alternate models that may falsify putative conclusions" (Johnson-Laird 1983, p. 144). Byrne and Johnson-Laird (1989) provide an example of inferences with a simple two-dimensional problem. Given the spatial relations of block objects for the first problem:

B is on the right of A,
C is on the left of B,
D is in front of C,
E is in front of A,

hence, landmark D is to the left of landmark E.

If you formed a model of the relationships as shown in Figure 4.11, you would conclude the inference statement is true. If you search for another model that expresses the same relationships as shown in Figure 4.12, you would conclude the inference statement is false.

Implicit inferences are the bases of "intuitive judgment and the comprehension of discourse." They tend to be "rapid, effortless, and outside conscious awareness" (Johnson-Laird 1983, p. 127). *Implicit* inferences depend on the automatic construction of a single mental model. Consider another two-dimensional problem again based on Byrne and Johnson-Laird (1989). Given the spatial relations of block objects for the second problem is:

A is on the right of B,
C is on the left of B,
D is in front of C,
E is in front of B,

hence, landmark D is to the left of landmark E.

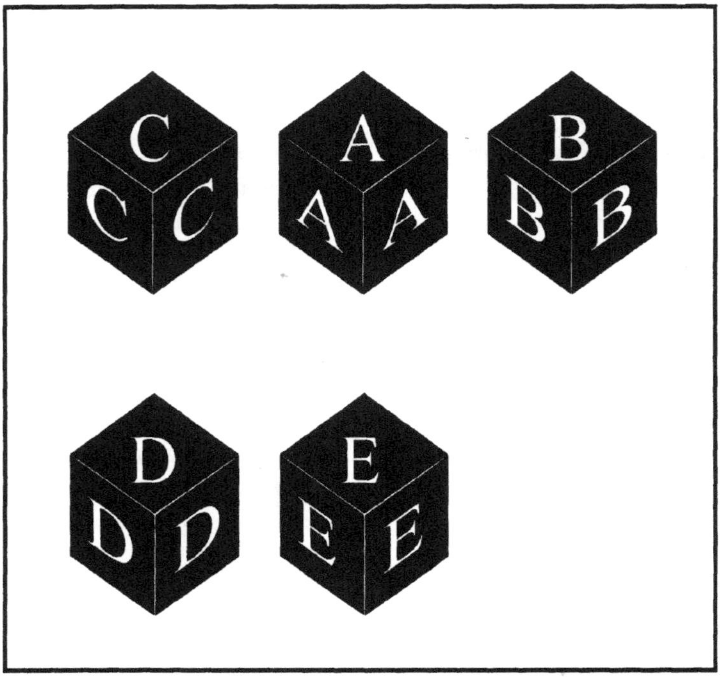

Figure 4.11: A mental model that represents the relationships among the block objects as described in the first problem that indicates the inference statement is true.

Figure 4.12: A mental model that represents the relationships among the block objects as described in the first problem that indicates the inference statement is false.

A model of the relationships would be like the one shown in Figure 4.13. No other model can satisfy the spatial relationships provided in the description.

It has been suggested by some researchers that describing environments in route terms (ahead, behind, left, and right) produces better mental models than describing environments in survey or map terms (north, south, east and west). Perrig and Kintsch

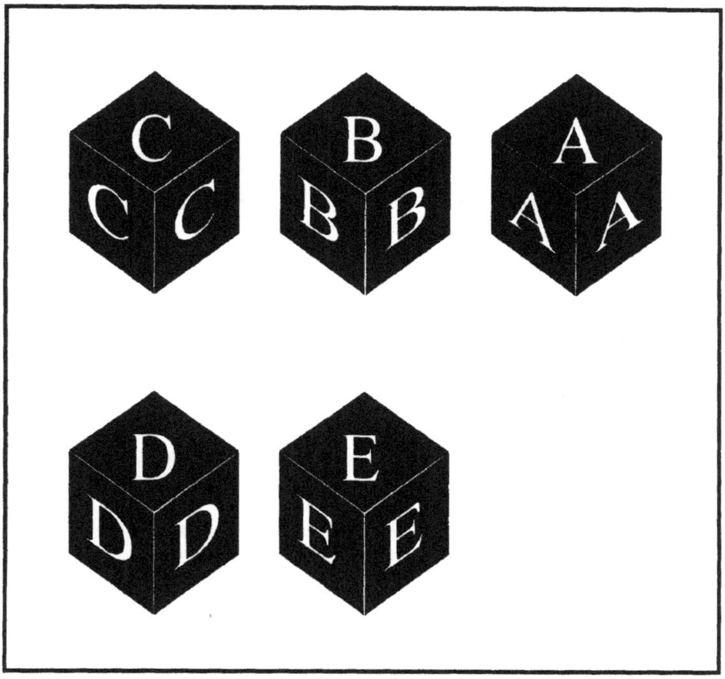

Figure 4.13: A mental model that represents the relationships among the block objects as described in the second problem that indicates the inference statement is true.

(1985) reported that a text comprehension was better for subjects studying route texts over survey texts. Ferguson and Hegarty (1994) reported that subjects who read route texts constructed more accurate sketch maps than subjects who read survey texts, but that these differences were eliminated if a map accompanied the texts. Texts that described an environment were read by their subjects for as long as they needed "to learn the information well enough to be able to describe the area to someone else" (Ferguson and Hegarty 1994, p. 457). Sketch maps of the learned environment were drawn by subjects and scored for accuracy. The study demonstrated that *anchor* features were recalled more accurately than secondary features (Couclelis, Golledge, Gale, and Tobler 1987). The *anchor* features were all linear features (a river, street, and highway) rather than point features. This is possibly because the *anchor* features were mentioned in the text more frequently than secondary features. It is also possible the mental models of some subjects may not have been completely encoded because some secondary features had location accuracies lower than 50 percent.

Mani and Johnson-Laird (1982) argued that readers of texts do not always encode mental models using the spatial information in a text. The contended mental models were encoded if text descriptions were spatially determinate, but were not encoded if text descriptions were not spatially determinate. They defined spatially determinate descriptions as texts that are sufficiently well organized and coherent to yield a single model and spatially indeterminate descriptions as texts so unorganized and incoherent that they are ambiguous. Incoherent descriptions prevent a clearly defined mental model from being created. Mental models effectively represent spatial environments because they are very flexible structures. One can use the encoded mental model for a variety of tasks because the model integrates visual and verbal information. This means a single mental model can create images for multiple views of an environment (Franklin, Tversky, and Coon 1992) and also be used to determine the route between two locations in an environment (Taylor and Tversky 1992a).

Taylor and Tversky (1992b) argued that mental models of environments should not be tied to a single perspective. This means the mental model will not be biased by the perspective of the initial learning experience. To test this notion, some subjects in their study read descriptions of naturalistic environments that were presented as route descriptions that provided an *internal* perspective with the viewer described as driving or walking around in the environment. Their route descriptions used egocentric terms such as *ahead*, *behind*, *left*, and *right* Other subjects read survey descriptions of the same environments that provided an *external* perspective with the environment having a fixed orientation and the viewer outside and above the environment. Survey descriptions used map terms such as *north*, *south*, *east*, and *west*. Both route and survey subjects verified True/False verbatim and inference statements made from both perspectives and drew sketch maps of the environments. Subjects could know the answer to verbatim statements directly from the text describing the environments. Subjects could not know the answers to inference statements directly known from statements in the text, but could infer the answers from information in the text.

Sketch maps of the environment were successfully drawn by both route and survey subjects. Responses to verbatim statements were faster and more accurate when the statements were presented in a perspective corresponding to the perspective in the text the subjects had read. This result indicated that verbatim statements were verified by subjects against a representation of the text. Subjects in both groups, however, were as fast and accurate verifying inference statements presented in either a route or survey perspective. The authors argued that subjects used a mental model to verify inference statements because they could not directly compare inference questions to a representation of the text. This evidence, that inference statements were responded to with equal efficiency even when a statement's perspective did not correspond to the perspective represented by the text, was used by the authors to argue that mental models have no perspective. Subjects who studied maps of the environments responded to similar statements and produced similar results. It would appear that people can construct mental models from either reading about environments or looking at maps.

Lloyd, Cammack, and Holliday (1995) performed experiments with human subjects that had them experience a hypothetical city in four different ways. The city (Figure 4.14) had ten landmarks that were buildings association with common activities

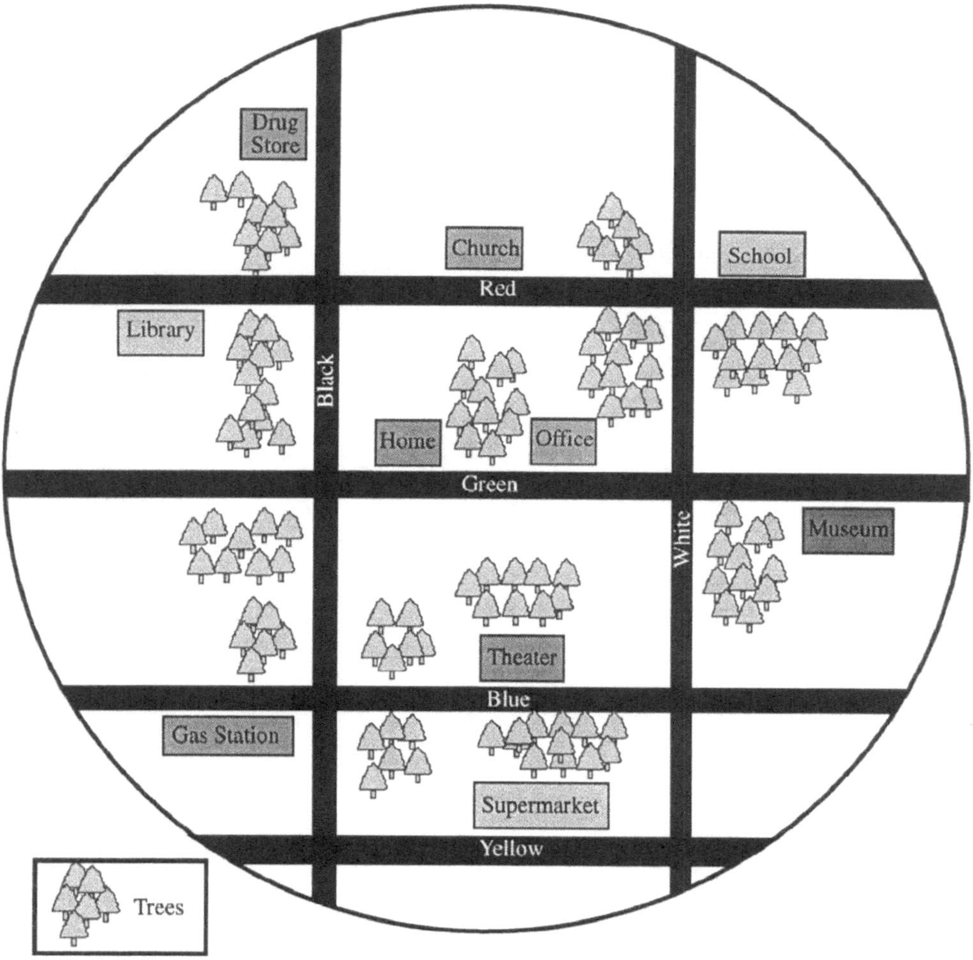

Figure 4.14: The hypothetical city used to investigate mental models. After Lloyd, Cammack, and Holliday 1995).

and six streets that were named for colors. Two groups of subjects had visual experiences to learn about the city. One group of subjects learned from an external perspective by experiences with maps that traced the routes between the home or office landmark and the other landmarks. A second group of subjects learned about the city from an internal perspective by viewing animations that simulated moving along the streets over the same routes. Two other groups of subjects had verbal experiences to learn about the city. One of these groups read descriptions of the same route from an external map perspective that used terms like north, south, east, and west. The other verbal groupsread descriptions presented in an internal route perspective using terms like ahead, behind, left, and right. After subjects learned the routes and passed a test to verify they had encoded knowledge of the city, they were given statements to verify about the relationships among the landmarks. Some of the statements were presented in the perspective they had experienced to learn the routes (map or route) and other were presented in the alternate perspective. At the end of the experiment all subjects drew sketch maps that were found not to be significantly different. This suggested that subjects could learn equally well from a variety of experiences that provided internal and external perspectives and visual and verbal information. Analysis of reaction times and accuracy for the verification task suggested that all subjects had encoded a cognitive map of the city that was a mental model. Subjects were able to verify statements with equal efficiency when the statements were presented in the perspective they had experienced when learning about the city or in the alternate perspective.

4.4. Conclusions

This chapter has discussed two theories that account for how spatial information is stored in memory. Object file theory (Kahneman, Treisman, and Gibbs 1992) explains how we keep temporary track of information about objects encountered in the environment or in a map animation. Johnson-Laird 's (1983) mental model theory explains how we integrate verbal and visual information into a model of the environment. The next chapter discuss theories of visual search that explain how we find objects in space.

CHAPTER 5

SPATIAL SEARCH PROCESSES

5.1. Introduction

Most spatial problems are solved by coordinating multiple tasks. Looking for something in a visual scene is a simple example frequently experienced by people. Lets say you are attending a football game and you know a friend is also at the game. You have an image of your friend encoded in memory and you are searching the stands for a match for that image. What variables make this example an easy or difficult problem? Some variables relate to what is in your memory. Can you construct a good image of what your friend might look like? Did you last see your friend at breakfast or in elementary school? Having accurate information about your friend's appearance would be a great help in the search. This is called *top-down information* because it is information in your memory. Does your friend have any unique features that would make her stand out from others in the crowd. Red hair would be more unique than brown hair and green hair would be even better. If she was seven feet tall and had green hair your search would be truly blessed. The information provided by the environment, the colors, shapes, sizes, etc., compete for your attention. This is called *bottom-up information* because it is information being perceived by your visual system. Your attention can be directed by focusing on important features. You might use size and shape features to focus attention on people who are female and away from those that are male. Your friend may attract your attention immediately if she has something unique about her appearance. You may have to scan each row in the stadium if she looks like most other fans at the football game.

5.2. Cognitive Theories of Search

This chapter is concerned with cognitive theories that explain how humans visually process information during a search process. The search may be needed to provide information for a larger task a person has undertaken. For example, one might need to search for a particular landmark during wayfinding or find the location of an important map symbol during map reading. Separate search processes may be initiated and completed multiple times during one of these larger tasks.

5.2.1. FEATURE INTEGRATION THEORY

Treisman's (1988) feature integration theory of attention (FIT) explains how spatial searches are conducted. Treisman's basic theory and other theories related to FIT provide a foundation for understanding how people search environments and maps (Duncan and Humphreys 1989; Treisman 1991; Wolfe, Cave, and Franzel 1989).

Geographers at this point in time have done few search studies and all have used maps and experimental research designs. This chapter will, therefore, focus on searching maps. The theoretical concepts discussed, however, should also apply to searching the immediate environment.

Map readers experience information on a number of separable dimensions (Garner 1974; Shortridge 1982; Dobson 1983). A *dimension* is "the complete range of variation that is separately analyzed by some functionally independent perceptual subsystem" (Treisman and Gelade 1980, p. 99). Examples conveniently divided into spatial properties (location, size and orientation) and object properties (color, texture, and shape) based on subsystems in the brain (Kosslyn and Koenig 1992). A *feature* is "a particular value on a dimension" (Treisman and Gelade 1980, p. 99). Examples of features would be top, small, horizontal, red, coarse, and round.

Targets that Pop Out of Maps

The earliest version of FIT made a simple distinction between a first stage in the visual search process that occurred before focused attention occurred and a second stage that required focused attention. It was argued that, during a preattentive stage, features are encoded "early, automatically, and in parallel across the visual field," while objects, which are constructed from features, are identified at a later stage in the search process using focused attention (Treisman and Gelade 1980, p. 98). Specialized modules for dimensions such as shape, orientation, color, and size are assumed to be automatically encoded (Figure 5.1). "Each module forms different feature maps for the different values on the dimension it codes - for example red, blue, and green within the colour module, vertical, diagonal, and horizontal with the orientation module" (Treisman 1988, p. 203). The separate feature maps allow the automatic detection of any target that has a unique feature on one of the maps. If the target had a triangular shape and all other objects had circular shapes, the target would "pop out" of the display (Figure 5.2). The search time will be independent of the number of distractor objects on the map because the target has a unique feature.

Researchers have subjects search displays with a varied number of distractors to demonstrate this effect. When the target pops out of the display (map), regressions using reaction time as the dependent variable and number of distractors as the independent variable theoretically should have a slope equal to zero (Figure 5.3). Spatially parallel searches are indicated by such a flat slope (Lloyd 1988). According to FIT, unique feature targets can be detected without focused attention during the preattentive stage of processing. One could know a target was on a map without knowing its spatial location. Treisman (1988, p. 203) argued "these separate [feature] maps allow the detection of targets with a unique sensory feature, simply from the presence of activity in the separate maps for that feature." Focused attention is needed to locate the target and conjoin its separate features, e.g., shape and color (Cohen and Rafal 1991). Information from the individual modular feature maps is automatically retrieved when the attention spotlight is focused on a location on the master map (Figure 5.1). The bottom-up

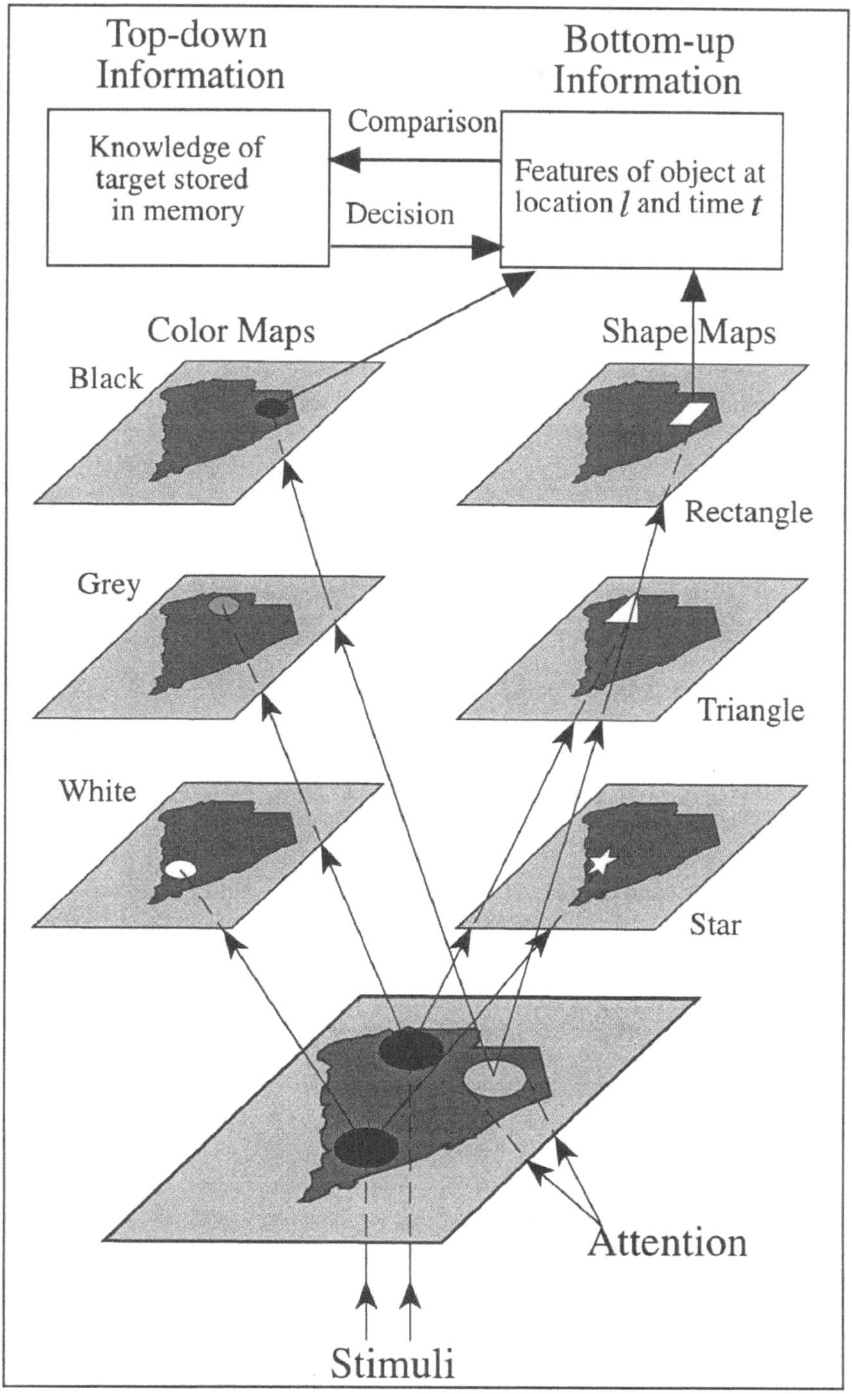

Figure 5.1: Feature maps for different dimensions like color and shape are available for use without focused attention. If a target object has a unique feature it will pop out of a display map. After Treisman (1988).

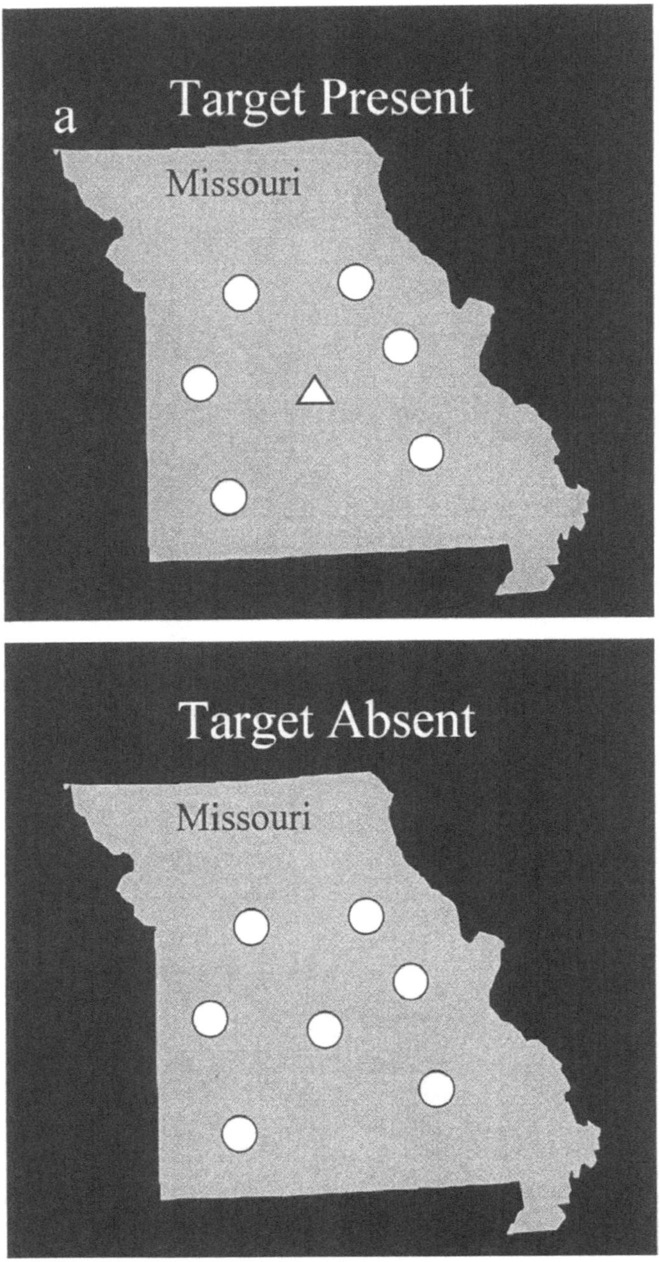

Figure 5.2: Maps for a feature search based on shape. Some maps would have the target (triangle) present (a) and some would have the target absent (b).

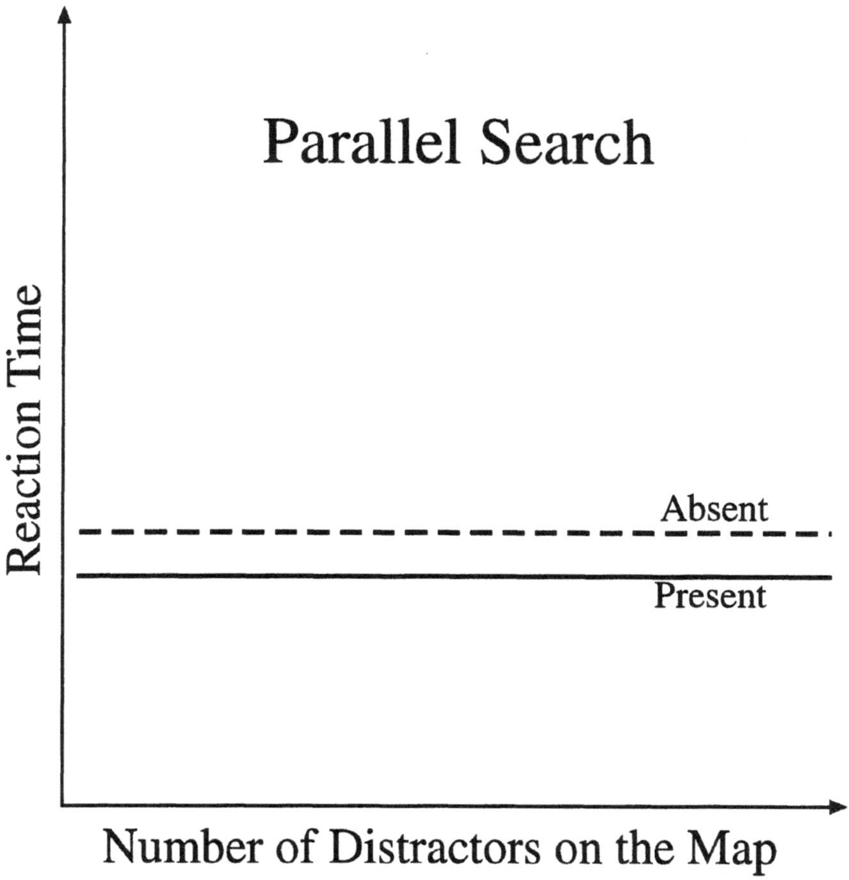

Figure 5.3: A feature search that can be parallel processed will have the target pop out of the map. For both target present and target absent responses regressions between reaction times and number of distractors will have flat slopes.

information about the object at that location can then be compared with the top-down information known about the target (Figure 5.1). A match makes us conscious of the target and its location. A difference makes us conscious of a distractor and its location. It also tells us to shift attention to a different location and continue the search.

FIT argues a target will not pop out of a display when the target shares features with distractors, i.e., a target does not have unique features. This is called a conjunctive search because the target is linked to the distractors through the shared features. An example of a conjunctive search defined by shape and color would be searching for a black triangle among black circles and white triangles (Figure 5.4). One could not be sure the target was present during the preattentive stage because objects at all locations are either black or a triangle. Neither the color feature maps nor the shape feature maps could offer unique information and focused attention on each location would be required to search for the black triangle. Regressions with reaction time as the dependent variable and the number of distractors as the independent variable should produce slopes

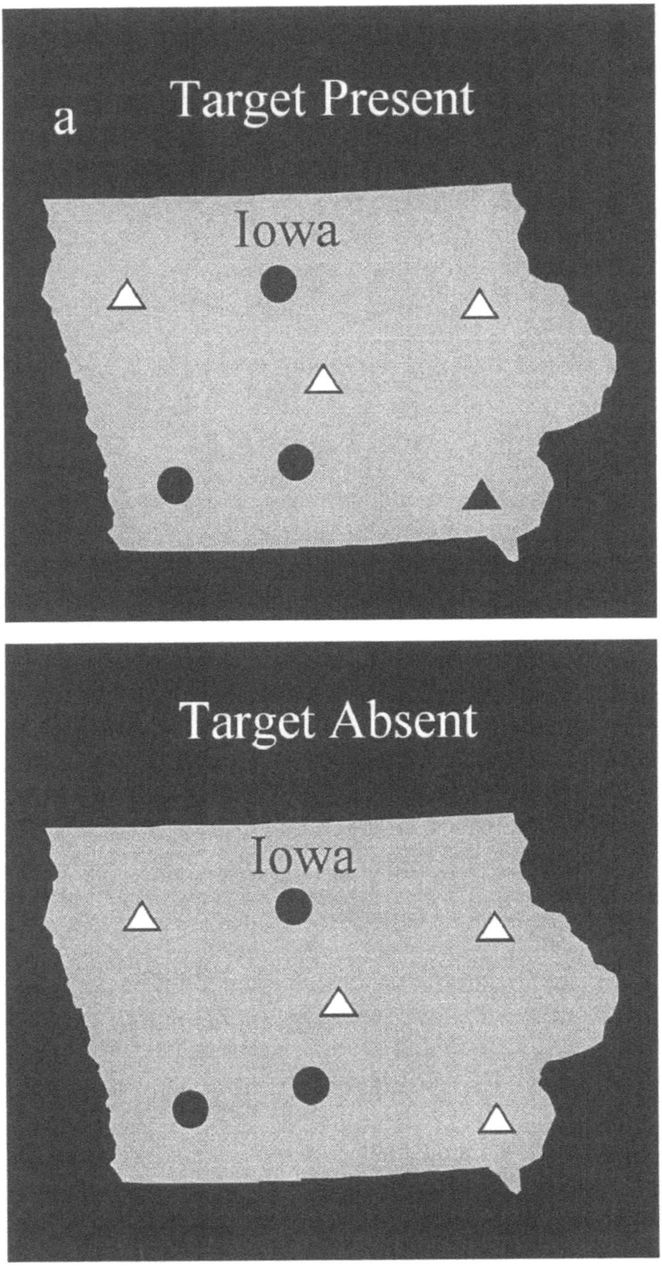

Figure 5.4: Maps for a conjunctive search based on color and shape. Some maps would have the target (black triangle) present (a) and some would have the target absent(b).

significantly different from zero (Figure 5.5). All the potential locations would need to be searched for an absent target and, on the average, half the locations for a target present search of a typical cartographic map. This type of search is called a serial self-terminating search because the searcher stops searching once the target is found (Lloyd 1988). Serial self-terminating searches have regression slopes for target present answers that are half as steep as regression slopes for target absent answers.

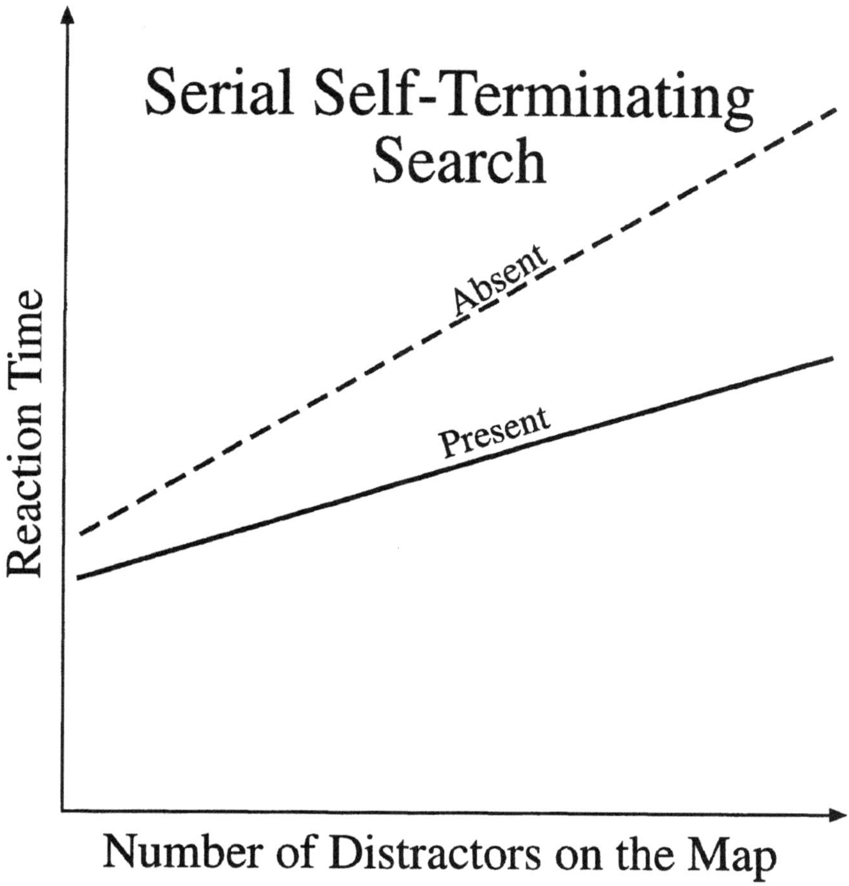

Figure 5.5: A conjunctive search that must be serially processed will not have the target pop out of the map. For both target present and target absent responses regressions between reaction times and number of distractors will have positive slopes. Slopes for target absent responses should be approximately twice as steep as for target present responses.

Like most theories, the original FIT has been modified to accommodate unusual search situations (Treisman and Gelade 1980). Not all searches for a single feature are parallel searches. The pop-out effect "for a target defined by a single distinctive feature is mediated by the unique activity it generates in the relevant feature map" (Treisman 1988, p. 205). The distractors might share relevant features to a degree with the target when the dimension is a continuous variable. Small differences among features may not be sufficient to differentiate between the target and all the distractors. This is true when both the target and distractors are lines differing only in length. A serial search process is necessary even though the target and distractors do not directly share features

(Treisman and Gormican 1988). Finding the smallest graduated circle on a map should be a serial search process (Figure 5.6a) unless the smallest circle is much smaller than any other circle (Figure 5.6b). FIT also predicts a serial search should occur when the

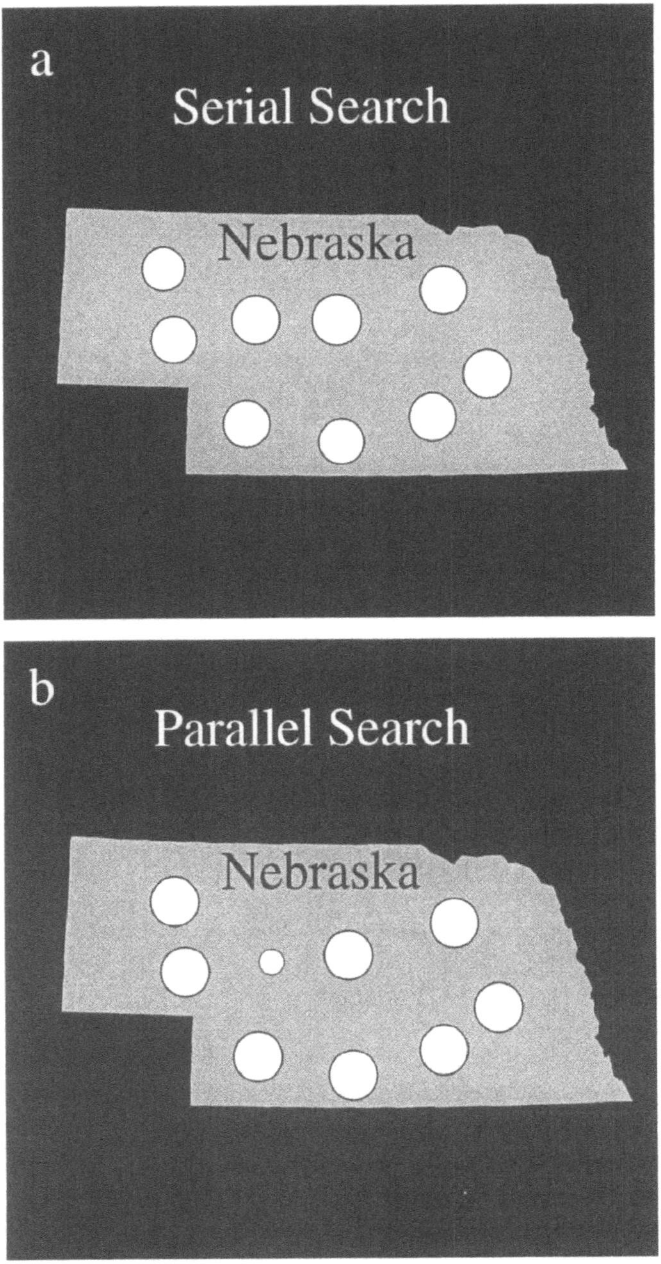

Figure 5.6: A search for the smallest circle may be relatively difficult if there is not much variation among circle sizes (a) or relatively easy if the difference between the smallest circle and other circles is large (b).

relevant feature is present on the distractors but absent from the target. Treisman and Souther (1985) showed that a circle with an intersecting line target will pop out of distractors that are just circles, but a serial search should be needed to detect a circle without an intersecting line among distractor circles with intersecting lines. This is because activity on a unique feature map signals the target is present, but attention is needed to detect that something is uniquely absent (Figure 5.7).

The original FIT predicted all conjunctive searches would be done with serial processes. Some experiments, however, have reported that conjunctive targets have popped out of displays (Nakayama and Silverman 1986a; 1986b). Parallel processing is apparently possible with some conjunctive features when the target and distractors are highly discriminable. Pop out effects have been reported for displays that paired binocular disparity (depth in a third dimension) and other dimensions, such as color, size, or spatial frequency and for targets defined by motion and color. Treisman (1991, p. 675) has argued "inhibition from distractor feature maps can be used to shut down activity in the master-map locations containing distractors with those unique features....search could be helped only in cases in which the target or the distractors activate a unique set of standard detectors in addition to any shared ones." This effect only occurs with highly discriminable features such as color and motion. Suppose the target was a black arrow that moved up and down and you were searching for this target among distractors that were black arrows that moved left and right and grey arrows that moved up and down(Figure 5.8). If the discriminability of the features were such that you could ignore the left-right movement and the grey color in the display, you would have suppressed all the objects except the target and it should pop out of the display.

The Complexity of Cartographic Maps

Real map-reading tasks and maps are usually more complex than the relatively simple tasks and target/distractor displays previously described. Map readers often may not have a specific target in mind when they begin searching a map. The information coming from the map (bottom-up information) has to terminate the search when no "top-down" information from memory defines the target of the search. Searching is stopped to consider something on a map that is unusual or interesting. When the search was initiated, this glorious object of our attention may have been completely unknown.

Multiple targets also add complexity to visual searches. Map readers frequently search simultaneously for several targets on a single map. Treisman (1988) reported on an experiment that had subjects look for one of three known targets or three targets simultaneously. Displays had either 4, 8, or 12 distractors and subjects did not know which of the targets might be in the display. The increase in reaction time was relatively small when the three targets and distractors were defined within a single dimension, e.g., searching for a white, dark grey or black target among light grey distractors (Figure 5.9a). Mean reaction time increased significantly when the three

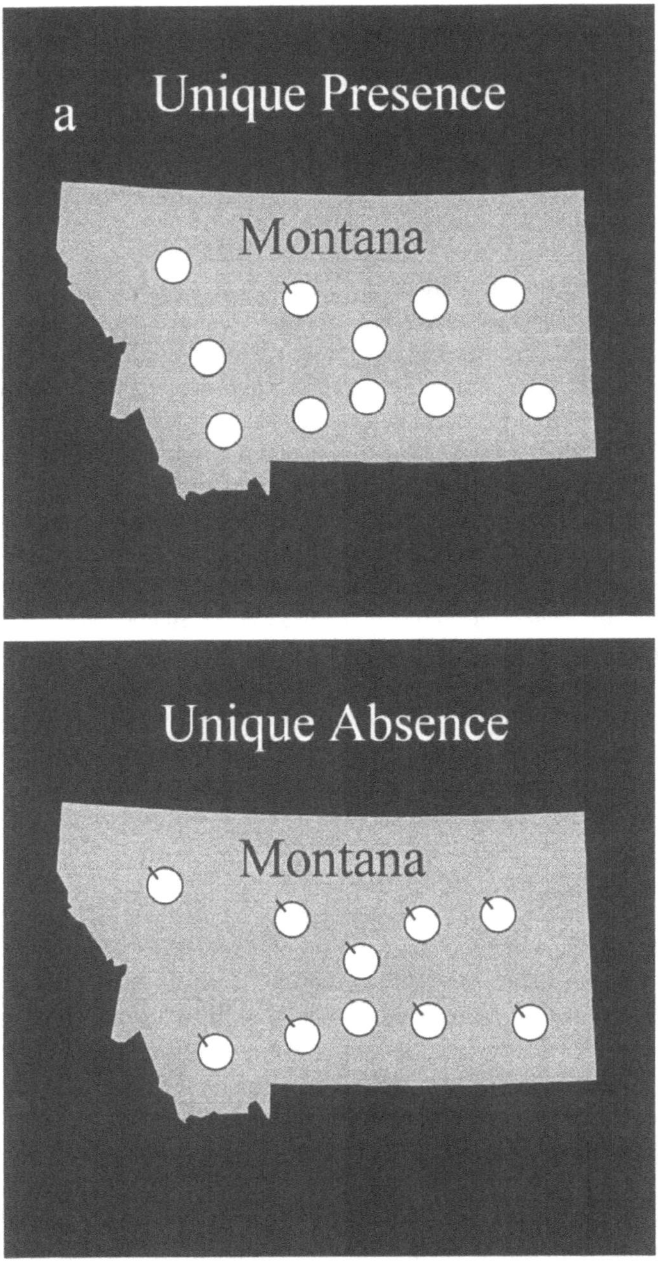

Figure 5.7: The search for a feature that is uniquely present, such as an intersecting line in a circle among circles (a), should be relatively easy while the search for a uniquely absent feature, such as a circle among circles with intersecting lines (b), should be relatively difficult.

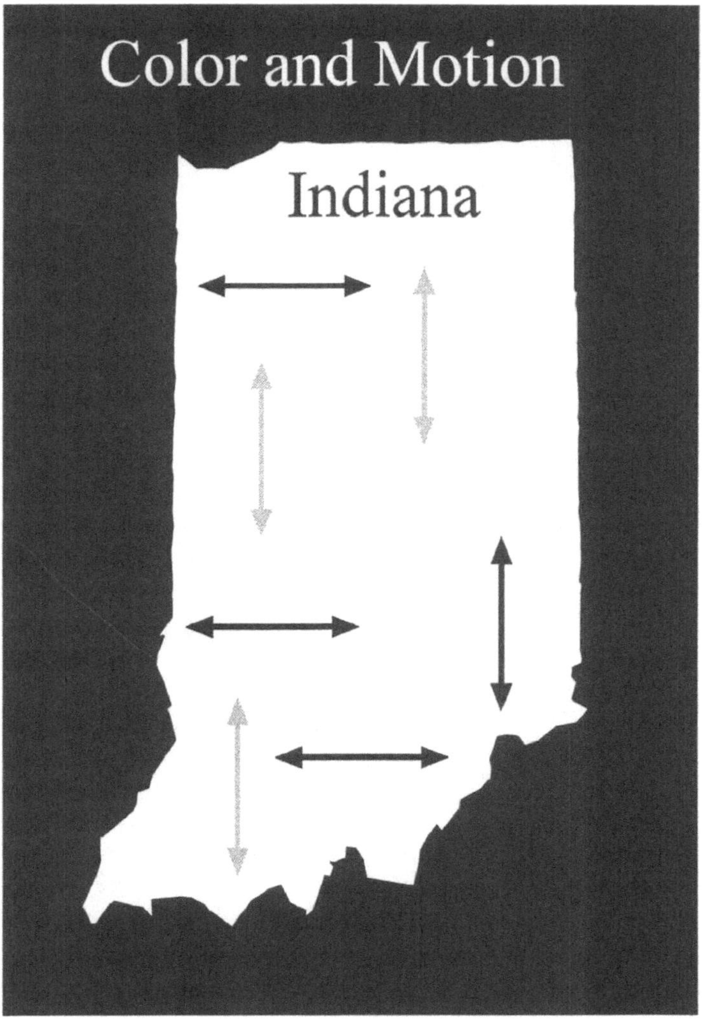

Figure 5.8: Some conjunctive searches based on features with high discriminability, such as color and motion, can produce pop-out effects.

targets were defined on different dimensions, e.g., black, vertical, or large target (Figure 5.9b). Parallel searches were done in both cases, i.e., reaction time was not related to the number of distractors in the display. "Thus the 'odd one out' pops out *within* a single, pre-specified dimensional module, but each different module may need to be separately checked to determine which of them contains it" (Treisman 1988, p. 207).

Given that different features are analyzed by independent modules some dimensions may be irrelevant to the search. The variation of distractors in shape and color should not affect the search when the search is for a large object because these modules need not be used. Variation within the size module, however, should slow the search. The fastest search should occur for a target among homogeneous distractors.

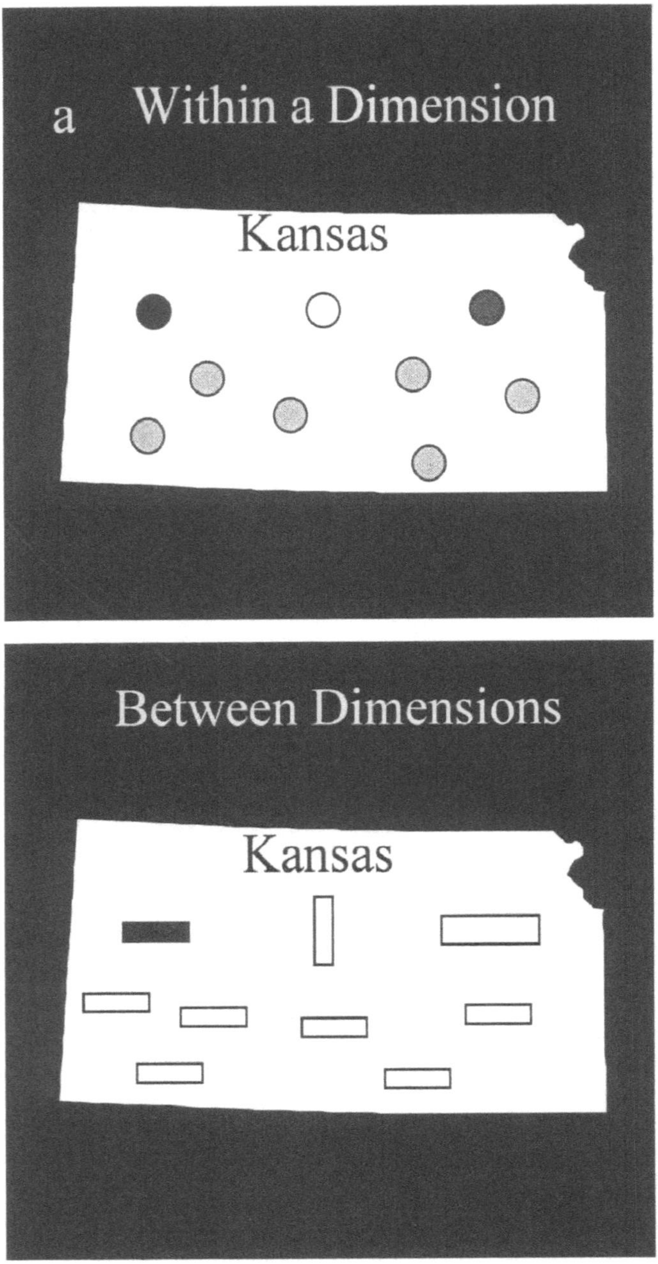

Figure 5.9: A search for one of several targets produces a pop-out effect when the search is within a dimension, such as searching for one of three colors (a), or when the search is among dimensions, such color, orientation, or size (b). The total time for searching within a dimension (a), however, is faster than searching among dimensions (b).

Here the target would vary by a fixed amount from all distractors, e.g., a large target among similar small distractors. As the heterogeneity of the distractors increases, i.e., more different sizes are assigned to distractors, the search should become more difficult. With many size feature maps activated the distractor features would contrast with each other and the target feature (Figure 5.1). Finding the target feature within the relevant module should be more difficult. The differences between the target feature and nearer distractor features are likely to decrease as the number of distractor features within a module increases. The detection of a specific-sized large target would be more difficult if other large objects were among the distractors.

Treisman (1988) compared three groups of subjects to test this notion with the color dimension. A single known target (a blue bar with a fixed size and orientation) among homogeneous distractors (green bars with the same fixed size and orientation) was searched for by a control group. The control group had relatively fast reaction times and flat slopes for the relationship between reaction time and number of distractors. The same blue bar was searched for among heterogeneous distractors that were green bars in three sizes and three orientations by an across-dimension group of subjects. Reaction times for this group were not significantly different from the control group. This indicated varying the features on irrelevant dimensions for the distractors had no impact on the search. The same blue bar was searched for among red, green, and white bars that had the same size and orientation as the target by a within-dimension group of subjects. Reaction times were significantly slower for this group and clearly indicated the interaction of processes engaging information on the map had significantly affected the efficiency of the spatial search.

5.2.2. ATTENTION ENGAGEMENT THEORY

Duncan and Humphreys (1989; 1992) have offered another theory that is very much related to FIT and search processes. They argued that there is no dichotomy between serial and parallel search modes and that the difficulty of a visual search varies continuously across tasks and conditions. The main concern of Attention Engagement Theory (AET) is with how easily visual attention can be directed to the relevant information in a display. AET has generated several new theoretical concepts by focusing on what causes the attention window to shift to particular spatial locations (Figure 5.1).

AET argues that parallel and serial searches are affected by the interaction of two critical stimulus variables and are considered as part of a continuum rather than a dichotomy. A parallel search is considered to be an extreme case in the continuum. Parallel searches still produce a pop-out effect with no significant relationship between reaction time and the number of distractors in the display.

AET defines a hypothetical three-dimensional *search surface* that measures and illustrates the theoretical difficulty of a search (Figure 5.10). The X-axis measures the

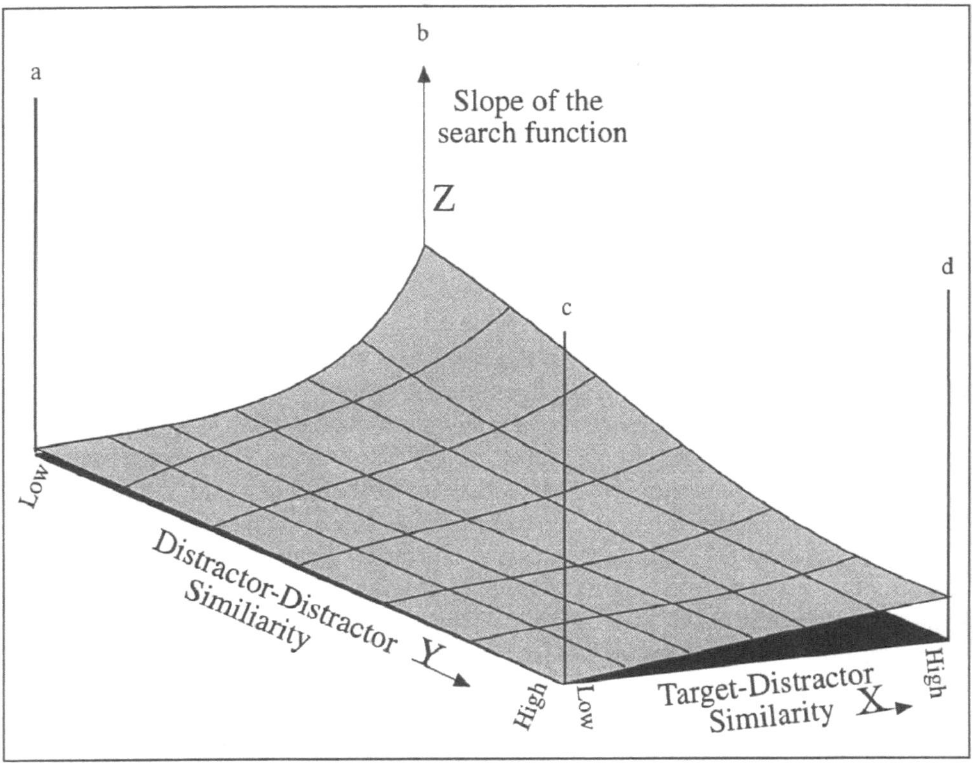

Figure 5.10: The search surface indicates the difficulty of a search (vertical axis) based on the similarity of the target of the search and the distractors and the similarity of distractors with other distractors. After Duncan and Humphreys (1989).

similarity between targets and distractors (Target-Distractor) and the Y-axis measures the similarity between distractors and other distractors (Distractor-Distractor). The slope of the regression line relating reaction time and the number of distractors is measured along the Z axis. Easy searches, when the target pops out of the display, have flat (zero) slopes and difficult searches, when serial processing is required, have slopes significantly steeper than zero. The interaction of the two similarity variables (Target-Distractors and Distractors-Distractors) determine the theoretical shape of the surface that measures the difficulty of the search. Duncan and Humphreys (1992) base three claims on the shape of the similarity surface. First, search functions are always flat if targets are sufficiently different from distractors regardless of the similarity among distractors (location *a* along the Y axis to location *c* in Figure 5.10). Second, there is a gradual increase in search slopes as Target-Distractor similarity is increased even when all distractors remain identical. Third, the most difficult search is produced when Target-Distractors similarity is high and Distractors-Distractors similarity is low as at location *b* in Figure 5.10.

A related study defined both targets and background distractor using only color. It was concluded that the interaction of the Target-Distractors similarity and Distractors-Distractors similarity produced large latency differences among the various conditions

(Moraglia, Maloney, Fekete, and Al-Basi 1989). Dramatic increases in response time were produced by introducing chromatic variation among the background items when target-background similarity was high. When target background similarity was low, however, the effects of background heterogeneity were minimal.

Two important concepts were used by Duncan and Humphreys (1989) to explain the relationship between stimuli similarity and the difficulty of visual search. By considering information as *weights* that *link* elements of the search process together they expressed their arguments in a neural network context. Once information from a display is coded into the visual buffer it must enter visual short-term memory (VSTM) where it can be used to make decisions. It was argued that entering VSTM is functionally equivalent to moving the attention window to some part of the visual buffer. This movement is considered a competitive process that is based on two issues (Duncan and Humphreys 1992 p. 579):

> First, each element in the input array gains or loses weight according to how well it matches an *attentional template*, an advance description of the information currently needed in behavior. Second, to the extent that different elements of the input array are linked by perceptual grouping, any change in weight for one is distributed to the other (*weight linkage*). Thus, the more strongly different parts of the visual input are linked, the greater is the tendency to select or reject them together.
>
> According to this model, the difficulty of a visual search depends on how easily any target in the array gains access to VSTM and hence (because selection is competitive) on the *difference* in selection weight between targets and nontargets.

The *attentional template* may not be the whole target but only selected features of the target. Knowing the target has a unique feature, e.g., its color is yellow, allows one to focus the *attentional template* to consider only that feature. Suppose the target and distractor stimuli varied in color, size, and shape. Limiting the *attentional template* only to a consideration of size and ignoring the color and shape information would obviously result in a more efficient search.

The similarity between the target and distractors increases as the number of shared features increases. As the number of shared features increases, the likelihood that the target will be visually grouped with distractors also increases. For example, if the target is a green triangle it might be visually grouped with green distractors that were squares or circles because of its shared color. Visually grouping targets and distractors together in a display links them together and makes the search process more difficult. Visually grouping distractors together, however, benefits the search for the same reason. When a distractor is rejected as the target, the weight of that distractor is reduced and this reduction of weight is distributed to other stimuli in the group. Duncan and Humphreys

(1989) call this effect *spreading suppression.* It allows distractors to be dismissed more efficiently by rejecting all members of a group simultaneously rather than rejecting each distractor separately.

D'Zmura (1991) considered searches with targets and distractors defined by colors that varied in hue and chromaticity. His color experiment supported the notion that opponent process mechanisms can be used to explain some pop-out effects. He reported orange targets would pop out of green or yellow distractors when subjects used a detection mechanism tuned to yellow. Orange targets would also pop out of yellow and blue distractors when subjects used a detection mechanism tuned to red. This use of focused attention to consider a key part of the target rather than the whole target is an example of Duncan and Humphreys' (1989) *attentional template* concept. D'Zmura (1991) also reported pop-out effects for orange targets within green-yellow and red-blue distractors, but not for orange targets within yellow and red distractors. Neither of these situations should have a pop-out effect because neither the red nor the yellow in orange is unique. The author argued that subjects used a higher order chromatic detection mechanism that was spectrally sensitive to the target's chromaticity, to detect the orange target among green-yellow and red-blue distractors. A simpler explanation might relate to what was unique in the first experiment but not in the second. The distractors in the first experiment were green-yellow and red-blue so they were unique in that part of them, the green or blue part, was not shared by the target. This unique information about the distractors could have been used to inhibit them and make the target pop out of the display. The yellow and the red distractors in the second experiment were not unique in any way compared with the orange target and, therefore, the target could only be detected using a serial process.

5.2.3. GUIDED SEARCH THEORY

The original FIT argued feature searches terminate during the initial parallel stage because the unique feature pops out of the display. No pop-out effect occurs for a conjunctive search because no unique feature exists and a second stage serial process is initiated to continue the search for the target. The serial search continues to search for the target by randomly moving the attention window from object to object until the target is identified. The original FIT assumed the two processes are independent of one another. AET eliminated the distinction between uniquely parallel and serial stages. It focused on how the *spreading suppression* of distractors affected the competition among objects to get attention and, there by, move the attention window to their location.

Guided Search Theory (GST) reconstructed FIT to preserve the distinction between parallel and serial stages in the search and made the competition for attention an essential element of the search process (Wolfe, Cave and Franzel 1989; Cave and Wolfe 1990). GST kept Treisman's notion of a preattentive parallel stage of processing followed by a serial stage, but two important conceptual changes were made that make GST different from FIT. GST argued it is illogical to consider parallel and serial stages as totally independent (Wolfe, Cave, and Franzel 1989). The information processed during the parallel stage may still be useful even if the target cannot be detected during the parallel stage (Egeth, Virzi, and Garbart 1984). GST offered a new account of the early stages of visual processing with the parallel stage, like FIT, representing the separate dimensions in the visual field as independent modules (Cave and Wolfe 1990). The locations on maps for individual dimensions, e.g., shape or color, have an activation level determined by both a bottom-up component and a top-down component (Figure 5.11 and 5.12).

Searching for a feature defined by a star shape among distractors defined by triangle shapes is illustrated in Figure 5.11. The display map at the bottom of the Figure 5.11 has the star-shaped target in the center of the map and eight triangular distractors. Other dimensions, such as color (all objects are white), and size (all objects have the same size), are the same for all objects. A conjunctive search is illustrated in Figure 5.12 using a black star-shaped target. The distractors are four black triangles and four white stars. Size is again the same for all objects.

Above the display maps in Figures 5.11 and 5.12 are individual maps that represent each dimension (color, size, and shape). One set of these maps represents the bottom-up activation of the dimensions and the other set represents the top-down activation. Each location in South Carolina that has an object on the display map is shaded according to that location's activation with lighter shades indicating higher activations.

The bottom-up activation does not depend on what is known about the target and will be the same no matter what is expected for a target. The visual pop-out effect is based on bottom-up activation and even an unknown target should pop out of a display (Treisman 1988). The sum of the differences between a location's value and the values at all other locations is used by GST to compute the activation for a particular location on a particular dimension. For example, the color of all the objects on the display map in Figure 5.11 is the same (white) so the bottom-up activation for color is zero for all locations. Bottom-up activation is also zero for the size dimension at all locations for the same reason. The bottom-up activation for shape would be nonzero for every location because the target location is different (star) from the distractor locations (triangles). The differences would sum to a high activation level for the target location because the difference between the target location and all other locations is large on the shape map. Each distractor would be different from the target, but not different from each other and, therefore, each of them would sum to a low level of activation. This would result in a clear contrast on the shape map, i.e., the target would "pop out."

The conjunction example (Figure 5.12) shows a different pattern of bottom-up activation. Here the target is a black star but distractors are white stars and black triangles. All locations have either black or star-shaped symbols. Although there is some bottom-up activation for all locations on the color and shape maps, the target location does not contrast with the distractor locations, i.e., no pop-out effect. The bottom-up activation on the size map is still zero because all locations still have the same size.

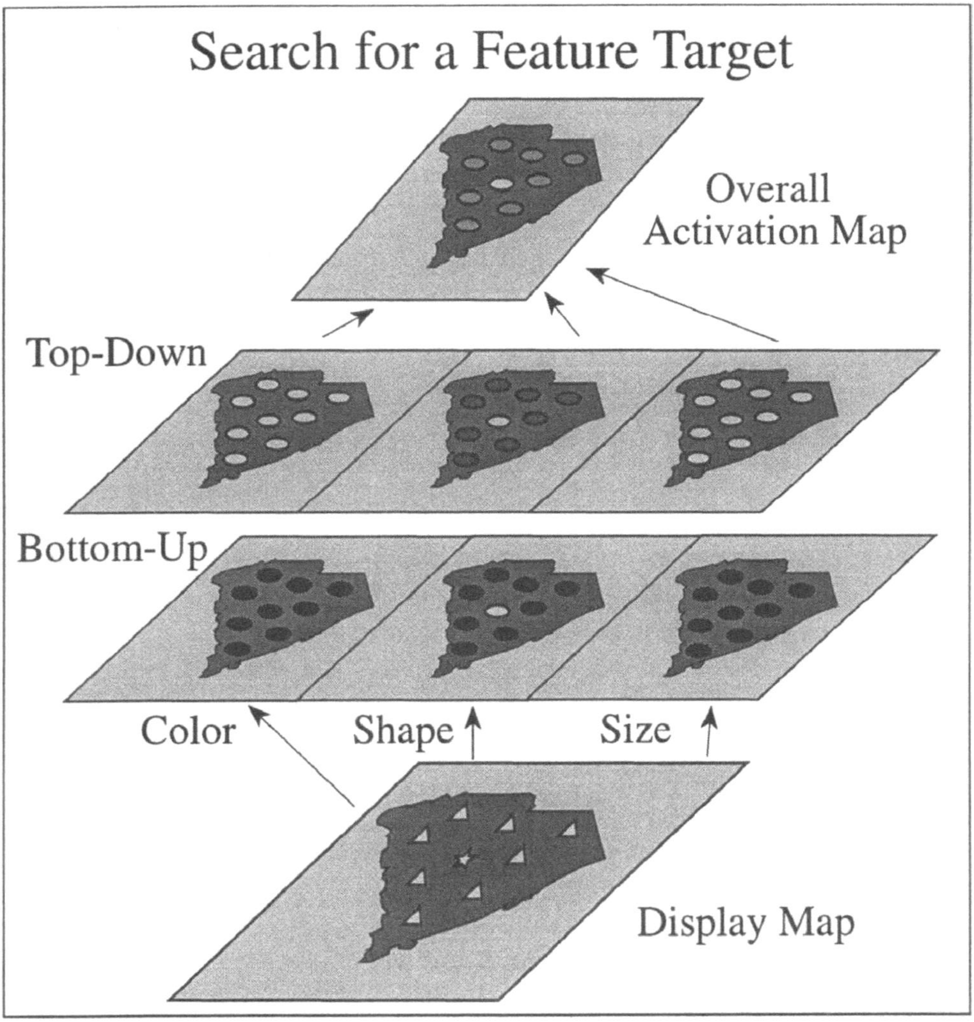

Figure 5.11: A feature search for a white star-shaped map symbol in the center of the display map. Bottom-up and top-down information for the locations combine to give the highest activation to the center location in the overall activation map making the symbol at that location it pop out of the map. After Cave and Wolfe (1990)

The top-down component of the process reinforces a feature search (Figure 5.5) but is more critical to a conjunctive search (Figure 5.12). Top-down activation at a location for each dimension is simply based on the similarity between the feature value at that location and the feature value of the target. This is because top-down activation depends on the searcher's knowledge of the target. Locations with the same color as the target will have high activation levels and those with different colors will have lower activations levels that depend on how similar that location's color is to the target's color. Top-down activation levels for size would, following similar logic, depend on the

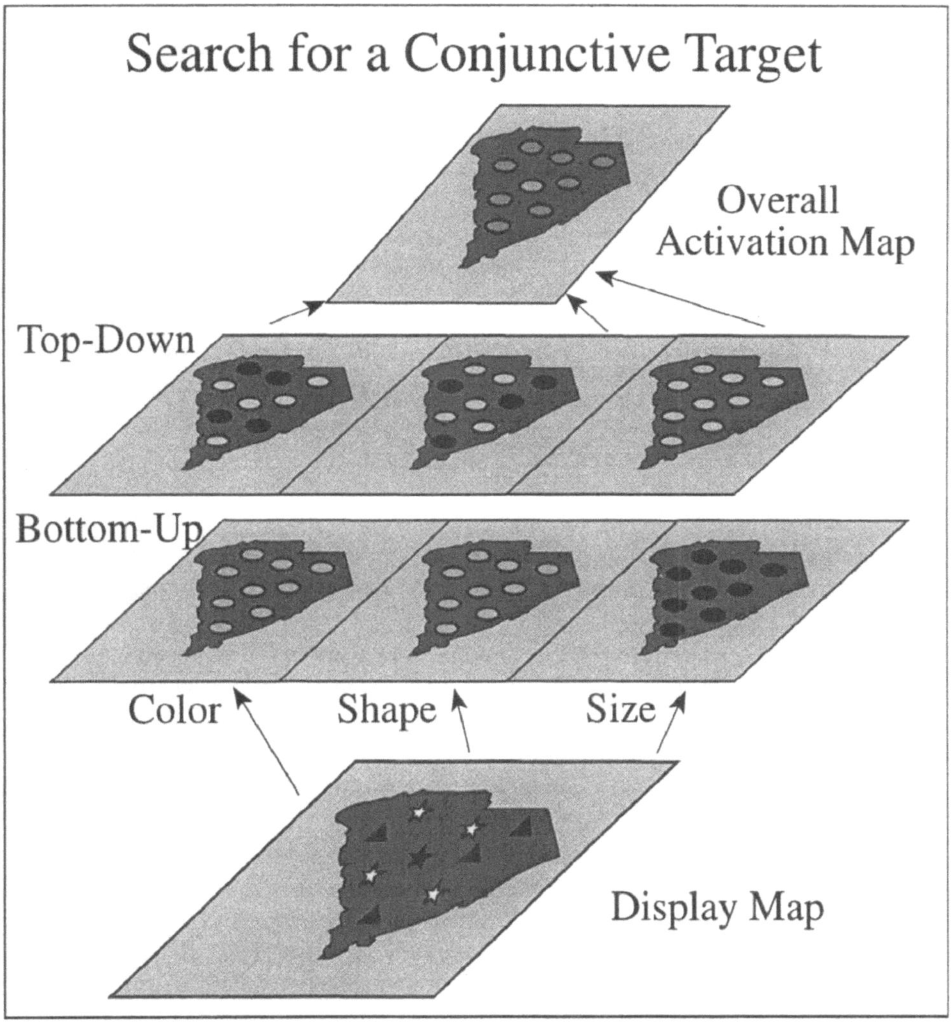

Figure 5.12: A conjunctive search for a black star-shaped map symbol in the center of the display map. Bottom-up and top-down information for the locations combine to give all locations about the same activation in the overall activation map making no symbol pop out of the map. After Cave and Wolfe (1990).

similarity of a location's size and size of the target. The top-down activation for color for the feature search example (Figure 5.11) has high activation at all locations because all locations match the target's color. The shape map, however, only has a high activation level for the target location because it is the only one to match the target's shape. The top-down activation values for the conjunctive search example (Figure 5.12) show a different pattern. The color map has high activation levels for the target location and other locations that share the black color. All other locations have low activation levels. The shape map follows similar logic but exhibits a contrasting pattern. Only the target is highly activated on both the color and shape maps.

Cave and Wolfe (1990) computed the activation levels for a computer simulation of the parallel search stage using the following equations:

$$BottomUp_i = \exp\left(\frac{\sum_{j=0}^{n} | f_i - f_j |}{n-1}\right) \tag{5.1}$$

$$TopDown_i = d\,| f_i - t | \tag{5.2}$$

$$Activation_i = BottomUp_i - TopDown_i \tag{5.3}$$

Activation$_i$ (Equation 5.3) represents the level of activation for location i on a map representing a particular dimension (Figures 5.11 and 5.12). The feature value for a particular dimension at location i is represented by f_i and t is the target value for that dimension. The bottom-up activation (Equation 5.1) for location i on a particular dimension is computed as the summed differences between the feature value at that location and all other locations (j represents the other locations) scaled by the number of locations (n). An exponential function is used to enhance larger activation levels so they will rise above the visual "noise". The top-down activation (Equation 5.2) for location i on a particular dimension is computed by multiplying a coefficient d times the absolute difference between location i's feature value and the target's feature value. Coefficient d measures the relative effectiveness of dimensions in separating targets from distractors. Color is often used to define targets and distractors because it appears to be more effective than many other dimensions (Bundesen and Pedersen 1983; Carter 1982; D'Zmura 1991; Egeth, Virzi, and Garbart 1984; Treisman and Sato 1991). Top-down activation is subtracted from bottom-up activation to compute overall activation (Equation 5.3). This form of the GTS model has the bottom-up component (Equation 5.1) acting in an excitatory role while the top-down component (Equation 5.2) acts in an inhibitory role. Cave and Wolfe (1990) argued that the model could be implemented using excitation or inhibition for either component. After the activation for each location i is computed for each dimensional map, the overall activation map (Figures 5.11 and 5.12) is computed by summing over all the dimensions. This overall map represents the result of the parallel stage of the search process and indicates the likelihood that the stimulus at each location is the target.

The information from the parallel stage can be used to direct attention to the most likely candidate. GST argues that no decisions are made during the preattentive parallel stage of processing. The actual decisions that accept or reject locations as being the target are done in the serial stage. Comparisons are made with the target and the candidate is accepted or rejected. The most likely candidate and the first one considered should be the target for a feature search (Figure 5.11) and for any conjunctive search producing a pop-out effect. Visual "noise" does not allow the target to be significantly more activated than the distractors for most conjunctive searches (Figure 5.12) and the target usually will not be first on the list of candidates. This could require the searcher to examine a number of candidates before finding the target. Although the information coming from the parallel stage is imperfect, it should improve the target's chances of being considered over a random search process (Wolfe, Cave, and Franzel 1989). It is assumed that the search is taking place over time and that transmission from the parallel stage is continually updated. "The noise in the transmission from parallel to serial stage is random; the signal is not. Therefore, the parallel processes might incorrectly guide the spotlight to a distractor at first. However, with the passage of time, the signal would emerge from the noise to guide the spotlight of attention to the target" (Wolfe, Cave, and Franzel 1989, p. 428).

GST makes a prediction concerning triple conjunctions that is qualitatively different from the prediction made by FIT. It predicts that more efficient searches should occur if more information from the parallel process can be used by the serial process. A feature that a distractor has that is not shared with the target could be used to inhibit that distractor, i.e., make it a less likely candidate as determined by the parallel process. This argument has been made by a number of other authors (Duncan and Humphreys 1989; Quinlan and Humphreys 1987; Treisman 1988; Treisman and Sato 1990). Having more unshared features should allow for more inhibition. GST, therefore, predicts that searches for some triple conjunctive targets should be more efficient than searches for standard conjunctive targets.

Wolfe, Cave, and Franzel (1989) reported results from an experiment that used a triple conjunction task and defined targets and distractors using size, color, and orientation. One condition of the experiment had distractors that shared one of the three features with the target and had two others that were not shared. Another condition had distractors that shared two of the three features with the target and had one unshared feature. A third condition was a simple conjunction where all features were shared between the target and the distractors. The shallowest slopes for the relationship between reaction time and the number of distractors were for the condition having distractors differ from the target on two dimensions. This supported the notion that information is passed from the parallel to the serial process and that this information can be used to inhibit distractors.

The three search theories discussed above have many similar arguments and some that are unique. Map readers perform the same types of feature and conjunctive searches discussed above, but do it within a map-reading context. A handful of geographic studies have begun to consider spatial searches of cartographic maps and how the nature of the map can affect the search.

5.3. Geographic Applications

Kilcoyne (1974) considered if searching a map for a target symbol was affected by the nature of the symbol. None of the spatial search theories previously discussed had been published when Kilcoyne did his map-reading study. Although these cognitive theories were not available to guide his research, his experimental design showed a remarkable intuition for the nature of the problem. He constructed maps of Illinois using pictorial and geometric symbols. A set of thirty pictorial symbols was collected from existing maps and a set of thirty geometric symbols was collected from standard 'Zip-a-tone' sheets used at the time to manually construct maps.. Two point symbol maps, one with 60 and one with 120 symbols, were constructed using each type of symbols. These four maps were used in a series of nine map-reading tasks that required subjects to answer questions related to verifying symbols were on maps, counting the number of symbols on maps, and comparing the number of symbols on maps. The time it took a subject to complete all the tasks and the number of correct answers were recorded as measures of efficiency. The basic conclusion of the study was that the map-reading tasks were completed faster and more accurately when the maps were constructed with pictographic symbols. Kilcoyne (1974) also concluded that it was possible to evaluate effective map reading using reaction time and accuracy as indices of efficiency.

Lloyd (1988) considered searching for pairs of pictographic map symbols on cartographic and cognitive maps. He wished to determine if searching for symbols previously encoded in memory was the same as searching for symbols currently being viewed on a map. The experiment with perception subjects used the following procedure. Subjects were first shown a pair of yellow pictographic symbols on a monitor. The symbols stayed in view long enough to be encoded into memory. A map was then presented that had an island figure surrounded by a background and a number of symbols on it (Figure 5.13). The map included the target symbols on half the trials and the symbols could be found in both the figure and the ground. The number of symbols varied from as few as four to as many as 24 on each trial. The subjects indicated if both symbols were on the maps as quickly and accurately as possible. The results for the perception subjects indicated they were using a serial self-terminating process to perform the task. Since shape was the only feature that could be used to detect the target symbols and targets were displayed among heterogeneous groups of distractors, this could be expected.

The memory subjects followed a similar procedure expect for the order of presentation. The memory subjects were first shown a map for a given trial that had a fixed number of symbols. The map was remained in view long enough for subjects to encode the symbols into memory. Two symbols were then presented and the subjects

Figure 5.13: Map with a figure and ground showing pictographic map symbols similar in style to those used by Lloyd (1988).

determined if they were on the previously viewed map. It was assumed that subjects had encoded a memory structure that contained the information on the map. The results for the memory subjects indicated they were able to perform the task using a parallel process. Reaction times were not significantly related to the number of symbols on the map. Neither group of subjects were affected by the relative location of the targets on island figure or the background. Subjects apparently focused on the yellow symbols as a dispersed figure on the map display. It was concluded that memory and perception processes were not functionally equivalent.

Nelson (1995) has evaluated Attention Engagement Theory in experiments that considered searching for color targets on choropleth maps. Her results indicated that differences between the target and distractors were the most important factor affecting the search process. She also reported that it was difficult to create maps that represented Duncan and Humphrey's hypothetically most difficult search condition (location *b* in Figure 5.10). This extreme case, with the target being very similar to the distractors, but the distractors being very dissimilar to each other, may be impossible to produce on real maps.

Brennan and Lloyd (1993) had subjects search choropleth maps for a target boundary among other boundaries. The boundaries were defined by color differences between regions on the maps. The target boundary was defined as two polygons with two specific colors on either side of the boundary. A map reader instructed to search for a target boundary with 10 percent grey in one polygon and 90 percent grey in another polygon would find this boundary between South Dakota and Minnesota on the map represented in Figure 5.14a. A subject presented with the task of finding the same target boundary on the map represented in Figure 5.14b would not find the target. The authors did an analysis of the reaction times and the accuracy of *present* or *absent* responses to test a number of hypotheses related to the search process being used by their subjects. Result supported the Cave and Wolfe's (1990) argument that both parallel and serial processes are important components of visual search processes. The location of the target boundary on trial maps had a significant effect on the efficiency of the search process. This indicated that subjects were serially processing the locations on the map in a systematic fashion and targets in some parts of the map were easier to detect than in other parts of the map. As the number of boundaries with one of the target colors (critical boundary) increased, reaction times also tended to increase, but as the number of boundaries without either boundary color (non-critical boundary) increased reaction times tended to decrease. The fastest response times were for maps with few critical boundaries and many non-critical boundaries. This result supported Duncan and Humphrey's (1989) argument that distractors with common characteristics can be suppressed as a category rather than individually to save processing time. In this case the common characteristic was the boundaries lacked one of the target colors. The results also indicated that some color combinations used to define boundaries produced faster response times than other combinations. The authors used opponent process theory to explain these results. This theory argues that opponent colors, i.e., red or green, blue or yellow, and black or white, can only be processed by the visual system one at a time. The receptors responsible for detecting a particular opponent pair can only detect one of the two colors at any moment in time. This suggests that boundaries defined by two colors that were not from the same opponent pair, e.g., red and blue, should be detected faster than boundaries defined by colors that were from the same opponent pair, e.g., red and green or blue and yellow. The results from Brennan and Lloyd's (1993) experiments supported this prediction of opponent process theory.

Lloyd (1996c) conducted search experiments with maps to determine whether four features commonly used to produce variation in map symbols varied in their ability to produce a pop-out effect. Color was found to produce results identical to the theoretical predictions illustrated for a parallel process in Figure 5.3. Other dimensions tested, shape, size, and orientation, showed some positive slope for the target absent trials although all indicated target present responses were parallel processed. The ability to enhance the pop-out effect for a feature search by encoding color and one or more additional dimensions as multiple unique features for the target indicated no improvement in response times over using color alone. Searches for conjunctive targets defined by color and a second variable indicated subjects were experiencing a pop-out effect for

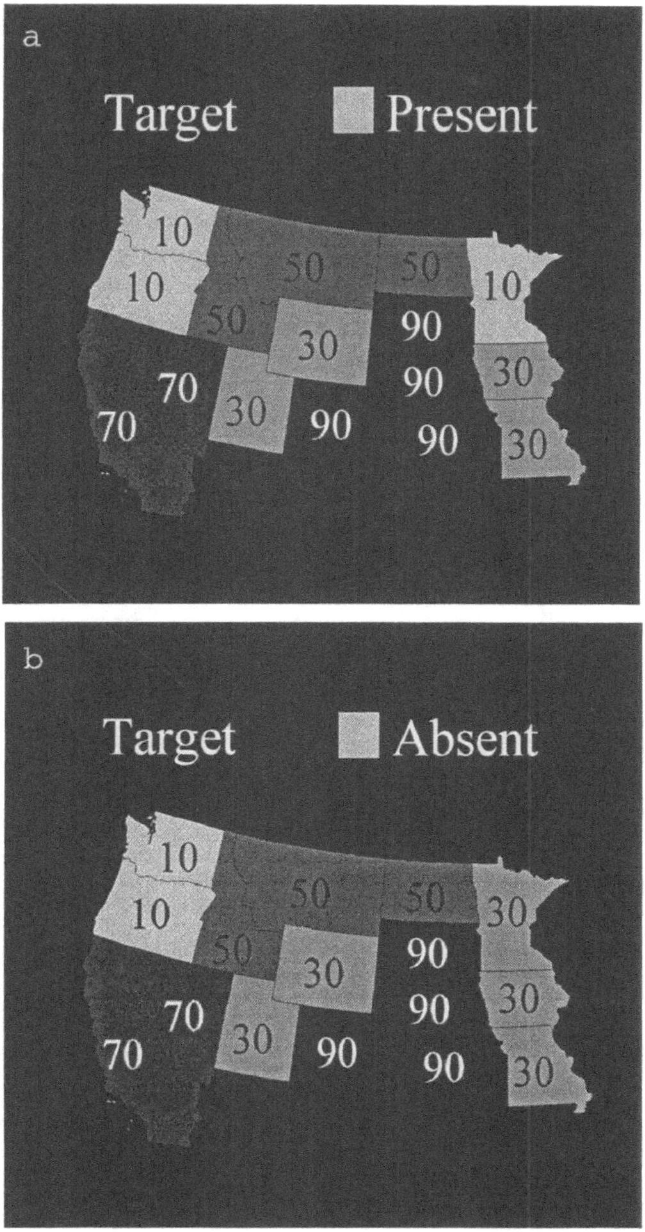

Figure 5.14: A target boundary defined as 90 percent and 10 percent grey is present as the boundary between South Dakota and Minnesota on map *a*, but no such target boundary can be found on map *b*.

conjunctive targets defined by color and shape. Target map symbols defined by other combinations, such as color and size and color and orientation, did not produce a pop-out effect for the subjects and had to be serially processed. It was argued that maps usually define a relatively large *visual field*, i.e., the space that could have a map symbol, that is too large to be searched without eye movements. Eye movements are necessary when a map reader's *field of vision* or the area around a fixation point from which a person can extract information, is smaller than the *visual field* defined by the cartographer. Given the relatively large visual fields typically defined for maps, color may offer some advantages over other dimensions that possibly could be used to define map symbols. Nagy, Sanchez, and Hughes (1990) have shown that the size of a person's *field of vision* is related to the relative discrimination possible for a particular dimension. If the *visual field* for color discrimination is relatively large compared to other dimensions, a search for a map symbol defined by color would require fewer eye movements than a search for a symbol defined by a less discriminating dimension.

5.4 Conclusions

This chapter has discussed theories of visual search and related them to searching for objects or boundaries during map reading. Feature Integration Theory (Treisman 1988), Attention Engagement Theory (Duncan and Humphreys 1989), and Guided Search Theory (Wolfe, Cave, and Franzel 1989) explain how information from a map is combined with information in a map reader's memory to find a target symbol among distractor symbols on a map. The use of neural networks to simulate the visual search process is discussed in Chapter 9. The next chapter considers how spatial information is learned.

CHAPTER 6

LEARNING GEOGRAPHIC INFORMATION

6.1. Introduction

Understanding intelligent spatial behavior is the ultimate goal for most geographic research concerned with human activities. Since human spatial behavior is frequently guided by visual interpretations of the environment or secondary representations of the environment, it is essential that geographers have an understanding of the visual processes used in decision making. This chapter is concerned with the processes used to learn geographic information. Discussions are centered on learning new information using visual processes and creating and storing spatial knowledge from initially learned information. The focus is on our ability to learn categories based on experiences and to make classification decisions that are fundamental for activities such as map reading and way finding (Golledge, Gale, Pellegrino, and Doherty 1992; Lloyd 1989; MacEachren 1992a). Most spatial behavior is the product of a sequence of events. Estes (1994) argued that intelligent behavior must be preceded by thinking that allows hypothesis testing and problem solving. Since such thinking requires information, it must be preceded by learning. Leaning information about the environment we live in must be a continuous process. For example, we learn categories from experiences and use them to make decisions. Each new experience, however, can potentially alter our understanding of the categories (Estes 1994). This means that decision-making processes that result in spatial behavior frequently combine previously learned information and newly acquired information and continuously change our understanding of the environment.

An important prerequisite for acquiring spatial knowledge is an ability to create categories by processing visual information. Nigrin (1993) argued that the ability to form useful categories as a response to continuously changing external stimuli is the most important prerequisite for autonomous intelligent behavior. We must learn to form meaningful categories because the world is too complex. If we could not simplify our sensory experience through classification, we would be overwhelmed by the intricacy of the details. This requires that we lean to detect common patterns in the information we process from the environment. All the experiences that produce similar patterns are grouped together and given a meaningful name. This category name is the connection to the information we have learned.

6.2. Learning Categories

Humans must sort through an enormous amount of sensory information to classify individual objects encountered in the environment into categories. The ability to perform specific classification tasks is thought to be learned through experiences (McClelland and Rumelhart 1988; Rumelhart and McClelland 1986). Our knowledge of particular categories evolves over time, with the errors we make gradually reaching a lower limit. Our earliest childhood experiences supply useful examples. A child starts with no categories and encounters many new objects in the environment. Suppose a young child had an initial experience with two objects. One had a brown color, was large, had a short tail, and made a loud sound (bark) and the other had a grey color, was small, had a long tail, and made a soft sound (purr). The child's mother or father or some other *expert* on animals might inform the child the first object was an exemplar of the dog category and the second object was an exemplar of the cat category. Each subsequent encounter with an animal provides the child additional experiences with the characteristics of the exemplars for animal categories. As many more exemplars of the dog and cat categories are experienced the patterns of the visual characteristics associated with the category names are reinforced. When the child identifies an animal incorrectly, the *expert* can correct the mistake to make learning the categories more efficient. Soon the child learns to classify dogs, cat, and other frequently experience animals with few mistakes. As we grow older we continue to experience new exemplars of old categories and to refine our knowledge of these categories. We use the old categories to make decisions and create new categories when we experience patterns that fit none of the current categories we have learned. The decisions we make that are based on patterns of visual information frequently appear to be intuitive judgments (Huttenlocher, Hedges and Duncan 1991; Kosslyn and Koenig 1992). Explanations of human decision making that seems to be intuitive rather than logical are particularly difficult because they are not easily modeled with standard statistical techniques (Lloyd 1994).

Although much of our learning continues to be through primary contact with exemplars of categories, physical events that vary greatly in space and time are difficult for one person to experience directly. The learning of climate categories is a good example of this spatial and temporal scale effect on the acquisition of environmental knowledge. Most people would not be familiar with the formal category names used by geographers to name climates. Since few people have had long-term primary experiences with more than a few climate categories, they have no particular incentive to make up informal category names. Living in one place for a long time would enable you to experience the seasonal variations of that location's climate. You are not, however, likely to be able to experience directly many places with the same degree of comprehension. Learning categories for patterns in the physical environment from primary experiences is also likely to be unsupervised and haphazard. Few students in the United States, for example, have had any primary experiences with climates related to tropical rainforest (Figure 6.1) or tundra (Figure 6.2) environments.

Secondary experiences with climates through formal educational experiences are more frequent and more organized. Climatic variables such as temperature and precipitation can be graphed and viewed to represent the experience one would have living at a location. A person looking at such graphical representations for many locations should be able to identify common patterns for locations that have similar variations in temperature and precipitation. For example, tropical rainforest climates have high temperatures that are relatively consistent from month to month and high levels of precipitation (Figures 6.1). Tundra climates have relatively cold temperatures with some seasonal variation and very little precipitation throughout the year (Figure 6.2).

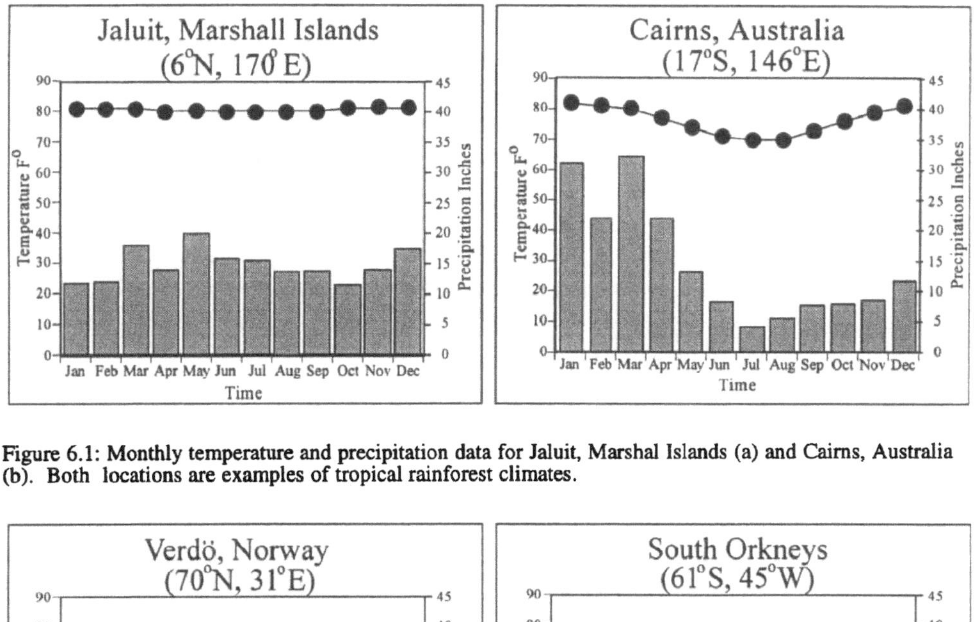

Figure 6.1: Monthly temperature and precipitation data for Jaluit, Marshal Islands (a) and Cairns, Australia (b). Both locations are examples of tropical rainforest climates.

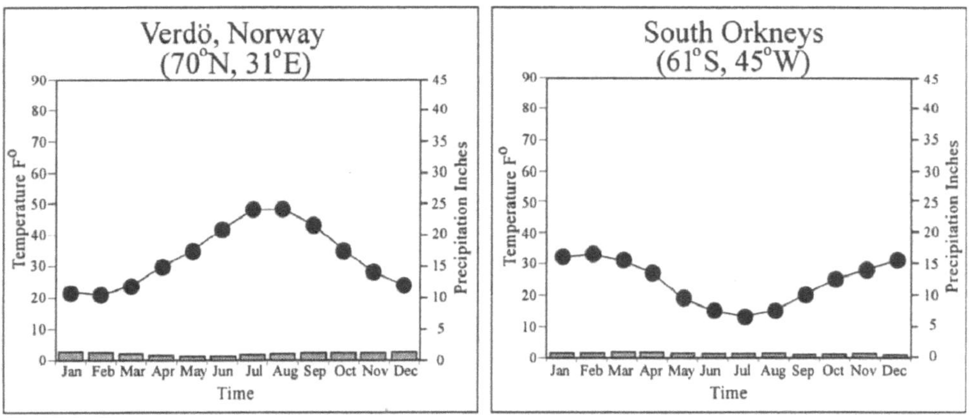

Figure 6.2: Monthly temperature and precipitation data for Verdö, Norway (a) and South Orkneys (b). Both locations are examples of tundra climates.

Locations with similar patterns could then be collected into one category and given a meaningful name. The supervision provided by an expert as part of a formal education process can also reinforce the learning of the categories. Precisely measured data, collected over many years for many locations, have provided formal definitions of climate categories that have specific characteristics and threshold values (Köppen 1931). For example, tropical rainforest climates (Af in the Köppen classification), have monthly average temperatures that exceed 64.4 ^{O}F and monthly average precipitations that exceed 2.4 inches (Figure 6.1a and 6.1b). The focus of this chapter, however, is not on how people memorize threshold characteristics or compute class membership with numbers. Its focus, instead, is on how people recognize visual patterns that cause a collection of exemplars to be considered part of one category.

6.3. Climate Categories

Lloyd and Carbone (1995) considered the learning process used by humans to encode climate categories (Montgomery, Weyel, and Petersen 1988; Murphy, 1976). Their student subjects were provided with temperature and precipitation data in a graphical form for a series of locations to learn climate categories. The authors also trained a pattern associator neural network to learn categories from the same patterns. They were than able to make a comparison between the performances of humans and the neural network. The authors hypothesized that both human subjects and neural networks could learn categories of climates through a supervised learning process. If the number of categorization errors decreased over a number of rounds of training (epochs), then learning was considered to be taking place (McClelland and Rumelhart 1989).

It was also hypothesized that the ability to learn specific climate categories and the rate of learning would be similar for humans and the neural networks. The ability to learn categories was measured by the error rates for specific climate categories at the end of the learning process. The rate of learning was measured as the decline in categorization errors made over N learning epochs (Figure 6.3). It was expected that humans and neural networks would find the same categories easy or difficult to learn. Since some real-world categories are separated by fuzzy boundaries and are, therefore, more difficult to learn, it was expected that error can reach a lower limit above zero for some categories (Kosko 1993; Kumler and Groop 1990). It was also anticipated that the weights learned by the neural network would correspond to critical distinctive visual characteristics of the climate categories such as seasonal variations in temperature or precipitation (Lloyd 1994; Rueckl, Cave, and Kosslyn 1989).

6.3.1. MESOTHERMAL CLIMATE CATEGORIES

Lloyd and Carbone (1995) selected four different mesothermal climate categories for their human subjects and neural networks to learn. According to the Köppen classification the climates would be labeled as humid subtropical (Cfa), marine west coast (Cfb), Mediterranean with warm summers (Csa), and Mediterranean with cool summers (Csb) (Lutgens and Tarbuck 1995). The differences among the categories were relatively subtle because the four classes were all considered mesothermal climates (the common C in their Köppen classifications). Basic differences relate to other letters in

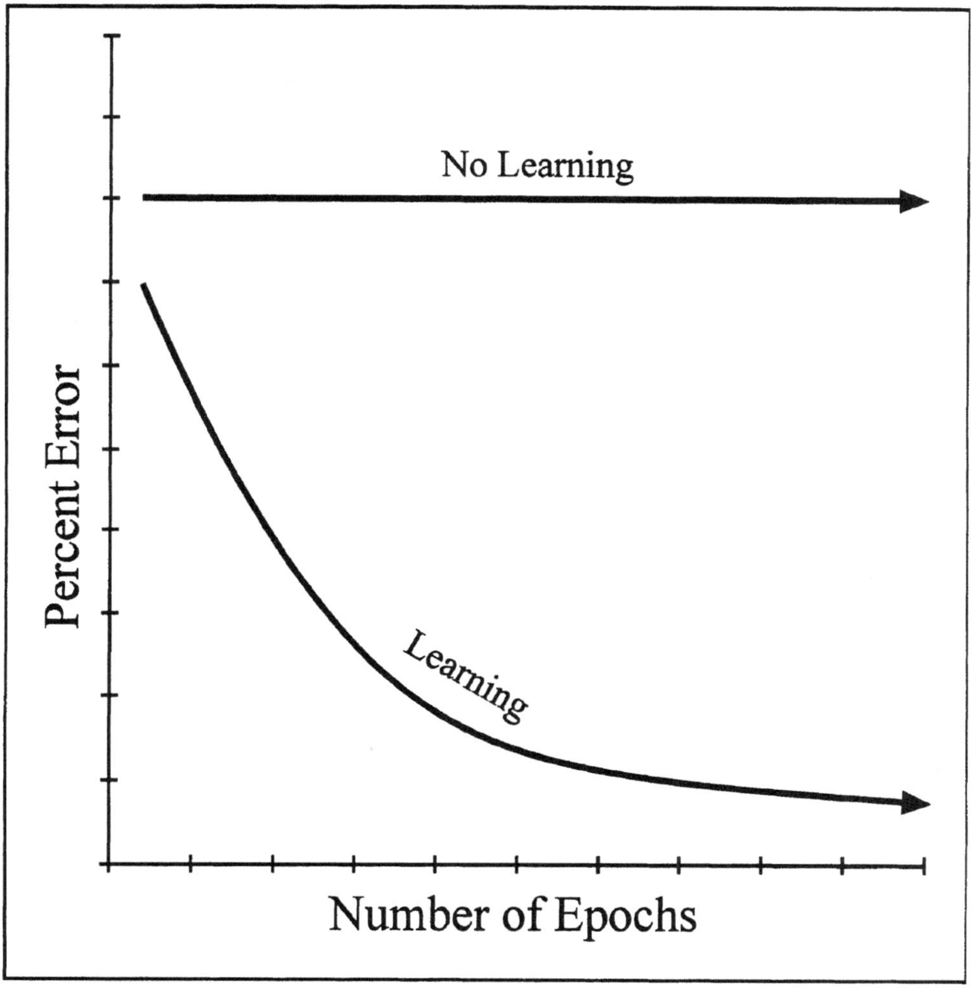

Figure 6.3: The top line implies that no learning has taken place because it has a flat slope. The bottom line implies that learning has taken place because percentage error has decreased at a decreasing rate over the learning epochs.

the Köppen classifications for temperature (a=warm or b=cool) and precipitation (f=evenly distributed or s=summer drought). Typical exemplars for the categories represent a category's basic visual characteristics (Figures 6.4, 6.5, 6.6, and 6.7). Average lines for both temperature and precipitation provided for each exemplar allowed the subjects to easily notice if the dot for a month was above or below average. This important binary distinction was also used to encode information (+ or -) to be used as inputs for training a neural network.

6.3.2. TRAINING HUMANS AND NEURAL NETWORKS

Lloyd and Carbone (1995) had human subjects view climate graphs that showed temperature and precipitation lines representing individual cities. After each graph was presented the subjects decided to assign it to one of four categories that were given

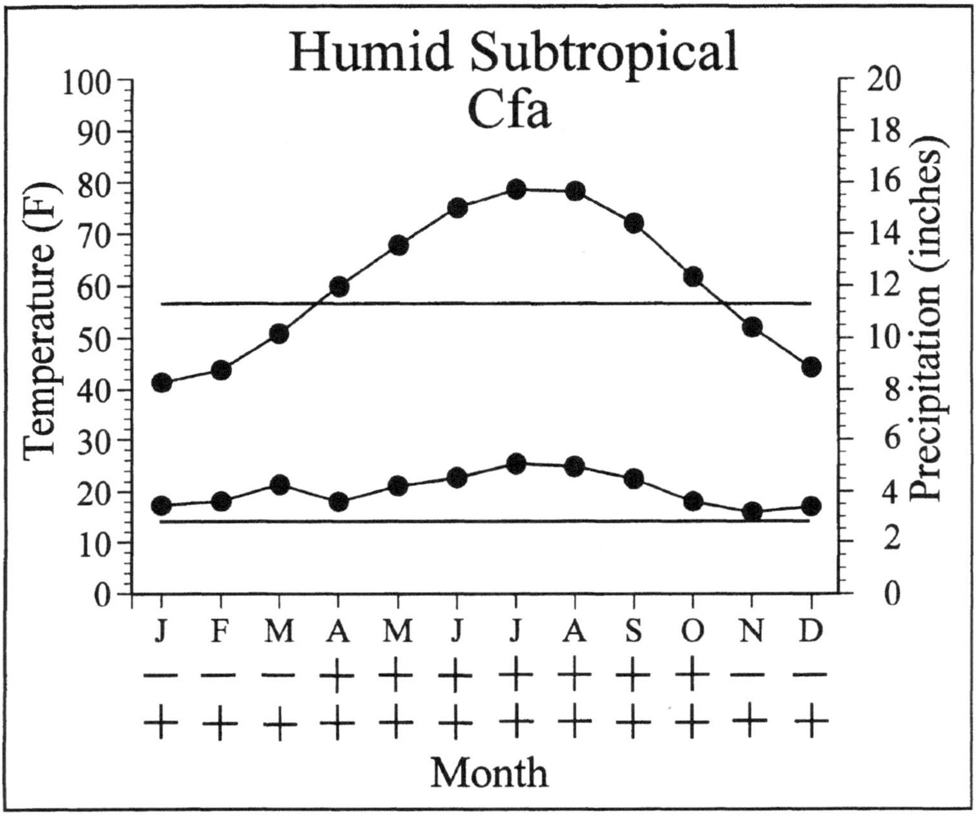

Figure 6.4: Monthly temperature and precipitation for a typical humid subtropical (Cfa) climate category. The upper lines and left axis of the plot relate to temperature and the bottom lines and right axis of plot relate to precipitation. The first row of signs indicates temperature is above (+) or below (-) average and the second row of signs indicates precipitation is above (+) or below (-) average. After Lloyd and Carbone (1995).

arbitrary letter names (A, B, N, and M). None of the subjects were familiar with the climate categories used by geographers and determined their decisions based on the visual impressions made by the lines representing temperature and precipitation. Subjects were given ten exemplars for each climate category for ten epochs of training. When subjects classified an exemplar into an incorrect category they were informed of their mistake and provided with the name of the correct category. This training method allowed subjects to reduce the number of error they made over the ten epochs of training as they learned the nature of the four categories (Figure 6.8). Error was initially very high for all four categories because subjects had no knowledge of the categories at the beginning of the learning process. Although the error decreased substantially for all categories, the Mediterranean with warm summers (Csa) category appeared to be more difficult to learn than the other categories (Figure 6.8).

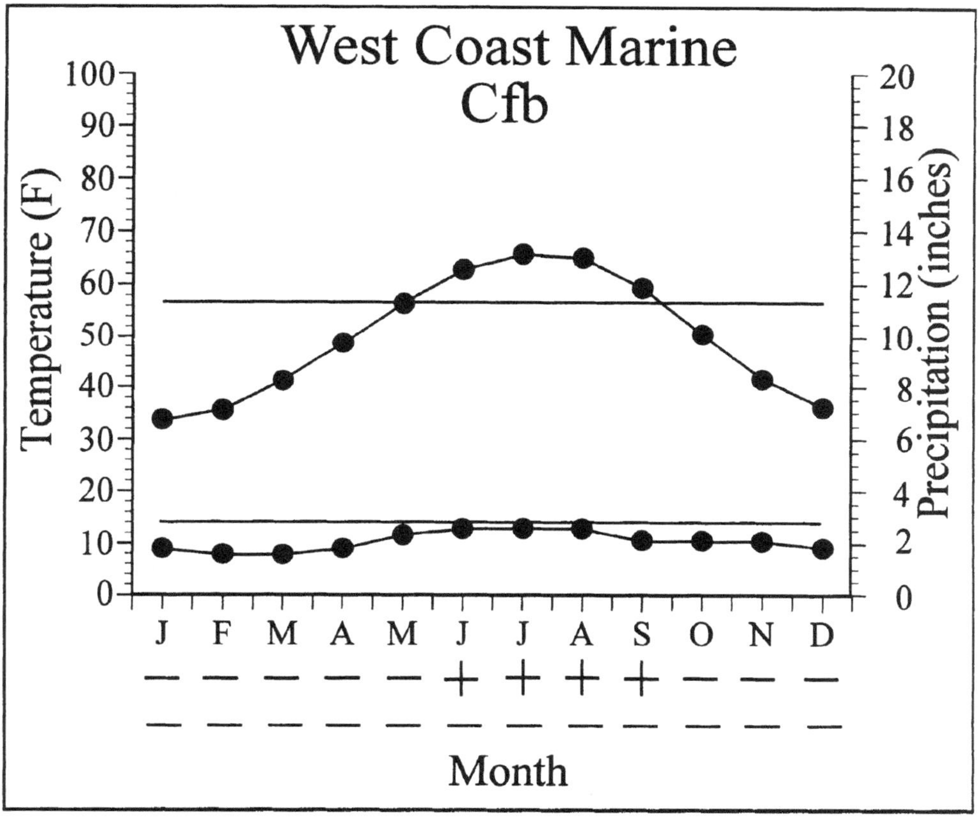

Figure 6.5: Monthly temperature and precipitation for a typical west coast marine (Cfb) climate category. The upper lines and left axis of the plot relate to temperature and the bottom lines and right axis of plot relate to precipitation. The first row of signs indicates temperature is above (+) or below (-) average and the second row of signs indicates precipitation is above (+) or below (-) average. After Lloyd and Carbone (1995).

A pattern associator neural network was designed that had input neurons that represented the temperature and precipitation for the 12 months of the year and the four climate categories as output neurons (Figure 6.9). All input neurons were connected to all output neurons and the 96 (12 X 2 X 4) connection weights were initialized to 0.0. The neural network was trained for ten learning epochs and the weights adjusted after each input pattern was passed through the network. The classification errors for the ten epochs suggested leaning was taking place (Figure 6.10). Unlike the learning curves for the climate categories for human subjects, the neural network had learned the categories equally well by the last epoch. On average, the neural network performed better over the ten epochs, but error rates were approximately the same for humans the neural network at the end of the learning process (Figure 6.11).

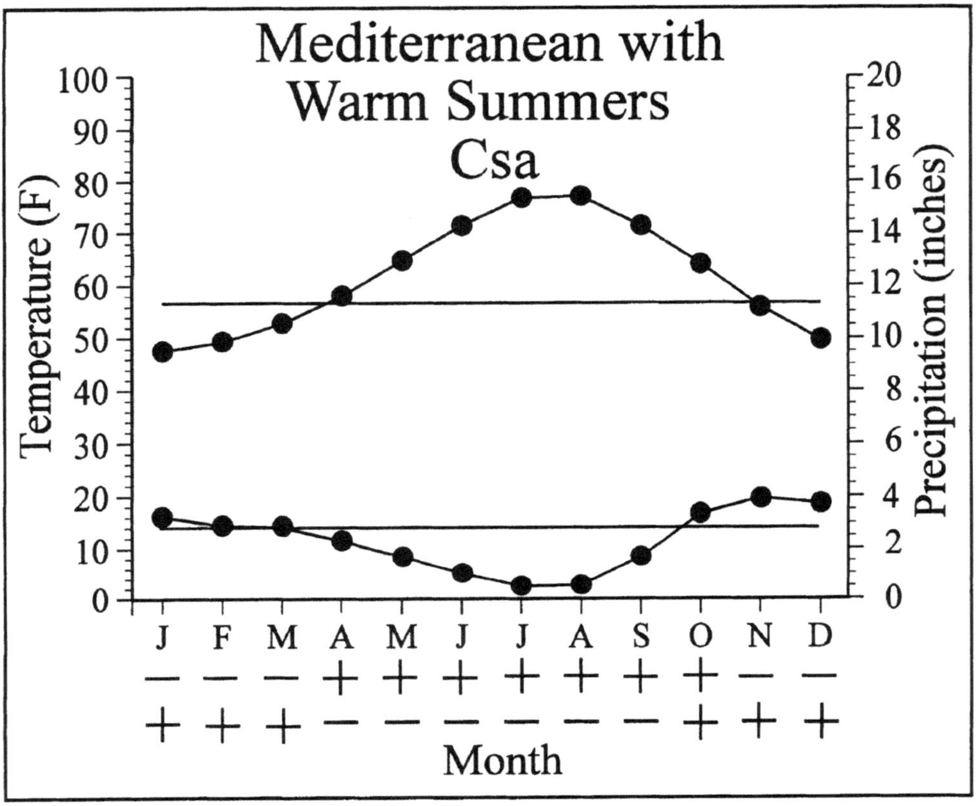

Figure 6.6: Monthly temperature and precipitation for typical Mediterranean with warm summers (Csa) climate category. The upper lines and left axis of the plot relate to temperature and the bottom lines and right axis of plot relate to precipitation. The first row of signs indicates temperature is above (+) or below (-) average and the second row of signs indicates precipitation is above (+) or below (-) average. After Lloyd and Carbone (1995).

A neural network stores the knowledge it has acquired from experiences in its connection weights. The weight learned by the neural network used by Lloyd and Carbone (1995) are represented in Figure 6.12. The graphs for the four output neurons represent their connection weights back to the 12 temperature and precipitation neurons. The bars extrude above the base for positive weights and below the base for negative weights. Positive weights turn the output neuron on and negative weights turn it off.

The humid subtropical category (Cfa) had its strongest weights associated with the summer precipitation neurons (Figure 6.12a). This should be expected since the humid subtropical category is the only category with precipitation values in the summer that were above average (Figure 6.4). The Cfa category also has a high positive weight for its connections to the April temperature input neuron and a moderately positive weight for its connection to the October temperature input neuron marking the longer summers associated with Cfa climates.

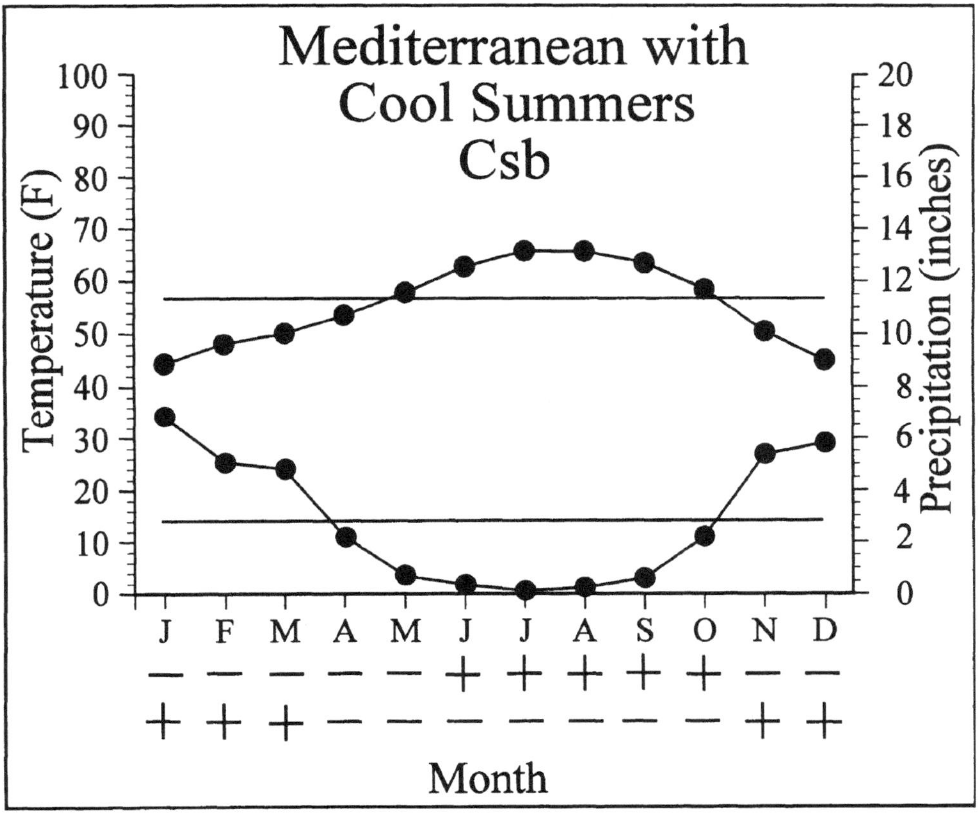

Figure 6.7: Monthly temperature and precipitation for typical Mediterranean with cool summers (Csb) climate category. The upper lines and left axis of the plot relate to temperature and the bottom lines and right axis of plot relate to precipitation. The first row of signs indicates temperature is above (+) or below (-) average and the second row of signs indicates precipitation is above (+) or below (-) average. After Lloyd and Carbone (1995).

The west coast marine category (Cfb) was defined by the neural network through its below average precipitation in the winter months (Figure 6.12b). This category was turned off by strong negative weight if the network experienced precipitation in the winter months. The neural network was able to learn that the west coast marine category was the only one not experiencing above average precipitation in the winter (Figure 6.5). The network also produced strong negative connections between the April and October temperature input neurons and the Cfb output neuron to document the shorter summer associate with the west coast marine category and to note that only Cfb category has below average temperatures in October (Figure 6.5).

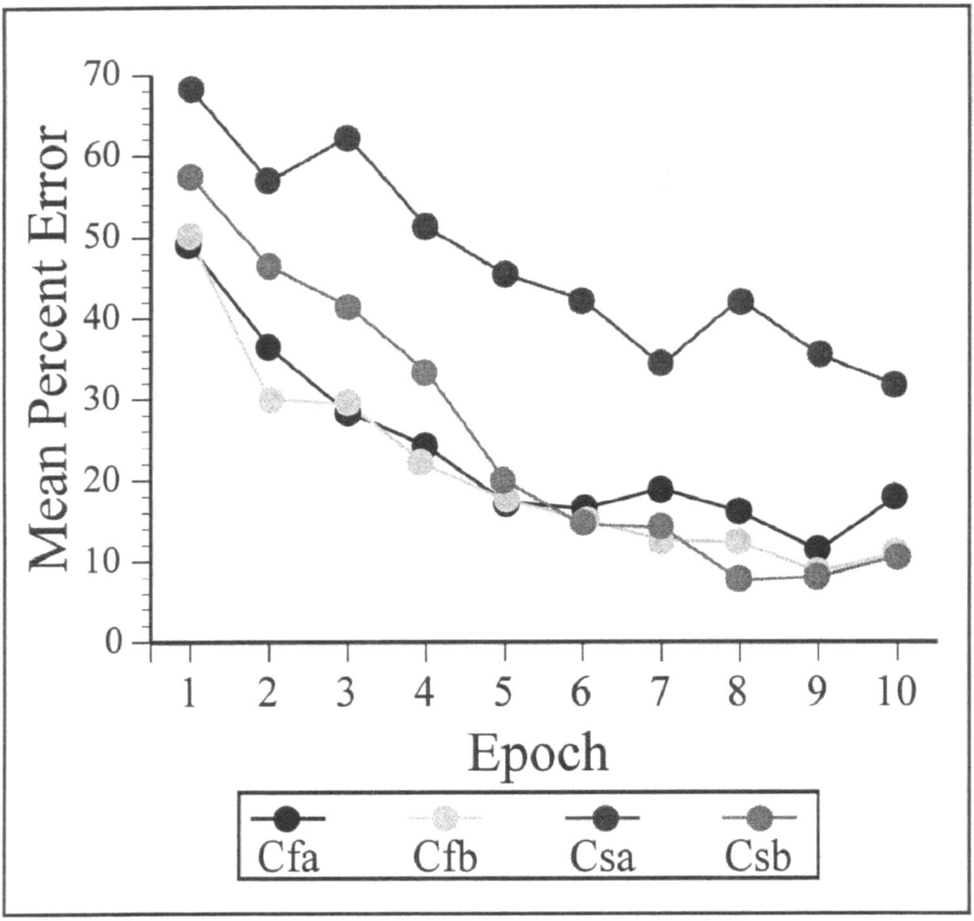

Figure 6.8: The mean percent error for human subjects for the four climate categories plotted over ten learning epochs. After Lloyd and Carbone (1995).

The neural network produced strong positive weights for the October and November temperature neurons and a moderately positive weight for the April temperature neuron for the Csa category to reflect the transition months of a long warm season(Figure 6.12c). This pattern resembles the temperature weight pattern for the Cfa category (Figure 6.4) except the end of the summer has the largest weights for the Csa category and the Cfa category has its largest weights at the beginning of summer. The Mediterranean with warm summers category (Csa) had relatively large negative weights associated with the summer and winter precipitation neurons and positive precipitation connections reflecting a transition period between above average and below average precipitation (Figure 6.12c). Although human subjects were not able to classify this category as successfully as the others (Figure 6.8), the neural network was able to learn the subtle patterns associated with this category by Epoch 10 (Figure 6.10).

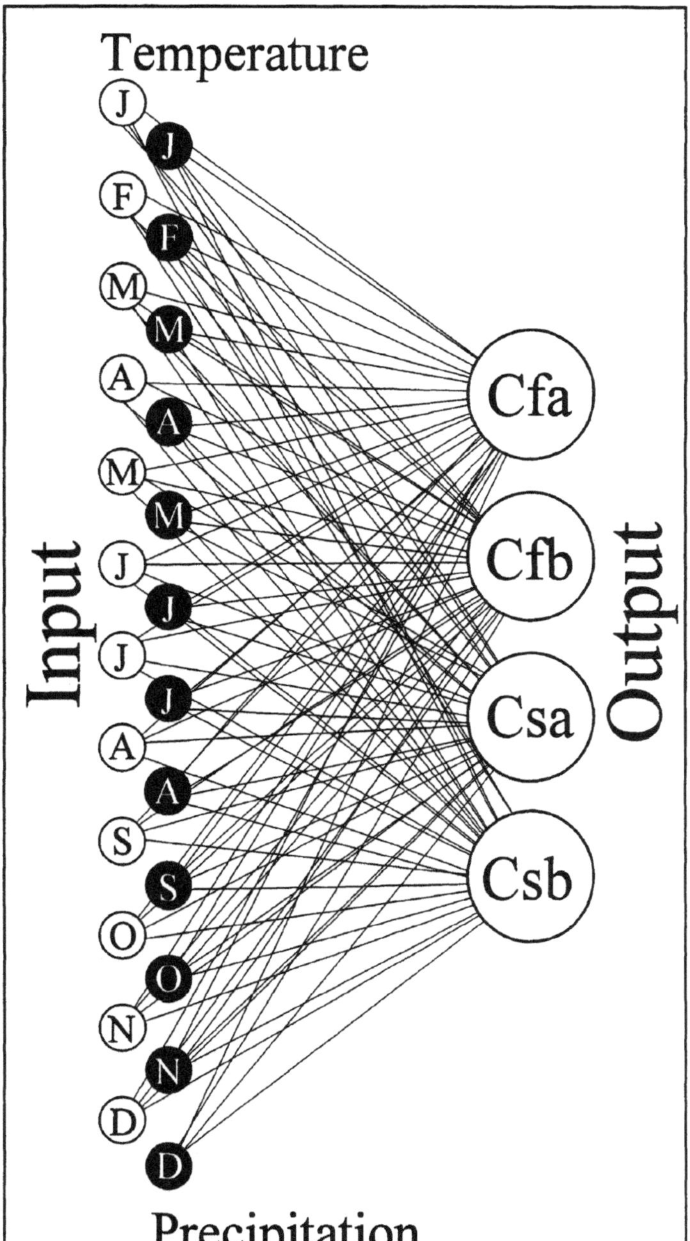

Figure 6.9: Pattern associator neural network with monthly temperature and Precipitation input neurons and four mesothermal climate categories as output neurons.

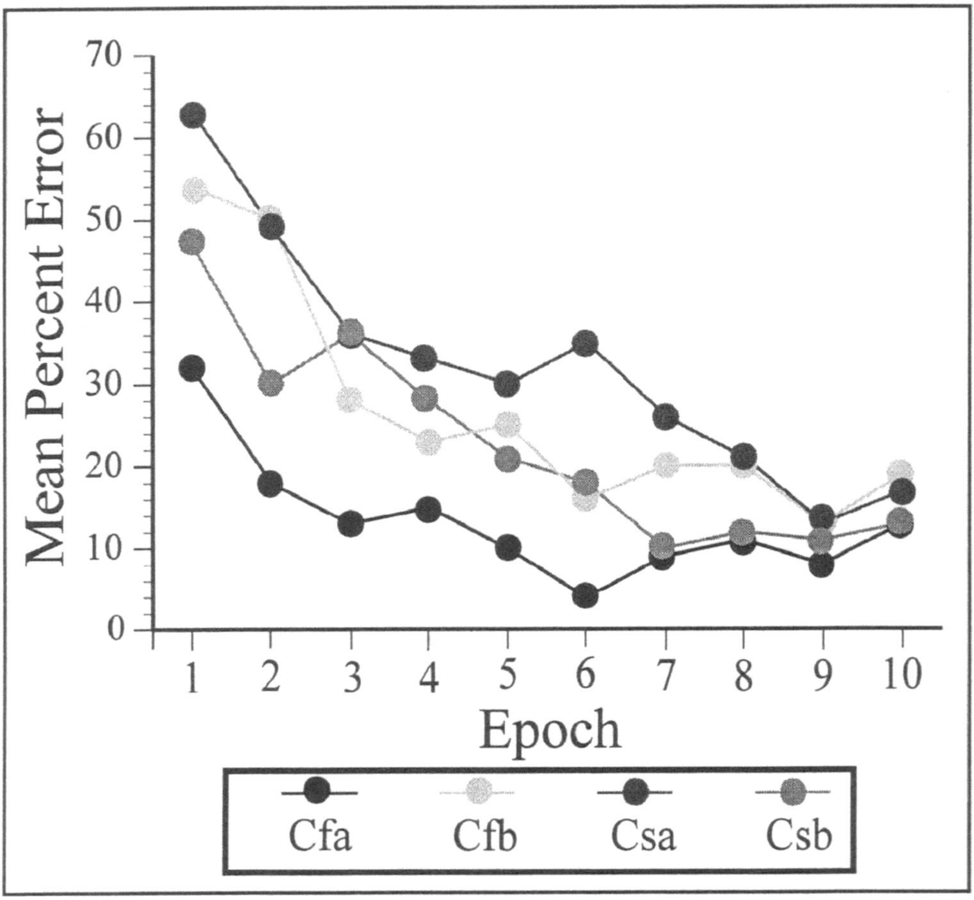

Figure 6.10: The mean percent error for the neural network for the four climate categories plotted over ten learning epochs. After Lloyd and Carbone (1995).

The Mediterranean with cool summers category (Csb) had a strong seasonal variation in precipitation and the neural network's weights reflect this (Figure 6.12d). The Csb output neuron is turned on by the large positive weights for winter precipitation input neurons and turned off by the strong negative weights for summer precipitation input neurons. None of the weights for the Csb category's temperature neurons were positive with the November weight being strongly negative. This made the Csb category distinct from the Csa category that had a large positive weight associated with its November temperature neuron.

The weights learned by the neural network reflected the visually distinctive features of the climate categories with the precipitation weights having stronger patterns than temperature weights (Figure 6.12). Weights for temperature neurons appeared to be

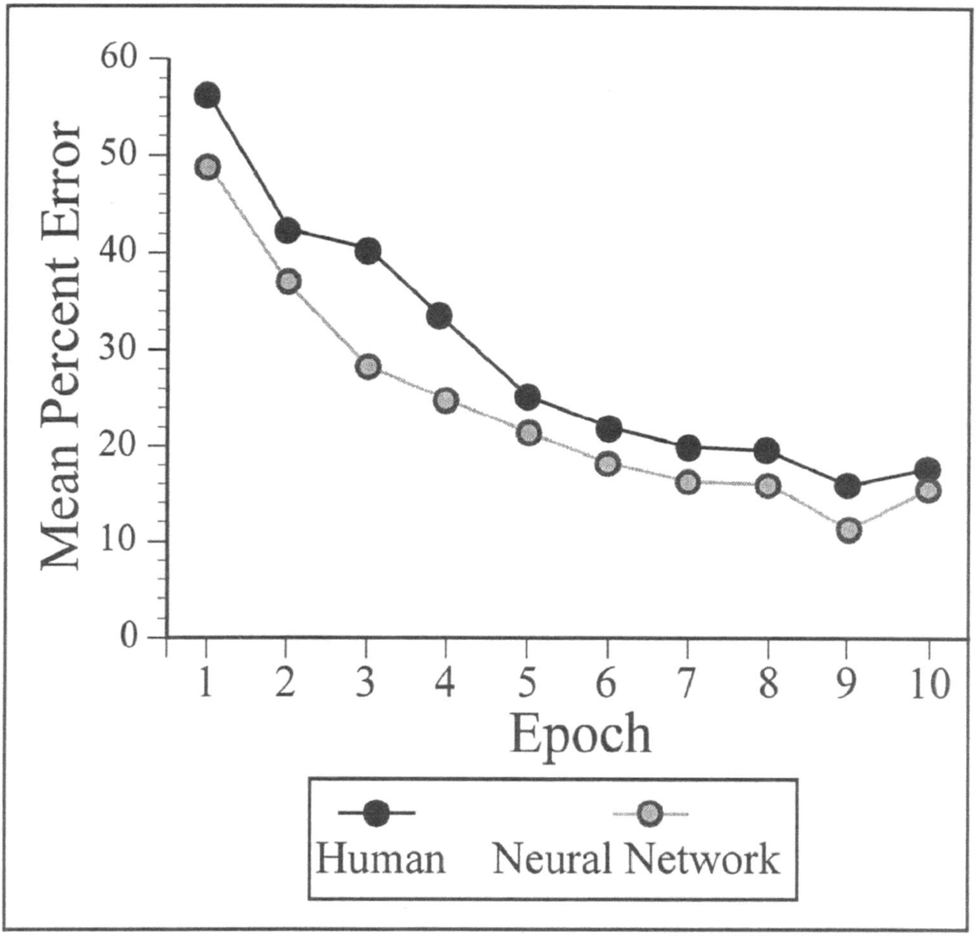

Figure 6.11: Comparison of learning curves aggregated over all categories for human subjects and the neural network. After Lloyd and Carbone (1995).

directed narrowly on the timing of the transition from above average to below average temperature. Precipitation weights highlighted broad differences in summer and winter rainfall patterns. This should be expected because all four categories were mesothermal climates. Extreme months for high and low temperature were about the same for the four mesothermal climates.

Lloyd and Carbone's (1995) simulations suggested both human subjects and the neural network could successfully classify the climates after ten epochs of learning and that the learning process was similar for humans and neural networks (Figure 6.11). Human subjects did appear to have more difficulty learning the Mediterranean with warm summers (Csa) category (Figures 6.8 and 6.10). The most encouraging result of this study, however, was that the neural network's connection weights reflected distinctive characteristics of the input patterns for the categories. Lloyd and Carbone (1995) argued that simulations done with neural network models can possibly give insights into the processes being used by human decision makers.

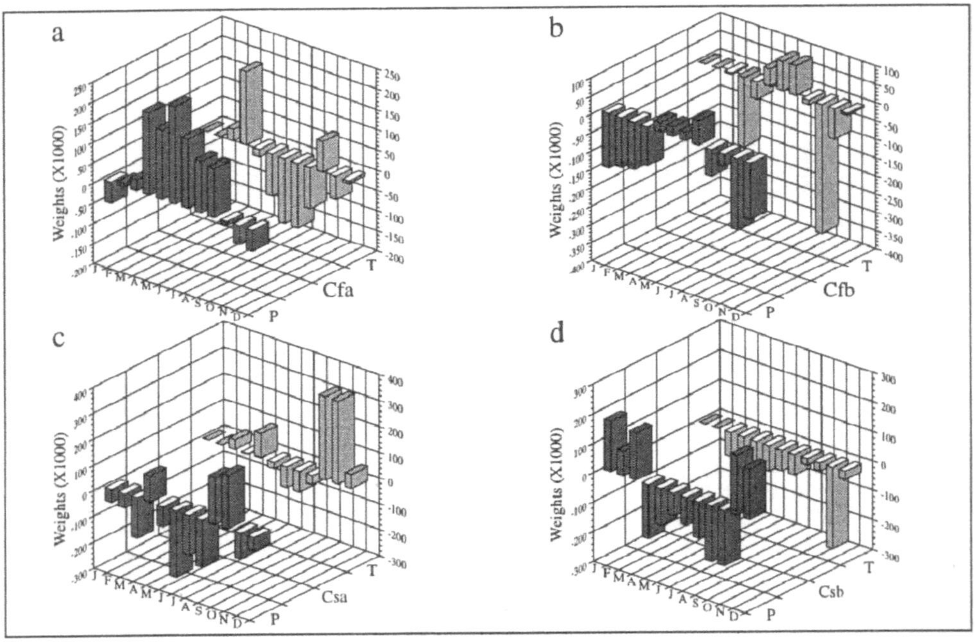

Figure 6.12: Final weights connecting input neurons to the four output neurons for Humid Subtropical (a), West Coast Marine (b), Mediterranean with warm summbers (c), and Mediterranean with cool summers (d) climates.

6.4. Learning higher-order categories

The remainder of this chapter considers how information about places is learned and organized in our memory. Experiences that we have with specific geographic locations allow us to generalize what we know about the geographic environment. For example, most people think of their home as a special place, but most of what you know about your home is generally true for other locations in the environment in the same category as your home. Our home is special because of the many experiences we have that forges an emotional attachment with our home that is not true for all places (Tuan 1974). There is, in fact, an entire hierarchy of geographic places that could be called home (Figure 6.13).

The woman in Figure 6.13 responding to the question "Where is your home?" must consider which scale is most appropriate for answering the man's question. If she was talking to a local man in a small village in a faraway country, she might respond with the name of a continent, e.g., "North America." If she was talking to a man walking in her neighborhood, she might respond by pointing out her house, e.g., "the blue house at the end of the street." A description of a house would not be an appropriate response to the man in the small village in a faraway country or a continent

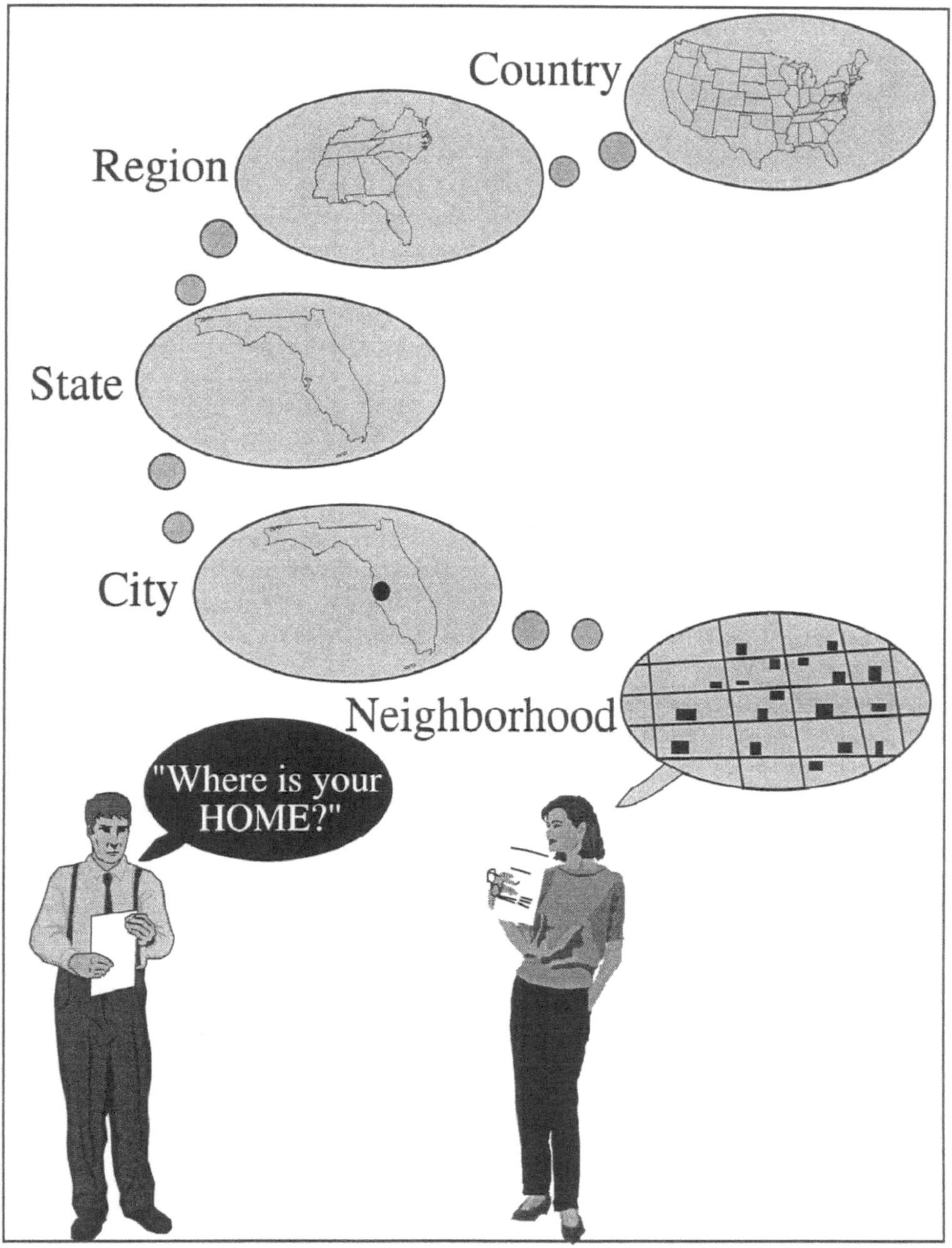

Figure 6.13: A person answering the question, Where is your home? could respond in a variety of ways. After Lloyd, Patton, and Cammack (1996).

an appropriate response for the man walking in her neighborhood.. The appropriate response considers the expected knowledge of the person asking the question and this is related to geographic scale. A person far away does not have the specific knowledge to appreciate a detailed description of the local environment that someone in your neighborhood might expect.

The organization of information on cartographic maps offers some hypotheses about how information might be organized in cognitive maps. As discussed in previous chapters, cognitive maps can be viewed in a number of ways. At one level the imagery versus conceptual-proposition debate focused attention on methods used to process and represent spatial information. At a more micro level, connectionist models consider information in our memories to be distributed throughout a system of neurons called a neural network. Within neural networks, information is stored as the strengths of connections between neurons and learning is considered as changes in the strengths of these connections (Gabriel and Moore 1990; Kosslyn and Koenig 1992; Lloyd 1994; Lloyd and Carbone 1995; Martindale 1991). The current discussion focuses attention at a more macro level. It is concerned with a fundamental conceptual structure supporting spatial information. It provides a view of cognitive mapping as more than just encoding and storing images or conceptual propositions about the environment. A significant part of the learning process generalizes the encoded information and creates an efficient hierarchical structure for storing environmental knowledge. Golledge (1993, p. 41) has argued "learning the language and unpacking the essence of the concepts (as well as providing many examples of their existence from the everyday environment) provides the tools for understanding the level of environmental knowledge that one develops through association and experience." Learning geographic concepts and levels of environmental knowledge is important because it is these concepts and this knowledge that is used to communicate about environments and make everyday spatial decisions.

6.5. The Organization of Geographic Information

Places on cartographic map are frequently organized in a spatially nested hierarchy. Higher-order geographic categories in the hierarchy act as containers for lower-order places and, in turn, these places are containers for other categories of places still lower in the hierarchy. For example, countries like the United States contain regions like New England, regions like New England contain states like Massachusetts, states like Massachusetts contain cities like Boston, and cities like Boston contain neighborhoods like Beacon Hill (Figure 6.14). The important levels of this nested hierarchy are clearly expressed on maps by cartographers with distinct symbols and boundary lines. A tree structure can express this type of relational hierarchy with container places connecting down the hierarchy with the specific places they contain (Figure 6.14). Note that each level in the hierarchy can be given a general category label for that class of place and that each node in the tree represented in Figure 6.14 depicts an actual place. This type of hierarchy has been suggested as a possible structure

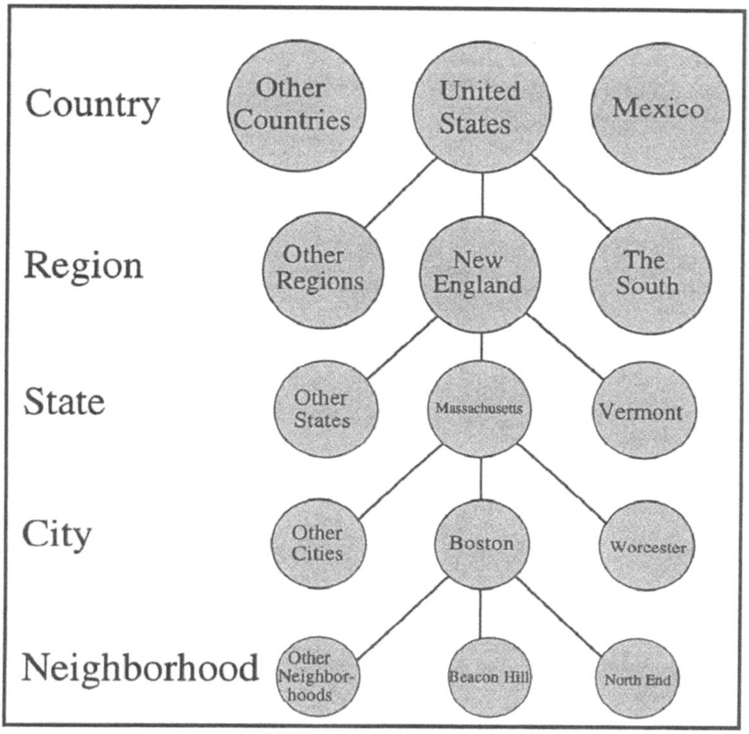

Figure 6.14: A nested hierarchy of geography locations.

for spatial knowledge by a number of researchers (Stevens and Coupe 1978; Eastman 1985).

It was suggested that this type of hierarchical structure is the underlying cause of metric distortions in judgments of spatial relationships (Stevens and Coupe 1978). It was also argued that it is efficient to encode explicitly the fact that Ohio is east of Colorado, but not to encode explicitly the fact that Cincinnati is east of Denver. The relationship between any city in Ohio and any city in Colorado can be inferred by the relationship between the higher-order containers (Figure 6.15). This encoding strategy generally will produce correct information, but also will produce errors for specific cases. To illustrate this point the authors performed an experiment that had subjects at the University of California, San Diego estimate the direction from origin cities to destination cities. Errors occurred when the direction between higher-order categories (California and Nevada) was different from the direction between lower-order categories (San Diego and Reno). Subjects living in San Diego made the same systematic error with their home town and Reno, Nevada that they made with other pairs of towns (Figure 6.16). Their estimates of the relationships between the lower-order places (cities) appeared to be influenced by knowledge about higher-order places (states).

This type of hierarchical structure is useful for expressing spatial relationships, but it has no way of expressing how we generalize information as we learn about our environment. Abstract conceptual knowledge that we acquire from learning cannot be stored in this type of structure since it has only one level of abstraction, i.e., all nodes represent specific geographic locations.

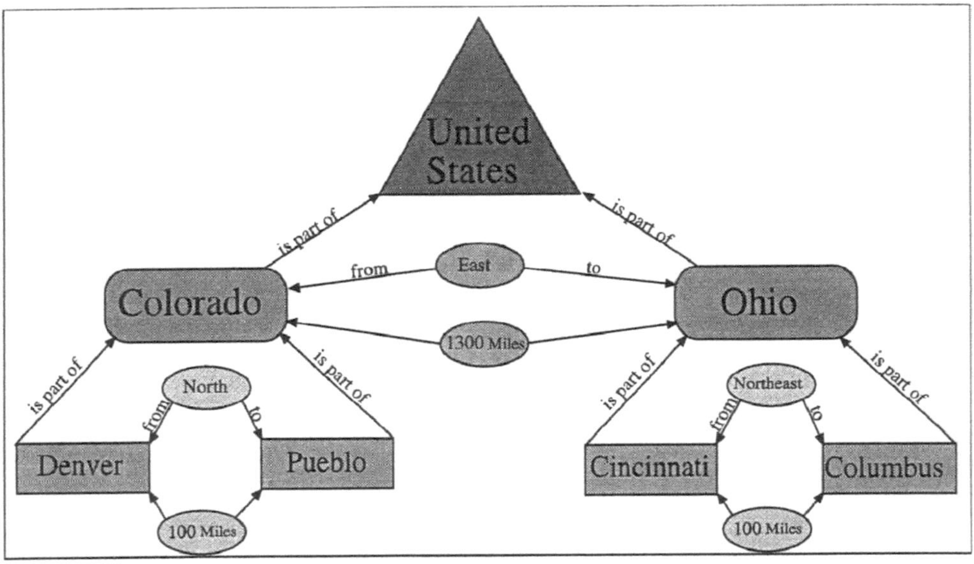

Figure 6.15: A hierarchical representation of spatial information in the United States. After Stevens and Coupe (1978).

Another hierarchy present on many cartographic maps has been discussed by Patton (1995). He argued that the highest level of abstraction on maps is the figure and ground distinguished by the cartographer (Dent 1972). Important objects are represented in the figure of the map and less important places are relegated to the ground. Cartographers steer the map reader's attention to objects that are important for the purpose of the map by manipulating what objects are in the figure and ground. This divides the map into two very abstract categories. Categories of geographic information at a less abstract level provide more information and are represented by point, line, and area symbols. A map might have dots to represent cities, lines to represent highways or rivers, and areas to represent political units. The symbolization selected by the cartographer provides categorical information at a middle level of abstraction. At the lowest level of abstraction, lettering is used to label symbols to identify them as specific places. Take as an example a simplified map of Eastern Ohio and Western Pennsylvania. At the highest level of abstraction, the areas on either side of the boundary between Ohio and Pennsylvania have been designated as the figure of interest and parts of the surrounding states of New York and West Virginia and Lake Erie have been assigned to the ground. Two states, represented as white and black areas, cities, represented white and black dots, and highways, represented as grey lines are shown as less abstract categories represented by the symbols. At this middle level it is easy to differentiate one state from the other or a highway from a city. Specific states, cities, and highways are identified with labels to bring them to a less abstract and more informative level. At this lowest level of abstraction we have more information and can differentiate Pittsburgh from Cleveland, the Pennsylvania Turnpike from Interstate 79,

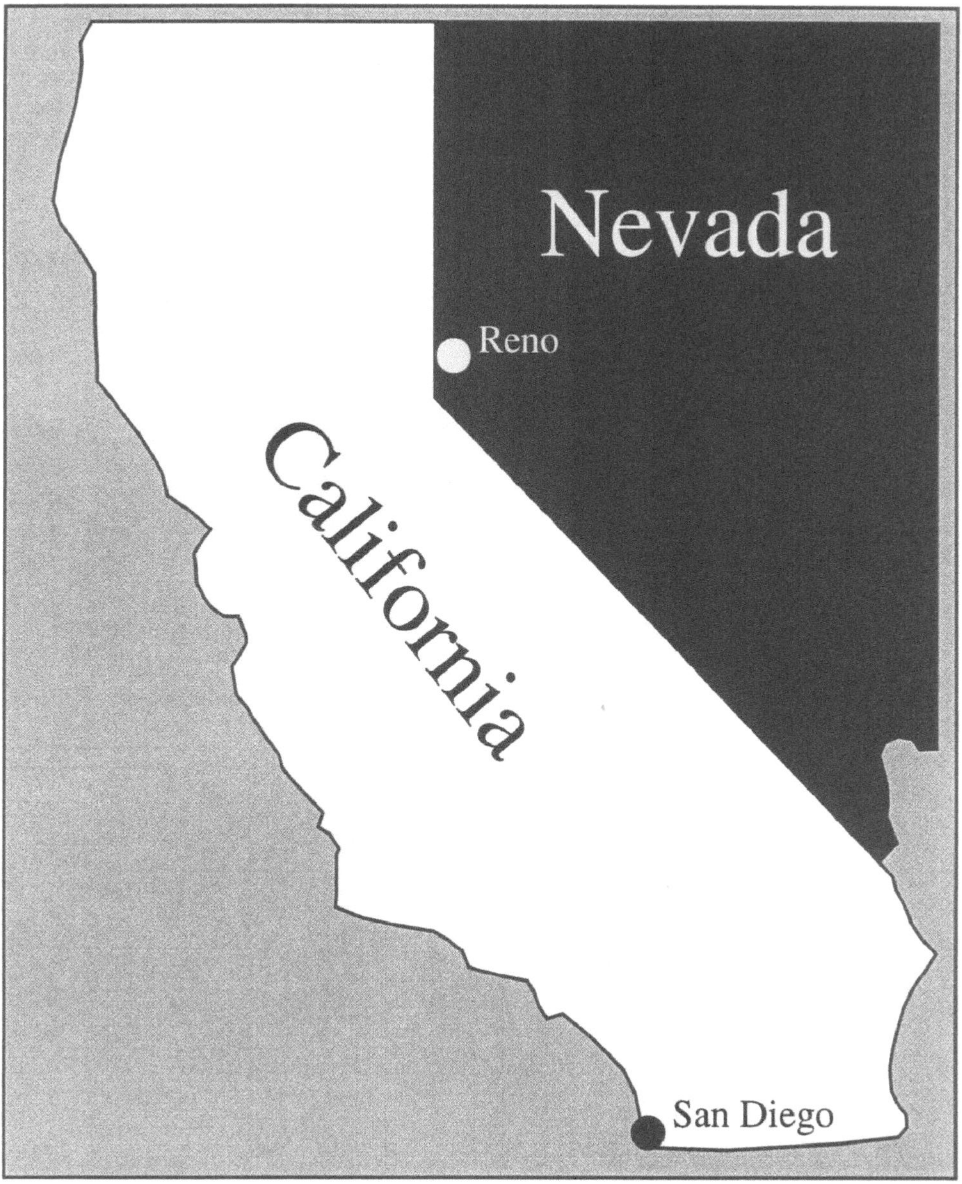

Figure 6.16: Reno is west of San Diego at a lower level in the hierarchy, but Nevada is east of California at a higher level in the hierarchy.

and Ohio from Pennsylvania (Figure 6.17). The other cities and highways remain unidentified and, therefore, exist only at the symbolic middle level. Other middle-level categories such as lakes or rivers have not been symbolized and, therefore, do not exist on the map at the lower levels of abstraction.

Another hypothesis for how information might be structured in our cognitive maps is suggested by this alternate view of how geographic categories are arranged on cartographic maps. Specific place nodes from Figure 6.15 are shown in Figure 6.18 as

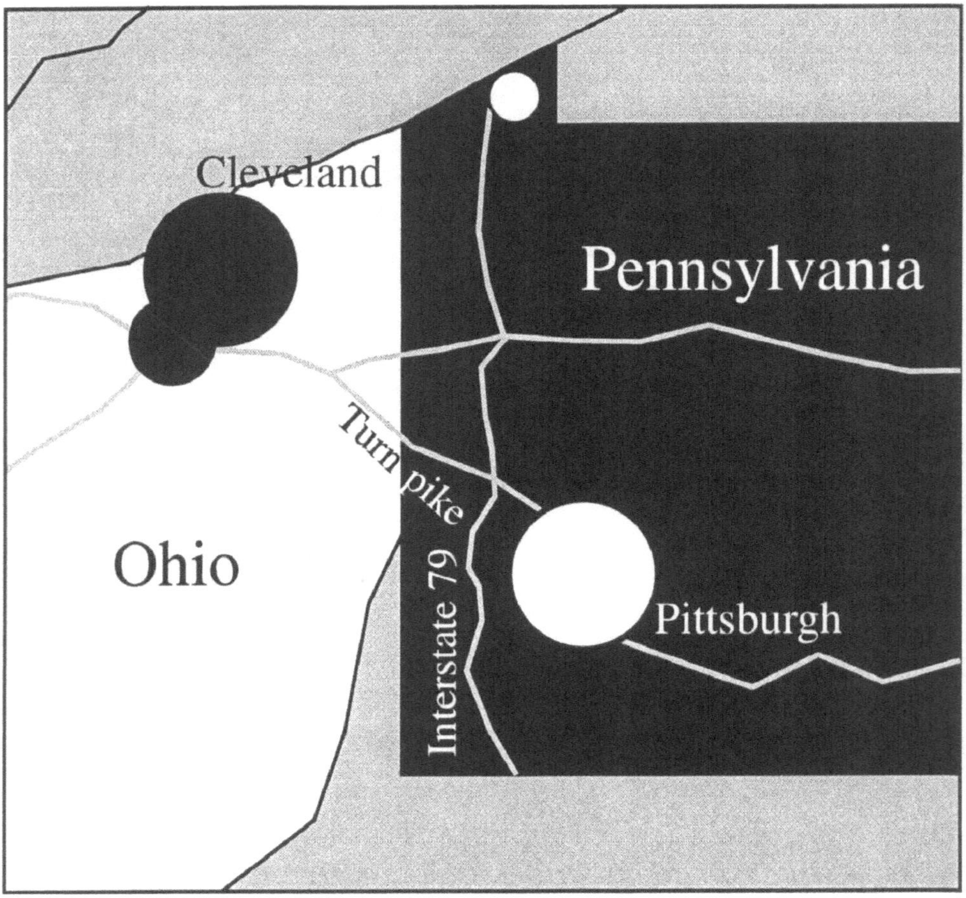

Figure 6.17: A simplified reference map of the boundary between Ohio and Pennsylvania. Map symbols represent basic-level categories.

the nodes in the lower layer of a three-level hierarchy. Note that the more abstract and less informative categories in the middle and upper level nodes are not specific places. The actual places shown at different levels in the nested hierarchy represented in Figure 6.15 are now the nodes in the subordinate level of the basic-level hierarchy. Each of subordinate nodes is connected to one node in the basic level of the hierarchy that represent abstract categories of places. Note the names of the nodes in the basic level of the hierarchy represented in Figure 6.18 are the same as the names of the levels of the nested hierarchy in Figure 6.15. We also learn about a category of place in the basic level of the hierarchy every time we experience a specific place in the subordinate level of the hierarchy. When we visit any country we learn about that country, but we also learn about the category country. The most abstract spatial concept, the category place, is at the highest level of this three-level hierarchy. Learning about specific places and using this information to learn about categories of places at the basic level also allows us to learn what the category place is like. The superordinate level might have other very abstract categories like persons, and things.

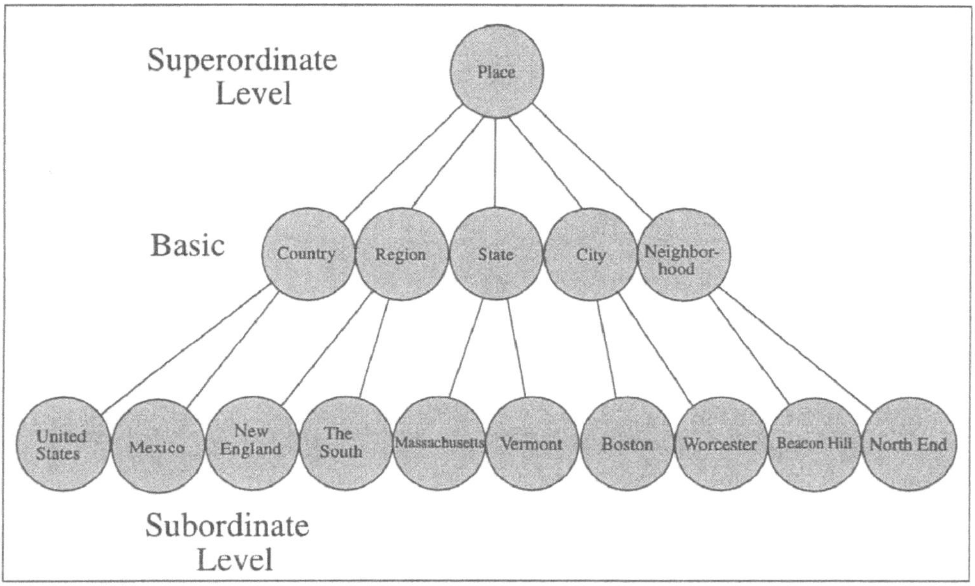

Figure 6.18: A basic-level hierarchy of geographic locations. After Lloyd, Patton, and Cammack (1996).

6.5.1. THE BASIC-LEVEL THEORY

People are interested in knowing where your home is located and what characteristics it has. Eleanor Rosch has argued people have three-level hierarchical memory structures that store information about objects (Rosch and Mervis 1975; Rosch, Mervis, Gray, Johnson, and Boyes-Braem 1976). She argued that the object that took you for your holiday in Nebraska would be called a vehicle at the superordinate level, an airplane at the basic level, and a Boeing 727 at the subordinate level. The subordinate level is the most informative and the superordinate level is the most distinctive (Figure 6.6). This implies the subordinate level holds more specific information and the superordinate level holds more general information. The basic level is a compromise between being informative and distinctive and the level most often used for naming objects. If you were asked how you got to Nebraska by another tourist, in most situations you would probably respond "by airplane" rather than "in a vehicle" or "on a Boeing 727." The response "vehicle" is too general and does not provide enough information. The specific type of airplane is not needed because the various types of airplanes share many characteristics and are, therefore, not that distinctive from one another. A variety of specific types of knowledge have been considered by studies testing basic-level theory. Topic of interest have included emotions (Shaver, Schwartz, Kirson, and O'Conner 1987), personality traits (John, Hampson, and Goldberg 1990), visual presentations (Murphy and Smith 1982; Murphy and Wisniewski 1989), events (Morris and Murphy 1990), and environmental scenes (Tversky and Hemenway 1983). The possible use of basic-level theory for the structuring of feature information in a GIS system has also been suggested (Usery 1993). Lloyd, Patton, and Cammack (1996) have also considered Rosch's basic-level theory for geographic categories and spatial information.

Objects that belong to natural categories were one of Rosch's earliest studies related to basic-level theory (Rosch, Mervis, Gray, Johnson, and Boyes-Braem 1976). Three-level taxonomies for familiar objects were considered as part of her study, e.g., fruit (superordinate level), apple (basic level), and Granny Smith apple (subordinate level). Subjects listed as many attributes for the objects as possible in 90 seconds after they were provided object names at all three levels in the taxonomy. She expected that an object named at the superordinate level as fruit would result in subjects being able to name relatively few attributes. This should occur because few attributes are appropriate for abstract categories like fruit. She also expected that an object named at the basic level as an apple would result in subjects being able to provide significantly more attributes than could be named for fruit at the superordinate level. This should occur because more information is known about the less abstract basic-level categories. It should be remembered that this information has been learned from experiences with specific objects. She also expected that an object named at the subordinate level as a Granny Smith Apple would not result in subjects listing significantly more attributes than could be provided for apple at the basic level. This should occur because specific objects at the subordinate level are only slightly less abstract than basic-level objects. They should, therefore, have only a few attributes not also associated with basic-level objects. In other words, we know a lot more information about apples than we do about fruit, but not much more about Granny Smith apple than about apples. The following conclusion was offered after using a variety of methods and stimuli to perform 12 different experiments:

> For categories of concrete objects, basic objects are the most general classes at which attributes are predictable, objects of the class are used in the same way, objects can be readily identified by shape, and at which classes can be imaged. Basic objects should generally be the most useful level of classification. Universally, basic object categories should be the basic classifications made during perception, the first learned and first named by children, and the most codable, most coded, and most necessary in the language of any people (Rosch, Mervis, Gray, Johnson, and Boyes-Braem 1976, p. 435).

Murphy and Smith (1982) studied why objects can be categorized faster at the basic level through a number of experiments that used pictures of artificial objects. Subjects were trained to associate pictures of artificial tools with artificial three letter names. They also learned to categorize the tools into superordinate categories based on the functions for the tools, into basic categories based on the shapes of the tools, and into subordinate categories based on differentiating features related to parts of the tools. Confounding effects, e.g., as the length of names of objects, familiarity of objects, and the salience of objects' features, were accounted for by the experiments. Results from Murphy and Smith's (1982) experiments supported the argument that objects can be categorized faster at the basic level (Rosch, Mervis, Gray, Johnson, and Boyes-Braem 1976) because basic-level objects have more distinctive attributes. People's preference to use basic-level categories in verbal communications might also be related to this characteristic of basic-level information.

People's preference for using basic-level categories to name objects was considered by Tversky and Hemenway (1983) to be a compromise between two important, but opposing, goals of categorization. It is important to use informative categories. Much additional information associated with that object is revealed when it is known that an object belongs to a particular category. This specific information related to the category was learned by experiencing objects that were members of that category. The more specific (less abstract) a category is, the more informative it is. Knowing your lunch will feature chicken cooked in a wine sauce is more informative than knowing your lunch will feature a dead bird. It is also important to minimize the number of different categories that need to be considered. As categories become more abstract, the number of categories diminishes. On the other hand, as categories become very specific they tend to have many irrelevant distinctions. While reading a report that terrorists have killed hundreds of innocent people with a car bomb, one is not likely to be concerned with the make of the car or the type of explosives used. Communicating the terrorists' methods in basic-level terms, i.e., car bomb, would be appropriate for a news article. Basic level categories are much more informative than superordinate categories, e.g., the terrorists used weapons, but more distinctive than subordinate categories, e.g., the terrorists used a 1972 Plymouth Duster and a barrel of 87 octane Texaco gasoline.

A number of experiments conducted by Tversky and Hemenway (1983) used geographic terms to study environmental scenes. Indoors and outdoors were the two superordinate categories for the experiments. Home, school, store and restaurant were basic- level categories for the indoor scene. Park, city, beach, and mountains were basic-level categories for the outdoor scene. Specific cases like elementary school under school or Midwestern city under city served as the subordinate categories. Subjects were presented both photographs and verbal descriptions of environmental scenes at three levels of abstraction. The subjects provided lists of attributes, activities, and parts appropriate for individual scenes. Scenes that represented superordinate categories (indoor and outdoor) shared very few attributes, activities, and parts in the subjects' lists.

This suggested these were very distinctive scenes. Many attributes, activities, and parts on the subjects' lists were shared by basic-level scenes (beach, mountains, home, school). This means basic-level scenes were less distinctive than superordinate categories, but more informative, i.e., subjects could list significantly more attributes, activities, and parts. Still more specific subordinate-level category scenes (elementary school and lake beach) did not share significantly more attributes, activities, and parts than basic-level scenes. The subordinate-level scenes were also only slightly more informative than the basic-level scenes since the number of activities, attributes, and parts listed for subordinate scenes was only a slight increase over the basic-level scenes. This demonstrated they were not much less distinctive than basic-level scenes. It was concluded, therefore, that scenes were like objects in that, "categories at the basic level are highly informative, but not at the expense of a multitude of categories and classifications" (Tversky and Hemenway 1983, p 140).

Of the various types of items listed by subjects, Tversky and Hemenway (1984) argued that parts proliferated in subjects' listings for basic-level categories, but few parts were listed for superordinate categories. They suggested parts are of particular interest because members of basic-level categories can be distinguished from one another by their parts, but members of subordinate categories share parts and differ from one another on other attributes. It was argued that both the perceptual salience and functional significance of a part appear to contribute to how strongly it is associated with an object. Unlike other attributes, names of parts frequently serve a dual role of informing at once to a particular appearance and to a particular function. Tversky and Hemenway (1984, p. 189) claimed that categories are learned by processing exemplars in a particular way. "Basic categories come first, and are based primarily on parts. Then, we form higher-order, superordinate groupings, that are typically based on function, not perception, where function is rather abstractly conceived. At the same time, we also subdivide basic level categories into more specific categories, on the basis of one (or very few) perceptual or functional features." Murphy (1991a) followed up on the notion that parts are critical for forming basic-level categories. Using artificial categories that could separate part and nonpart information, he concluded that basic-level structures could be found in categories that did not have parts in common. See also Tversky and Hemenway (1991) and Murphy (1991b).

It has been hypothesized that as a person becomes an expert in a particular domain of knowledge, the additional knowledge is added to subordinate level categories rather than basic or superordinate level categories (Tanaka and Taylor 1991). Bird and dog experts were asked to list features for superordinate, basic and subordinate categories that were related to their domain of expertise. If a feature was not listed at a higher level of abstraction, it was defined as a new feature for a particular level in the hierarchy. *Has a beak* would be a new feature for the basic-level category bird because it would not be listed for the superordinate-level category animal. Has a beak would not be considered new if listed for the subordinate-level category sparrow because it was also appropriate for the basic-level category bird. If expert subjects added expert knowledge primarily to the subordinate level, Tanaka and Taylor (1991) expected their experts to list as many new features for subordinate-level categories as they did for basic-level categories. For the same categories in the expert's knowledge domain, nonexpert subjects should list more new features at the basic level than at the subordinate level. Both of these hypotheses were confirmed by comparing lists of features made by experts and nonexperts. Other experiments indicated that experts used subordinate-level names as frequently as basic-level names to identify objects in their domain of expertise. A dog expert might identify a picture as a German Shepard while a nonexpert might identify the same picture as a dog. Experts were also found to be as fast to verify category membership at the subordinate level as at the basic level. Evaluating these results, Tanaka and Taylor (1991, p. 481) concluded "human categorization appears to be continually reshaped and altered by learning and experience."

Basic-level categories are important because they contain much more information than superordinate-level categories. The knowledge that something called X is a place does not supply much specific information. This is because the superordinate level only stores information that applies to all places. Few characteristics, activities or behaviors are associated with all places and few parts are included in all places. Knowing that something called X is a state, city or another basic level category provides a person with much more information than knowing it is a place. The characteristics, activities, or parts connected with all countries, regions, states, cities, or neighborhoods are stored at the basic level. Knowledge stored at the subordinate level, i.e., knowing that the something called X is Massachusetts, does not provide much more information beyond knowing that something called X is a state. This is true because what is known about states at the basic level is also known for specific states at the subordinate level. There is some specific information that just applies to Massachusetts or Vermont, but it is not significantly more than what is known about states. This implies subordinate level categories should be slightly more informative, but also less distinctive than the basic-level categories. This makes basic-level categories very useful because they are both sufficiently informative and distinctive.

6.5.2. BASIC-LEVEL GEOGRAPHIC CATEGORIES

For geographic categories, the superordinate level of the three-level hierarchy is occupied by the category *place* (Figure 6.18). The *place* category is the most general geographic category and knowing that a name refers to a place makes the thing referred to distinctive (Rosch, Mervis, Gray, Johnson, and Boyes-Braem 1976). The amount of information true for all places should be relatively small. *Place* is the only superordinate category relevant to the discussion, but many other nonspatial categories occupy similar position in other information domains, e.g., animal, plant, vehicle, or clothing. Members of the place category share some characteristics with other members of the place category. Places do not share many characteristics with members of other superordinate categories, e.g., members of the weapon, vehicle, or clothing categories (Rosch and Mervis 1975). No one is likely to confuse a *place* with a *weapon* because categories are very distinctive from one another at the superordinate level of categorization. Knowing that a name refers to a place distinguishes the thing referred to from members of other superordinate categories, but it does not provide much specific information about the individual location. Superordinate categories are distinctive but not informative.

At the basic level of the hierarchy there are many more geographic categories that can provide answers to the question "Where is your home?" (Figures 6.13, 6.14, and 6.18). Any member of these geographic categories could potentially be called your home. More specific information is known about members of these basic-level categories. The geographic basic-level categories are, however, less distinctive from each other than superordinate categories are from each other (Figure 6.18). The names of basic-level geographic categories are familiar to everyone and frequently used to communicate about space. The focus here is on the most familiar basic-level geographical categories (country, region, state, city, and neighborhood) that define spatial locations at different geographic scales. These are also the categories that form nested spatial hierarchies based on the container principle.

Below the basic-level geographic categories are subordinate categories (Figure 6.18). Knowledge about specific locations should be found at the subordinate level. Specific home countries (United State or Mexico), regions (New England or The South), states (Massachusetts or Vermont), cities (Boston or Worcester), and neighborhoods (Beacon Hill or North End) can be found at the subordinate level. Everyone has some information about many specific locations, including a home. The maximum information would be for a location that has always been your home. The minimum information might be just a name on a map identifying a city named X in a state named Y in a country named Z.

Basic-level theory asserts that we acquire information about geographic locations and efficiently store it in a hierarchical structure in our memories. This is an efficient process because we do not have to store redundant information. If you read in a geography text that "a place has location," then anything that you know to be a place will have this characteristic (Getis, Getis, and Feldman 1994, p. 4).

6.5.3. TWO HYPOTHESES

There would appear to be two plausible hypotheses concerning the storage of categorical geographic information in human memory. One hypothesis postulates that knowledge we have about geographic locations should be structured within a geographic hierarchy based on a container principle (Figure 6.14 and 6.15). This notion suggests that part of what we know about higher levels in the geographic hierarchy should also apply to lower levels. For example, part of what is known about a country (United States) also applies to a region in the country (The South), part of what is known about a region (New England) also applies to a state in the region (Massachusetts), and part of what known about a state (Colorado) also applies to a city in the state (Denver). This seems reasonable because the lower level geographic unit is contained in the higher level geographic unit.

The second hypothesis postulates that knowledge will be stored in a hierarchical structure based on abstractness (Figure 6.18). The basic-level theory makes the prediction that differences in amounts of information stored at the superordinate and basic levels will be significant. This means we should have less knowledge related to the abstract category *place* and much more knowledge related to the basic-level geographic categories, *country*, *region*, *state*, *city*, and *neighborhood*. The basic-level theory also makes the prediction that differences in amounts of information stored at the basic and subordinate levels will not be significant. This means we should have about the same amount of information stored about specific countries and the basic-level category *country*. The same should be true for regions, states, cities, and neighborhoods.

Lloyd, Patton, and Cammack (1995,1996) tested these hypotheses in two studies. Data were gathered from students at the universities in North and South Carolina using methods previously used to investigate basic-level hierarchies (Murphy and Smith 1982; Rosch and Mervis 1975; Tversky and Hemenway 1983). Subjects listed as many characteristics, activities, or parts that they associated with a particular geographic term in a fixed amount of time. Different groups of subjects responded for the superordinate category *place*, the basic-level categories *country*, *region*, *state*, *city*, and *neighborhood*, and subordinate categories *home county*, *home region*, *home state*, *home city*, and *home neighborhood*. Subjects responding to *place* listed characteristics, activities and parts that were appropriate for all places. The basic-level groups made their lists thinking of the basic-level geographic terms. Subjects in the subordinate-level groups made their lists thinking of their individual homes.

Amount of information

Data were analyzed to determine if the amount of information listed for the various geographic categories reflected the predictions made by basic-level theory (Lloyd, Patton, and Cammack 1996). The predictions that a substantial increase in information would occur between the superordinate and basic levels and relatively little increase in information would occur between the basic and subordinate levels were supported for the geographic categories (Figure 6.19). Parts information appeared to be important for geographic categories. The basic and subordinate level had more parts listed than characteristics or activities. This supported Tversky and Hemenway (1984) argument that parts are particularly important information categories for visual objects. Although the number of parts listed was generally larger than the number of characteristics or activities, this was particularly true for the neighborhood category at the lower end of the geographic hierarchy (Figure 6.20). This possibly reflects the intensity and frequency of contact we have with neighborhoods compared with other geographic categories. When information was aggregated over the three categories, the large increase in information between the superordinate and all five basic-level categories was obvious while differences between the five basic and subordinate categories were small (Figure 6.21). The fundamental expectation regarding the amount of information stored at each level of the hierarchy was strongly supported for geographic categories (Lloyd, Patton, and Cammack 1996).

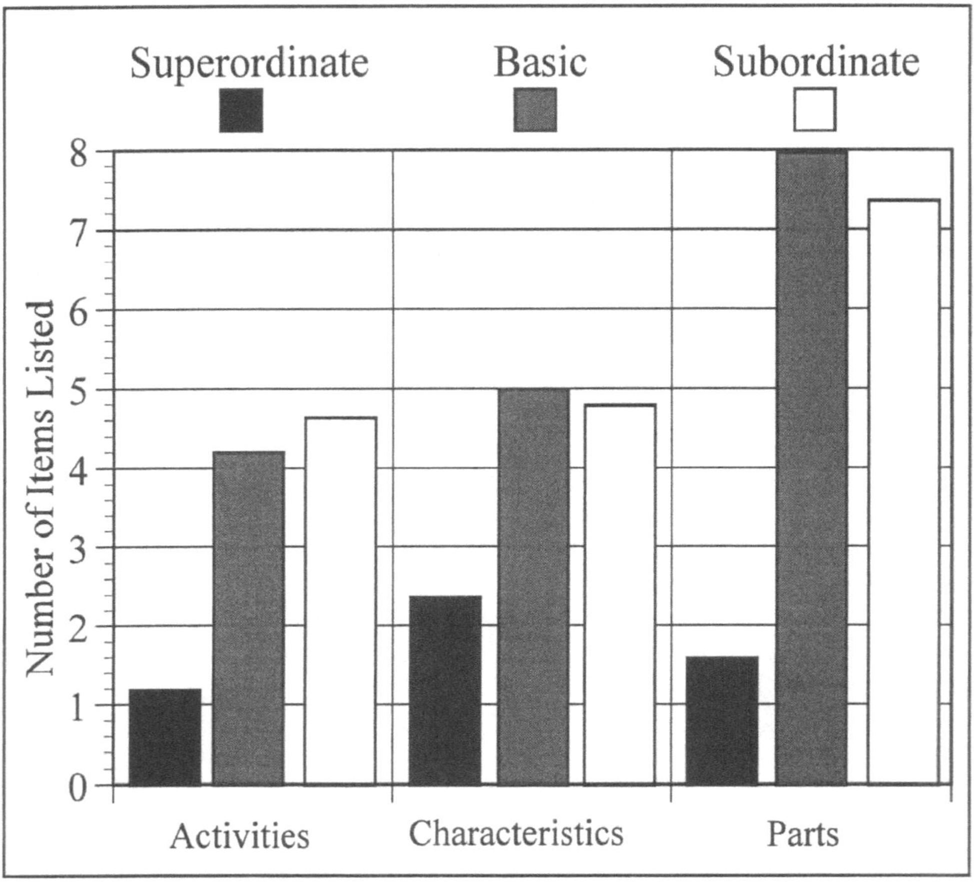

Figure 6.19: Mean number of activities, characteristics, and parts listed for superordinate, basic, and subordinate categories. After Lloyd, Patton, and Cammack 1996.

If we know that a name refers to a city or a country, we then know that all the generalized information we have acquired and stored about cities or countries applies to this location. Most of the spatial information we acquire should be stored in one of the basic-level categories. We will usually, of course, also know some additional information about the place that is unique for that particular location.

Information Content

Using the same data set, Lloyd, Patton, and Cammack (1995) focused their attention on the similarity of the information in the same 11 geographic categories. They considered if both the geographic (Figure 6.14) and basic-level hierarchies (Figure 6.18) were reflected in the content of the information stored in memory related to the.

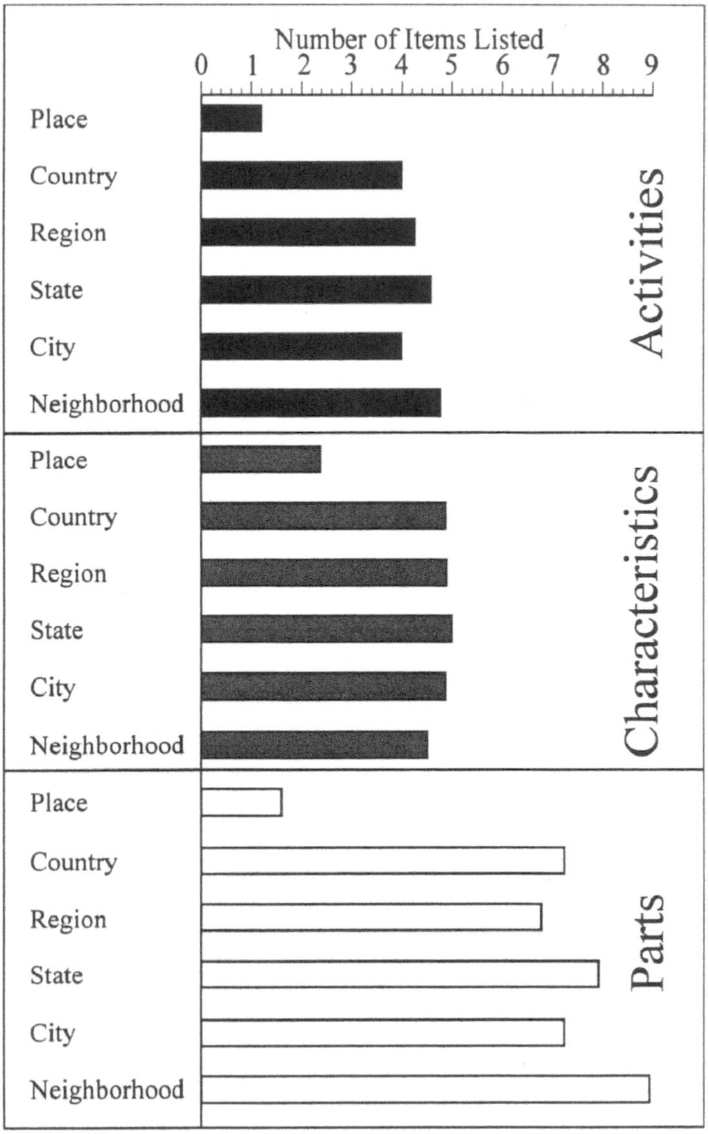

Figure 6.20: Mean number of activities, characteristics, and parts listed a by category name. Except for the place category, basic and subordinate categories are aggregated. After Lloyd, Patton, and Cammack 1996.

geographic categories.

A dissimilarity index was computed for each of the 55 pairs that could be made from the 11 geographic categories (Figure 6.16). This was done separately for the three types of information (activities, characteristics, and parts). The dissimilarity index was based on Tversky's (1977) contrast model that argues that the similarity (or dissimilarity) of two objects is based on both their common and distinctive features. Their implementation of Tversky's contrast model measured the dissimilarity of any two category lists with the following equation:

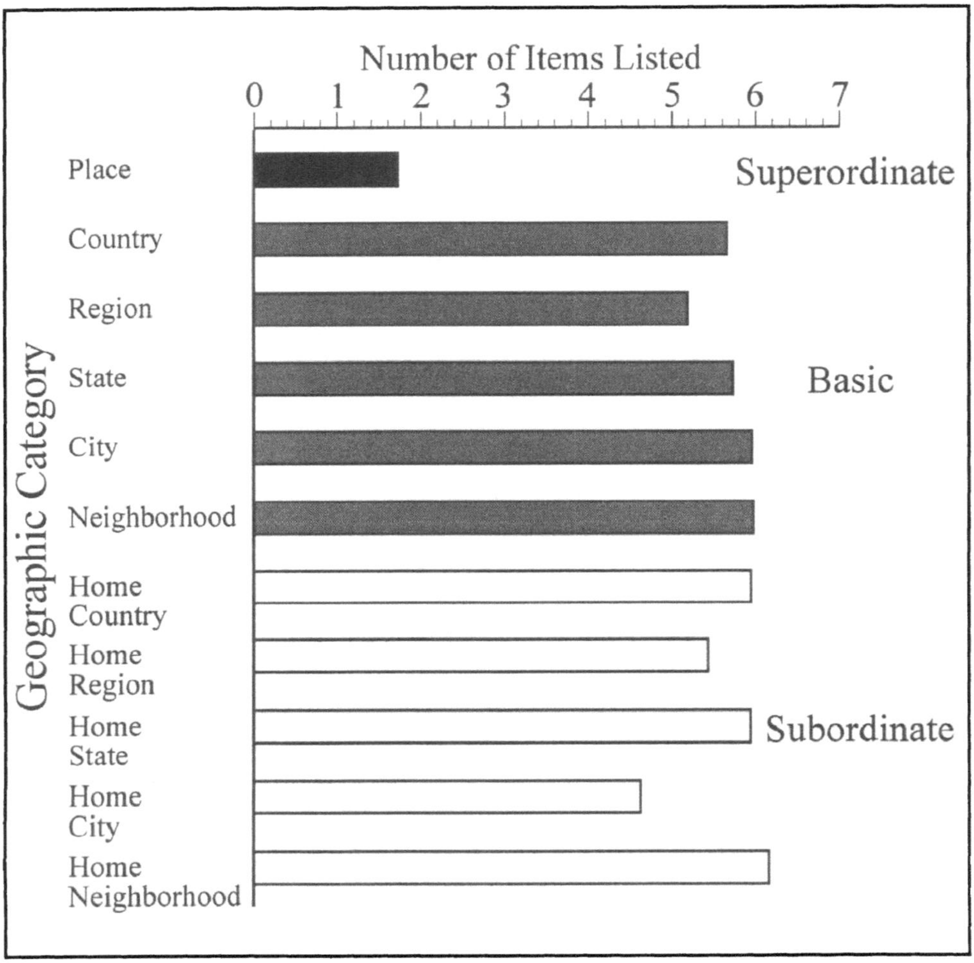

Figure 6.21: Mean number of items listed for the 11 geographic categories. After Lloyd, Patton, and Cammack 1996.

$$\text{Dissimilarity}_{i,j} = D_i + D_j - C_{i,j} \tag{6.1}$$

Where $\text{Dissimilarity}_{i,j}$ equals an estimate of the separation of categories i and j in the cognitive structure representing spatial information in our memory,
D_i equals the proportion of distinctive (unique) information associated with category i,
D_j equals the proportion of distinctive (unique) information associated with category j,
and $C_{i,j}$ equals the proportion of common information shared by categories i and j.

The dissimilarities for the three types of information were analyzed using an individual difference multidimensional scaling to recover a two-dimensional space (Figure 6.22). The final solution accounted for 85.7 percent of the total variance with

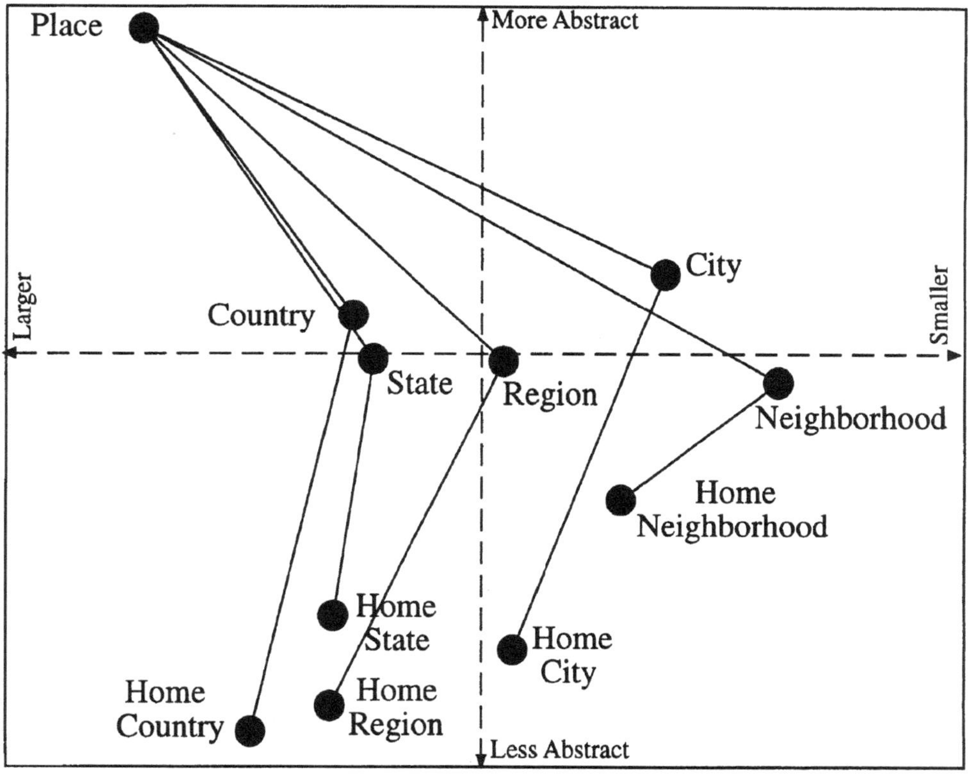

Figure 6.22: Two-dimensional space expressing the structure of the attributes, characteristics, and parts information associated with the 11 geographic categories. After Lloyd, Patton, and Cammack 1995.

the characteristics information having the lowest percentage explained (80.1) and the parts information having the highest percentage explained (91.3). The distribution of the 11 categories along the vertical dimension of the plot supported the prediction that abstractness would be one of the underlying dimensions (Figure 6.22). The distribution along the vertical axis supported the notion that the dimension measured abstractness with categories containing general information nearer the top of the space and categories containing specific information nearer the bottom of the space. The most abstract category, Place, was at the top of the space, the basic-level categories were near the center of the space, and the superordinate categories were at the bottom of the space.

The distribution of the 11 categories along the horizontal axis supported the prediction that an underlying dimension would reflect the geographic hierarchy suggested by the container principle (Figure 6.14). Larger spatial units (countries) generally were on the left side of the horizontal dimension and smaller units (neighborhoods) were on the right side (Figure 6.22). This was generally true for both the basic-level categories and the subordinate-level categories. The one exception to this generalization was that the State and Region categories switch positions within the basic-level categories. It was also interesting that place was not only on the abstract end of the vertical dimension, but also on the large side of the horizontal dimension.

An important conclusion made from these results was that the information stored in our memory related to spatial categories appears to reflect two hierarchical storage theories. Rosch's basic-level theory was supported by an analysis of the amount of information in the categories (Lloyd, Patton, and Cammack 1996), and by the analysis of the content of the information in the categories (Lloyd, Patton, and Cammack 1995). This would indicate that when we learn some information about our Home City at the subordinate level of the basic-level hierarchy we also learn something about the basic-level City category. It also implies that much of the basic-level information known about the City category is also true of any city. The analysis of the dissimilarity of the information in the spatial categories considered both the distinctive and common information in the categories and suggested a geographic hierarchy based on a container principle was also an important dimension in the cognitive structure containing spatial information. This indicated that containers should partially reflect the character of their contents and the contents should partially reflect the character of their container. These connections between categories in our memory allow us to store information efficiently and impose order on a complex environment.

6.6. Conclusions

This chapter has considered issues related to learning spatial information. Learning can be as simple as associating a visual pattern with a category name. Neural network models and humans learn climate categories by experiencing exemplars that are patterns of monthly temperature and precipitation information. The pattern of connection weights in the neural network indicates what was learned. As we experience individual exemplars we also can create more abstract categories and store this information in basic-level categories. This illustrates that spatial knowledge is not just observed information that has been stored, but generalized information that is learned by experience. The next chapter continues the discussion of learning by considering spatial prototypes.

CHAPTER 7

SPATIAL PROTOTYPES

7.1. Introduction

The purpose of this chapter is to explore how experiences with maps are translated into knowledge. The discussion will center on the hypothesis that a person experiencing a sequence of similar objects, e.g., map symbols that represent a category, encodes a cognitive structure in his or her memory. This cognitive structure will be called a prototype. This chapter uses the term prototype as it was defined by Eleanor Rosch's (1973, p. 328) to mean the "central tendencies of categories." The terms *category*, *exemplar*, and *prototype* are intended in the sense defined by Rosch. For example, a *category* of maps would be a distinctive class of maps, e.g., reference maps, azimuthal projections, or sketch maps. An *exemplar* for these categories would be an individual map that was determined to be an instance of one of the categories. The *prototype* of a category represents the central tendencies of the characteristics of all the exemplars of a category. Exemplars similar to the prototype are considered as most typical of the category even though it is possible for no individual exemplar to be an exact duplicate of the prototype.

There may be many different exemplars that are valid members of a category. We are frequently required to identify objects that are examples of a general category of object. Although we have not seen this exact example of the type of object, we can easily classify it by comparing it to a prototypical image of the object we have stored in memory (Nosofsky 1987,1988). We can usually identify a person we have just met as being male or female based on a variety of physical characteristics we expect the typical male or female to have. As the many exemplars of a category are experienced over time, knowledge about the nature of the category is encoded into memory. A particular exemplar may not have the exact same characteristics of other members of the category and still be a valid member of that category. A penguin is considered to be an exemplar of the bird category even though some of its characteristics make it an unusual member of the category. Frequently experienced exemplars are considered centrally associated with the category. The most typical or central member of the category is the category's prototype. For example, look at the letters in Figure 7.1. Which one is different?

If you selected the letter in the center box you probably based your choice on shape. This letter is a category 'L' letter and the others are all category 'R' letters. You probably also noticed that none of the letters have shape characteristics that are exactly the same. It was, however, easy for you to assign the center letter to the 'L' category and the other letters to the 'R' category. You are able to do this easily because you have experienced many different 'R's and have developed a prototype for an 'R' that gives you the essence of what an 'R' should look like. Kosslyn, Holtzman, Farah, and Gazzangia (1985) argued "literate adults have seen letters so many thousands (millions?) of times, varying in size, font, weight, and so on, we have come to abstract out and store a 'prototypical image' of each letter." The prototype 'R' is an internal representation in

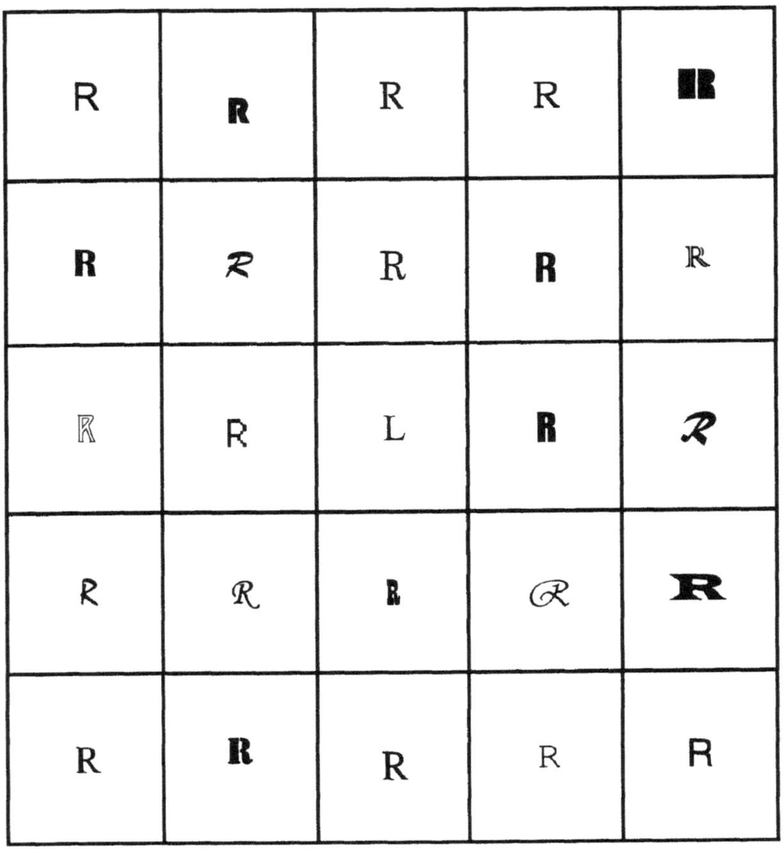

Figure 7.1: Letters represented for two categories in different fonts.

your memory. These prototypes are stored in memory as images of segments and descriptions of how the various parts should be arranged to form the general shape (Kosslyn, Reiser, Farah, and Fliegel 1983). Expressed verbally, it might be that an 'R' is a vertical line (more or less). Attached on the right side of the line, to the top half (more or less), is an open circular (more or less) line. Another line extend from where the bottom of the circular line meets the vertical line down and to the right. A candidate letter can be compared to your prototype 'R' in memory to determine if it is a member of that category. You can even identify the candidate as an 'R' when it is in a font you may not have previously experienced. This suggests that you are not just matching the candidate 'R' to every 'R' you have previously seen.

It is also possible to encode prototypes of maps into memory. Most Americans have viewed many different maps of the United States that showed the individual states. Adults have seen many versions of common maps drawn at different scales, using different map projections, and with various amounts of detail included in the lines forming the states. The facts that someone familiar with such maps can draw a sketch map, no matter how accurate, indicates some information has been encoded and stored in memory. People are also able to recognize a new version of a map that is depicted at a scale, with an amount of detail, or with a map projection they have never encountered before. The prototype theory would argue that the details of individual exemplars are not stored as individual cognitive maps. Instead a prototype is encoded and updated from the various individual exemplars that have been encountered over time (Hampton 1988). The sketch map one draws should not be a drawing of any individual map seen by the individual, but an abstraction of all the exemplar maps, i.e., the prototype. New maps are identified by comparing them to the prototype encoded in memory. For example, Americans have seen many maps of the state of Texas that represented the state in a variety of ways. Suppose a random sample of Americans were ask to imagine an outline map of Texas and to draw this imagined object on paper. What maps would be imagined? What influences would have created the representation of Texas internalized by individuals? A prototype of Texas may not have all the details encoded, but the essence of the shape and important characteristics, e.g., the panhandle, should be encoded (Figure 7.2). A prototype of Texas or any other object that is stored in memory is an abstraction that represents a category. It can be called an abstraction because it captures those features that are typical of the individual exemplar objects in the category. It represents the general patterns of the category rather than the specific details of all the exemplars. The two sketch maps were drawn by college students who were not geography majors (Figure 7.2b and 7.2c). The sketches represent the typical details one stores about the visual appearance of Texas through casual experiences with maps of the United States. Individuals will, of course, vary in their ability to express what they can imagine in a sketch. Figure 7.2d is a simplified version of Texas produced by encoding fewer details. It was created by imposing a grid on the original map of Texas (Figure 7.2a) and encoding points where the original outline entered and exited a grid cell (Figure 7.3). The essence of the shape can be expressed by the panhandle in the North, the point in the South, and the protrusions in the East and West. Both sketch maps represent these essential parts of Texas.

7.2. Category Prototypes

Humans learn prototypes for categories and then use these theoretical cognitive structures to make decisions and judgments (Johnson 1987; Lakoff 1986; Rosch and Lloyd 1978). Questions related to learning prototypes have centered on how the cognitive structure is encoded. The learning of spatial prototypes should be influence by both *spatial* and *temporal* properties of objects. Spatial properties of objects are thought to be encoded and processed by subsystems within the dorsal system in the brain (Kosslyn and Koenig 1992). Specific subsystems have been identified that encode location, size, and orientation. Non-spatial properties, such as shape, color, and texture, are thought to be encoded and processed by a separate ventral system.

Encoding studies have focused on theories such as *family resemblance* (Rosch and Mervis 1975) and *basic-level* categories in a hierarchy (Rosch, Mervis, Gray, Johnson, and Boyes-Brian 1976; Tversky and Hemenway 1983). Other studies have considered specific objects like semantic and spatial reference points (Rosch 1975a, 1975b) or features like colors and shapes (Rosch 1973, 1975c).

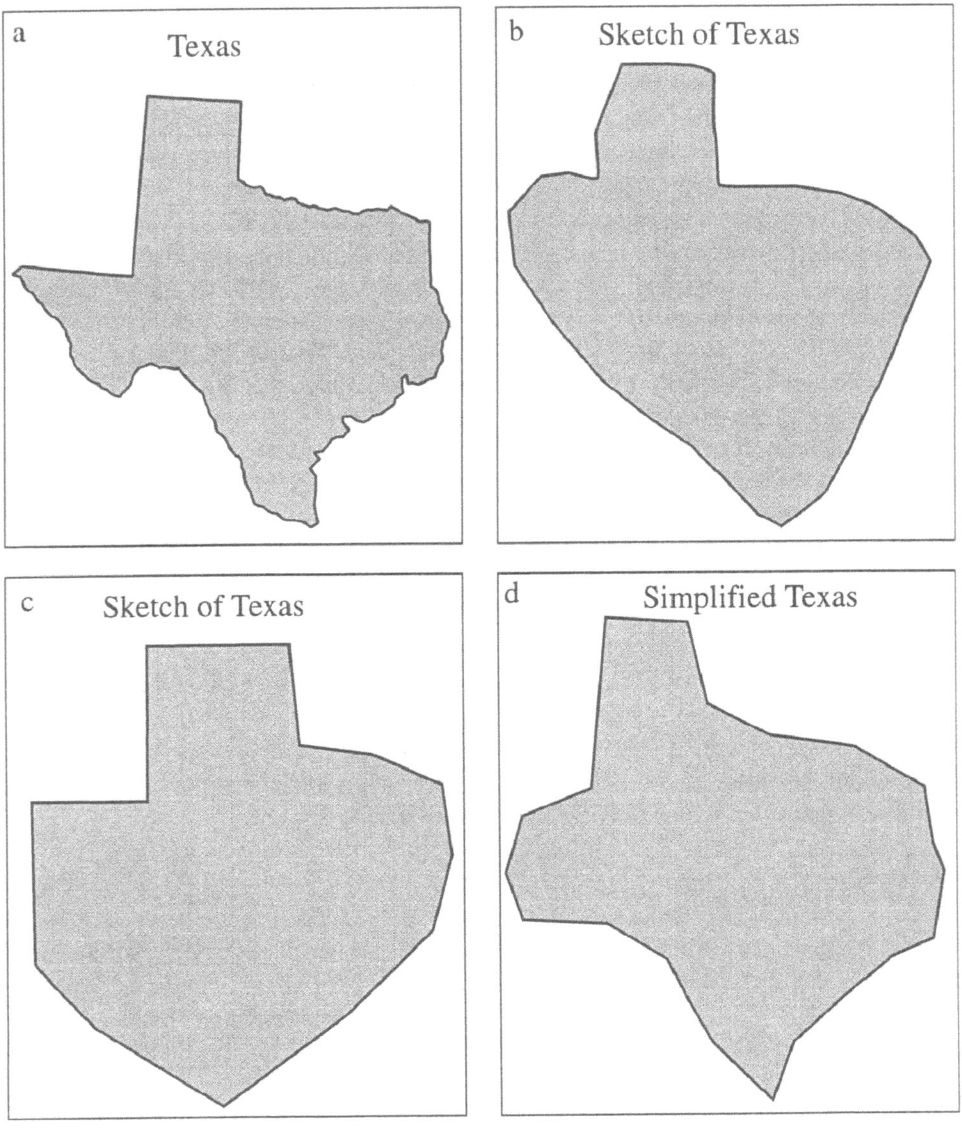

Figure 7.2: Texas from a map of the United States showing the details of the boundary *(a)*, two sketch maps produced by college students *(b* and *c)*, and a simplified version of the original map (*d*).

Questions related to using prototypes have centered on how the cognitive structure is decoded. A number of studies have considered how prototypes are used to make judgments. These include investigations of the decision rules used to compare exemplars and prototypes (Ashby and Gott 1988), how the knowledge about a category is generalized (Gelman 1988) and how people judge the typicality and membership of items in conjunctive categories made from two simpler categories (Gelman 1988). Other studies have been concerned with how decisions are made that classify new stimuli into existing categories (Malt 1989; Medin, Altom, and Murphy 1984; and Smith and Zarate 1992). Huttenlocher, Hedges, and Duncan (1991) investigated the impact of prototypes on our memory of spatial locations.

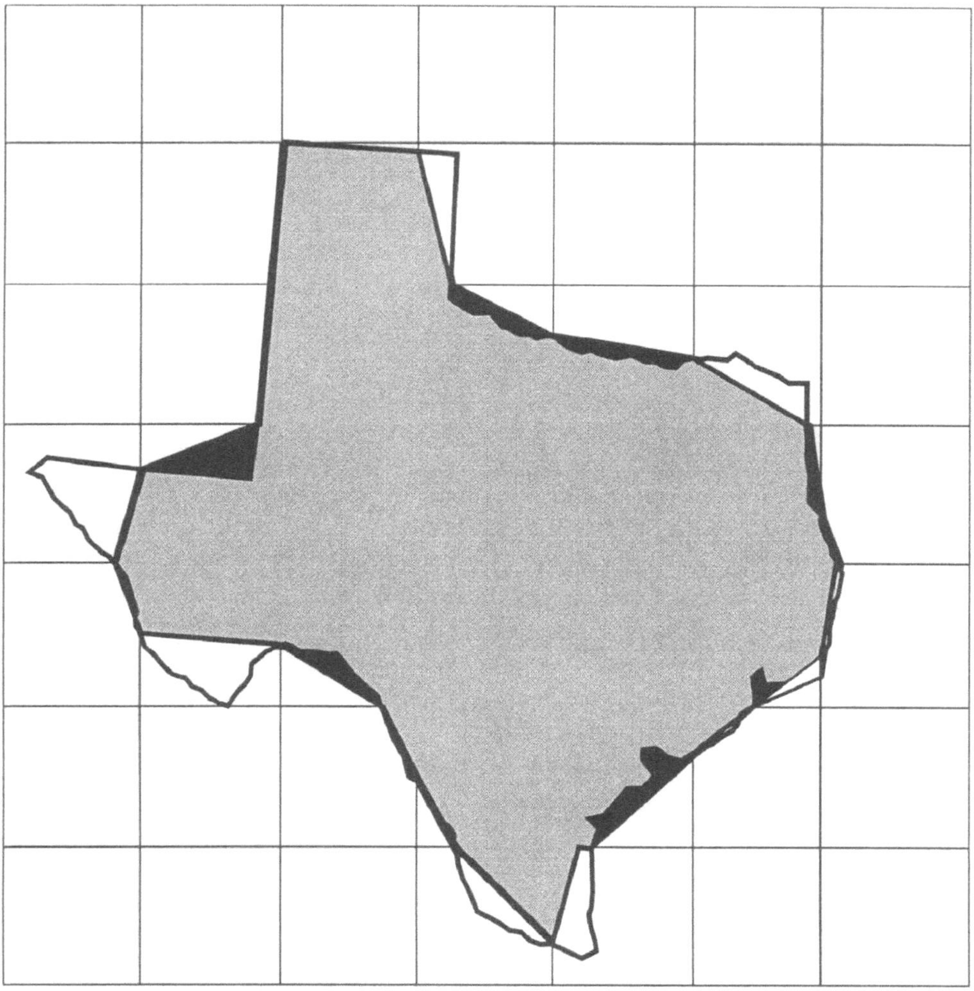

Figure 7.3: The simplified map in Figure 7.2d was produced by imposing a grid on the original version of Texas (Figure 7.2a) and recoding the outline as the points where the boundary entered and exited a cell of the grid. The grey polygon represents the simplified Texas and the black and white polygons represent discarded details of the original boundary.

The properties of color were considered in some of the earliest studies of prototypes. It was argued that the number of basic color terms is limited because different languages distinguish basic color names that refer to the same locations in color space (Berlin and Kay 1969). Rosch (1973) considered colors to be good examples of *natural prototypes*. It was argued that prototype colors attracted attention more readily and were more easily remembered than nonprototype colors of the same category (Heider 1971,1972). Because they constituted focal points in color space and served as reference points Rosch (1975a) called these prototypes *focal colors*. She demonstrated that cultures lacking names for these prototypical colors can still learn focal colors more readily than nonfocal colors (Rosch 1973).

Focal colors (prototypes) frequently are selected as the best example of a color category. McNiff (1991) considered if this was also true for compound categories frequently found on maps, such as the blue for water or the green for forests. She asked a group of geographers to select the best specific color to represent the water, forest, and urban categories on maps. Focal blue, focal green and focal red were not selected to represent these landuse categories. This suggested specific nonfocal colors also can be learned and function as prototypes when they are frequently used to symbolize objects like landuse categories on maps.

Our learning of the prototype for a category is directly influenced by how frequently we encounter specific exemplars of the category. Consider, for example, a hierarchy in which subordinate categories are grouped into a superordinate category. The superordinate bird category has robins, ducks, and penguins as subordinate categories. Most respondents in the United States thought of the prototype of the bird category when asked to give an example of the bird category. The exemplar "robin" was named more frequently than "penguin" when people were asked to name a bird (Lakoff 1986; Rosch 1975C; Tanaka and Taylor 1991). A robin is more central to the bird class because it shares more characteristics with other members of the class even though both robins and penguins are both members of the bird category. A robin has two legs, two wings, a beak, feathers, lays eggs, can fly and so on. Penguins are less typical because they cannot fly. There swimming skills and association with extremely cold climates also distinguish them from the typical bird. Because their attributes most nearly resemble the attributes of other members of their category and do not resemble the attributes of members of other categories, prototypes exhibit a strong *family resemblance* (Rosch and Mervis 1975).

7.2.1. FAMILY RESEMBLANCE

A prototype appears not to precede the category (Rosch and Mervis 1975). They are formed by learning processes that focus on the characteristics of items in the category. The notion of family resemblance argues that members of a category need not have exactly the same characteristics for them to be considered part of the same category (Wittgenstein (1953). This is in contrast to the notion that categories are logical bounded classes with a set of critical features. Categories based on critical features have sharp boundaries since an object is only in the class if it has the set of critical features. This mandates that an object must have a critical feature to be considered part of Category A (Figure 7.4). Categories based on family resemblance have fuzzier boundaries because members of the class share a set of features, but individual members may have only some of the features. Although most of the members of a class will have a feature, some will not (Figure 7.5).

Prototypes for Categories of Map Symbols

We can learn prototypes for categories of map symbols in the same way we learn a prototype for the bird category. Any encounter with a map symbol belonging to a particular category provides an experience that influences the formation of the prototype for that category in our memory. Two categories of map symbols have been

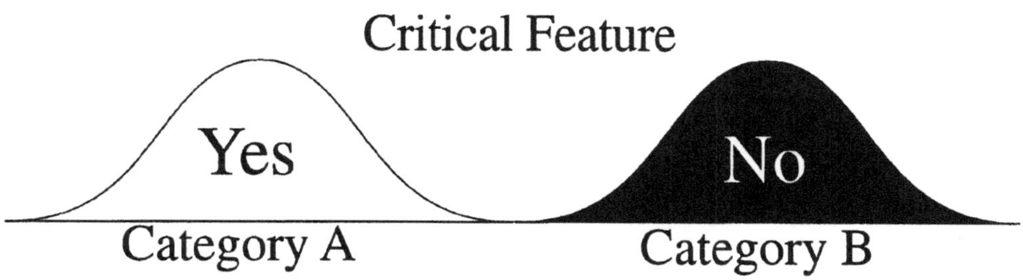

Figure 7.4: Categories A and B represented with a sharp boundary. No exemplars in Category B have a critical feature of the exemplars of Category A.

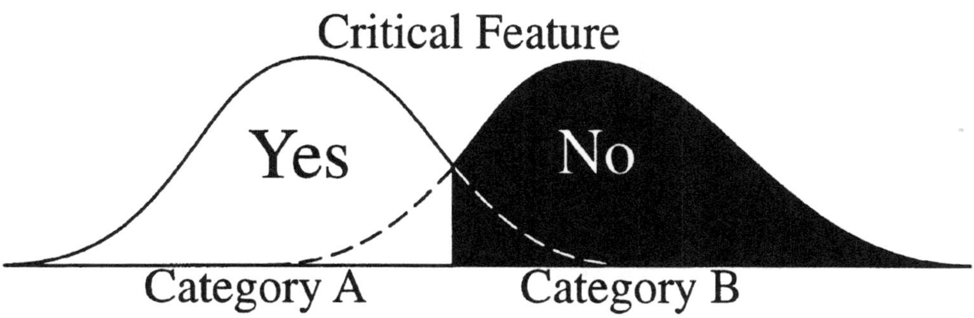

Figure 7.5: Categories A and B represented with a fuzzy boundary. Some exemplars in Category B have a critical feature of most exemplars of Category A.

developed to illustrate the learning of category prototypes based on family resemblance (Figure 7.6). Chernoff (1973) first suggested using facial expressions to encode multivariate symbols for maps (Muehrcke and Muehrcke 1992; Nelson and Gilmartin 1996). The symbols use here are circular faces that have features that can differ on six dimensions. Such faces seemed particularly appropriate for a discussion of categories based on family resemblance. The face for any symbol can take on either of two digital values (0 or 1) for each of the six dimensions. For example, the prototype face for

Category A has been encoded as all 0's (Figure 7.6). This translates to a face with straight hair, ears that are close to the head, eyes that look downward, a nose that bends to the right, a mouth that frowns, and a beard. The prototype face for Category B has been encoded as all 1's (Figure 7.6). This translates to a face with curly hair, ears that stick out, eyes that look upward, a nose that bends to the left, a mouth that smiles, and a mustache. If such faces were being used to construct a map, the six dimensions could be assigned some concrete meaning for the purpose of the map. For example, a frown might indicate lower than average income and a smile higher than average income.

With six dimensions to work with, 2^6 or 64 possible faces could be created to define a population. Exemplars were selected from this population of potential map symbols that had a family resemblance for two categories defined as Category A and Category B. The prototype for Category A (Figure 7.6) was encoded as (0 0 0 0 0 0) so each exemplar for Category A had four 0 codes and two 1 codes (Figure 7.7). The

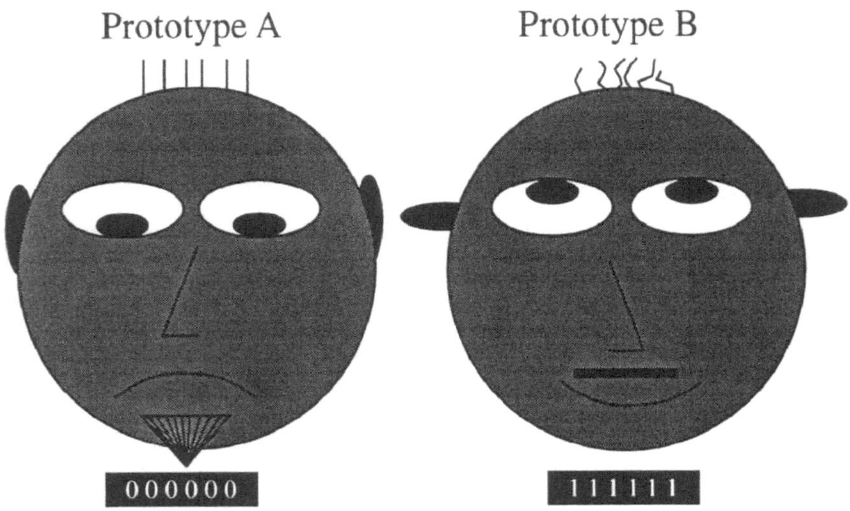

Code	Head Hair	Ears	Eyes	Nose	Mouth	Face Hair
0	Straight	In	Down	Right	Frown	Beard
1	Curly	Out	Up	Left	Smile	Mustache

Figure 7.6: Prototype Chernoff faces for Categories A and B.

prototype for Category B was encoded as (1 1 1 1 1 1) so each exemplar for Category B had four 1 codes and two 0 codes (Figure 7.8). Note that the prototypes are not included as 1 of the 15 exemplars. Also note that some members of Category A have a 1 code for each of the six dimensions and some members of Category B have a 0 code for each of the six dimensions. This means 2/3 of the exemplars of a category have the family characteristic, e.g., straight hair for Category A, but 1/3 of the exemplars have the alternate code that is a characteristic of the other family, e.g., curly hair for a member of Category A.

Learning the Categories

A simple pattern associator neural network model was used to simulate the learning of the map symbol categories. The model can also be used to illustrate how a prototype for a category can be learned even though it does not exist as an exemplar and has never been directly experienced. The model had six input neurons that represented the six dimensions used to create the map symbols and two output neurons that represented the two categories (Figure 7.9). The weights for the model were initially

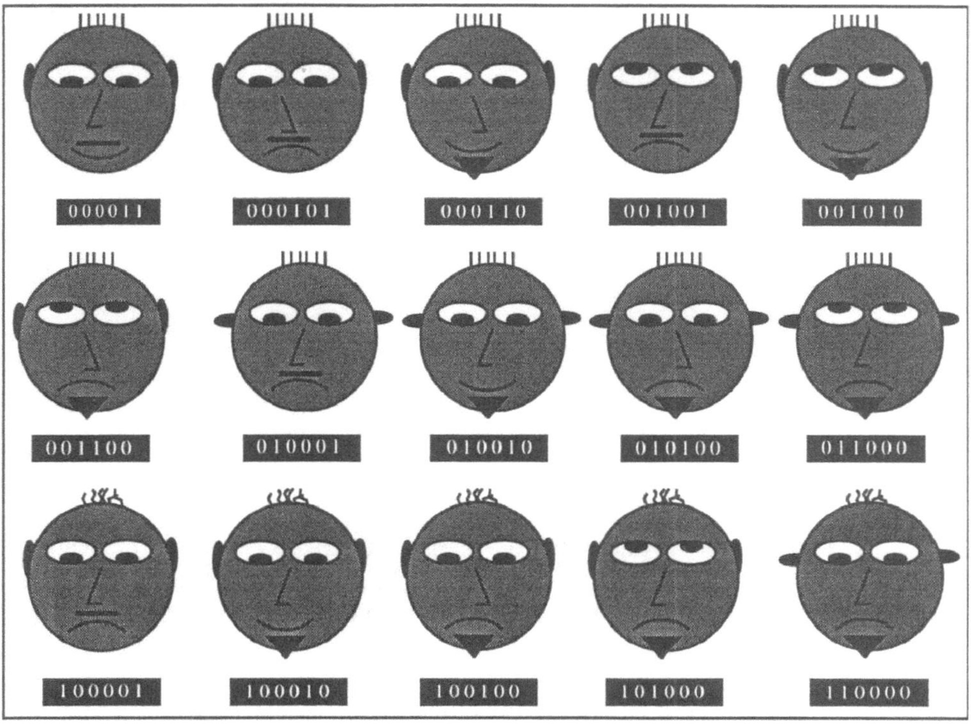

Figure 7.7: The 15 faces that were exemplars for Category A.

set to 0.0. The model was trained by inputting the digital codes for the exemplars and having the model output an activation for Category A and Category B. Weights were adjusted after each trial based on the differences between the activations of the output neurons and the expected activations. When an exemplar for Category A was presented to the model, the expected activations were 1.0 for the Category A output neuron and 0.0 for the Category B output neuron. The reverses of these responses were expected when an exemplar for Category B was presented to the model. An epoch of learning consisted on all thirty exemplars being presented to the model in a random order. The model was trained for 145 epochs or until it could categorize all 30 exemplars correctly. The learning curve for the model's training suggested the usual nonlinear decrease of error with number of training epochs (Figure 7.10).

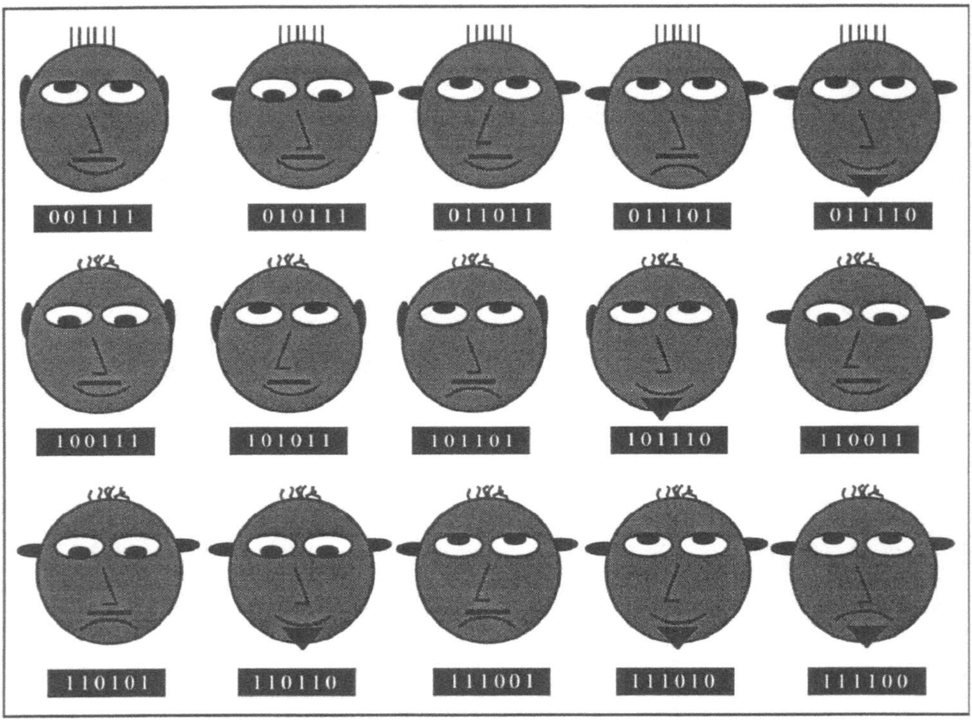

Figure 7.8: The 15 faces that were exemplars for Category B.

The weights for the final model are marked on the lines connecting input and output neurons in Figure 7.9. The weights indicate what the model has learned from its experiences with the exemplars. Note the connections between all the input neurons and the Category A neuron are dashed lines indicating a negative weight. Also note that the connections between all the input neurons and the Category B neuron are solid lines indicating a positive weight. The model has learned to produce higher activations for the Category A output neuron when inputs are mostly 0's and higher activations for the Category B output neuron when inputs are mostly 1's.

To illustrate that the model learned the prototypes for the categories even though none of the exemplars had the exact characteristics of the prototypes (Figure 7.6), the response of the model for the first ten learning epochs was monitored for both the exemplars and the prototypes. Since the model had no initial knowledge, it would not be expected to respond correctly for the early epochs of learning (Figure 7.10). Over the first 10 epochs the model showed a gradual improvement in its ability to correctly

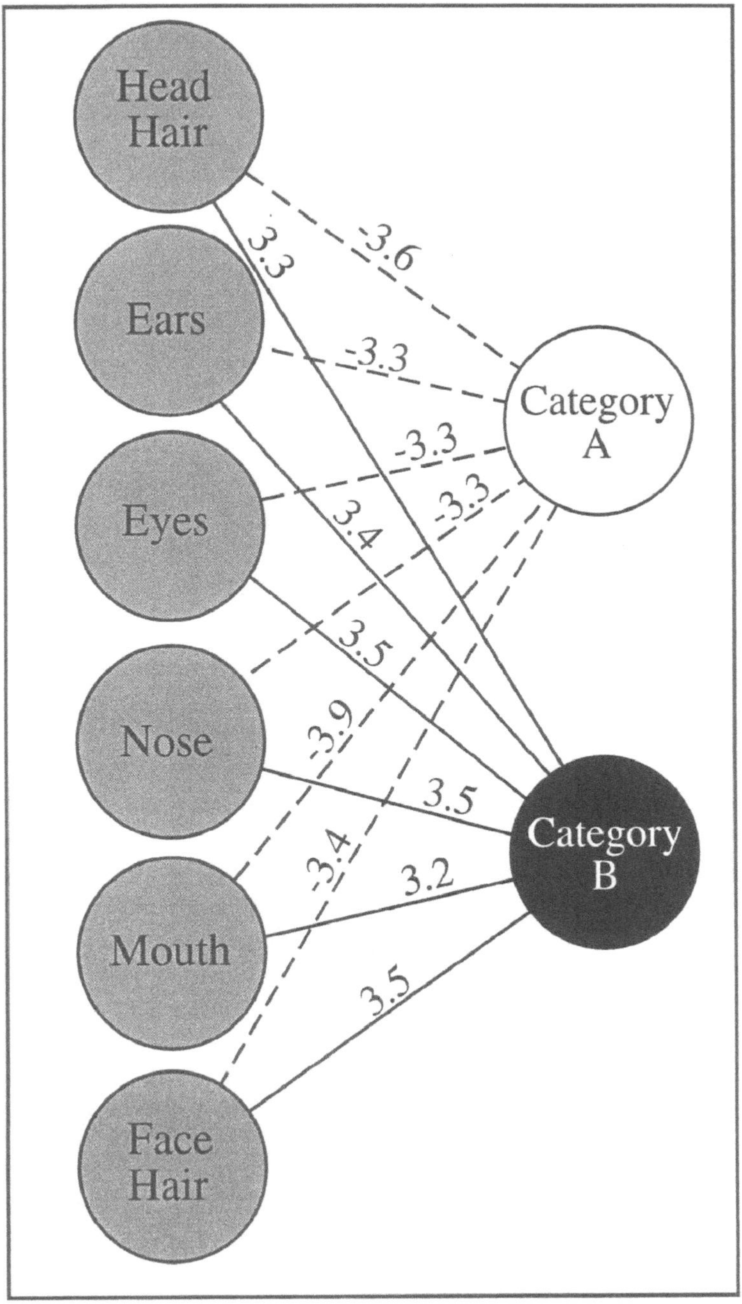

Figure 7.9: The pattern associator neural network that learned the categories of Chernoff faces.

classify the exemplars. The model was, however, able to learn enough in three epochs of learning to consistently classify the prototypes correctly (Figure 7.11). This was true even though the model was never trained with the prototypes. The basic patterns represented by the prototypes was easily extracted from the exemplars through the family resemblance of the exemplars.

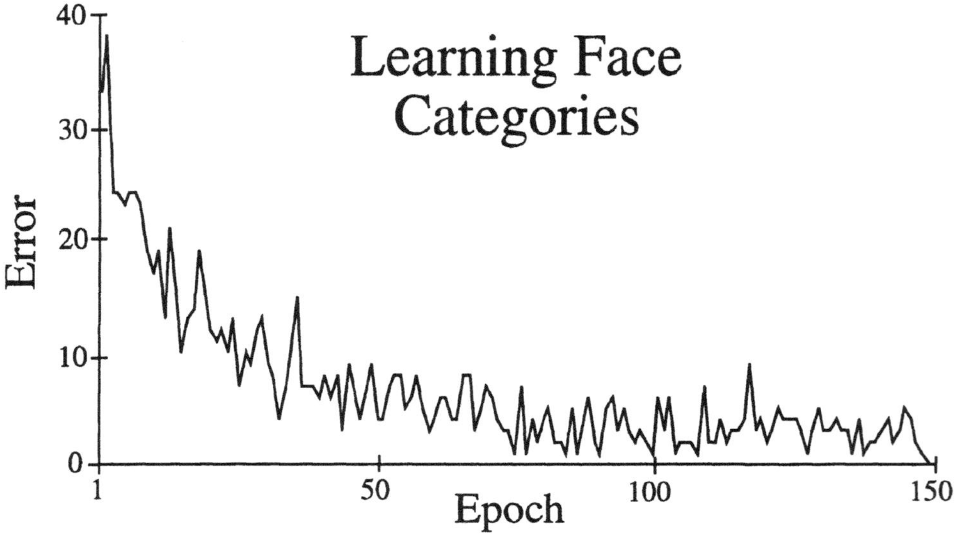

Figure 7.10: The learning curve for the pattern associator model.

7.2.2. NONLINEAR PATTERNS

The family resemblance in the above patterns was easily learned by a pattern associator neural network because the categories were linearly separable. This notion can be easily visualized for exemplars defined on three dimensions, but it applies to any number of dimensions. For three dimensions with binary (0 or 1) conditions, exemplars for categories could be defined that were located at corners of a cube. In this example the three dimensions are size, shape, and color (Figure 7.12). Size is defined as small (0) or large (1), shape is defined as a circle (0) or a star (1), and color is defined as white (0) or black (1). Some arbitrary collections of exemplars could be defined that exist as categories (white versus black corners) on either side of a linear partition of the space (Figure 7.13a). The category marked with black would be a small white circle, a small black star, and a large black circle (Figure 7.12) and the category marked with white would be a small white star, a large black star, and a large white circle. For the first category a small black circle would be the prototype and for the second category a large white star. These two categories and their prototypes should be easily learned by a pattern associator network. Other collections of exemplars could be defined for the same dimensions that cannot be separated by any linear partition of the space (Figure 7.13b). In this case the category marked with black would be a small black circle, a small black star, and a large white circle. The category marked with white would be a small white star, a large white star, and a large black circle. Pattern associator neural network

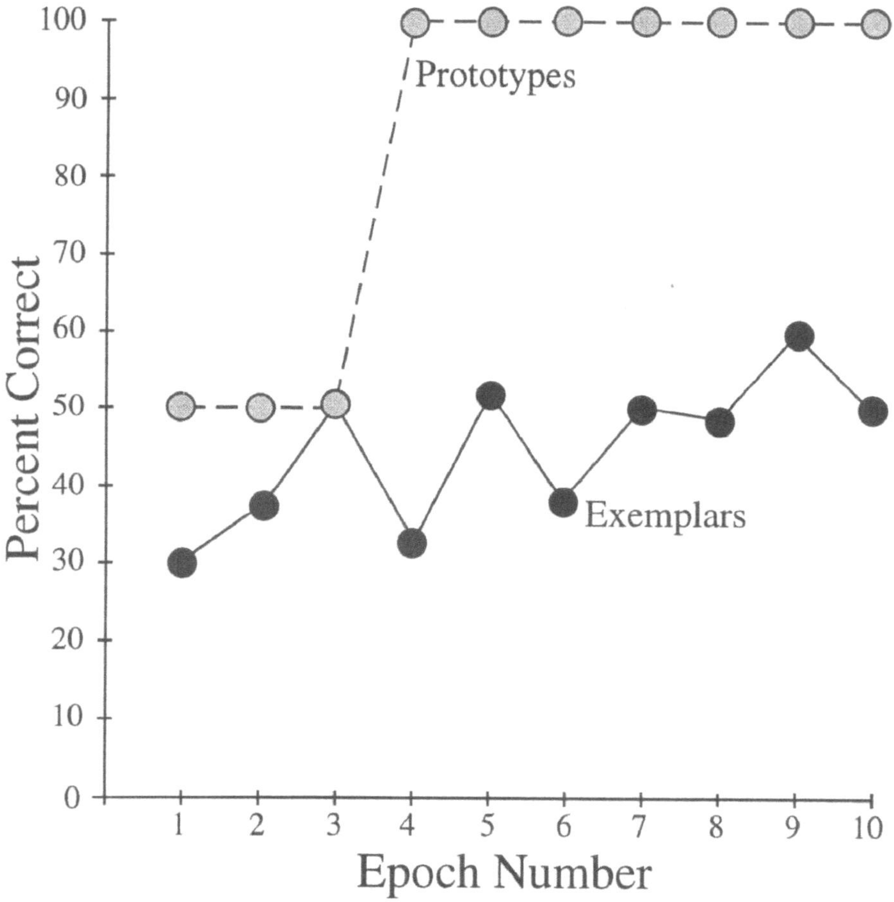

Figure 7.11: The percent correct for the 30 exemplars and the 2 prototypes over the first 10 epochs of learning.

models will find it more difficult to learn these categories because they can only be separated by a non-linear partition of the space.

Medin and Schwanenflugel (1981) designed an experiment with human subjects that produced linearly separable and non-linearly separable categories within a four-dimensional space. It was predicted that the human subjects would find the linearly separable categories harder to learn than the non-linearly separable categories and this was confirmed for the human subjects. Since pattern associator models should theoretically find it more difficult to learn non-linearly separable categories than linearly separable categories, Gluck (1991) reasoned another type of neural network was needed to simulate the experimental results reported by Medin and Schwanenflugel (1981) for human subjects. He suggested a configural-cue neural network for leaning non-linearly separable categories. This type of network can learn non-linear variations in the input patterns by encoding neurons that represent interactions between the dimensions. These neurons function like interaction effects in an analysis of variance. Consider binary information represented in a three-dimensional space to visualize the effect of configured neurons (Figure 7.14). Gluck (1991) argued that single neurons in a neural network function to activate faces of the cube. This provides very general knowledge about one part of the space. Double neurons activate edges that provide more specific information and triple neurons activate corners of the cube to activation even more specific information.

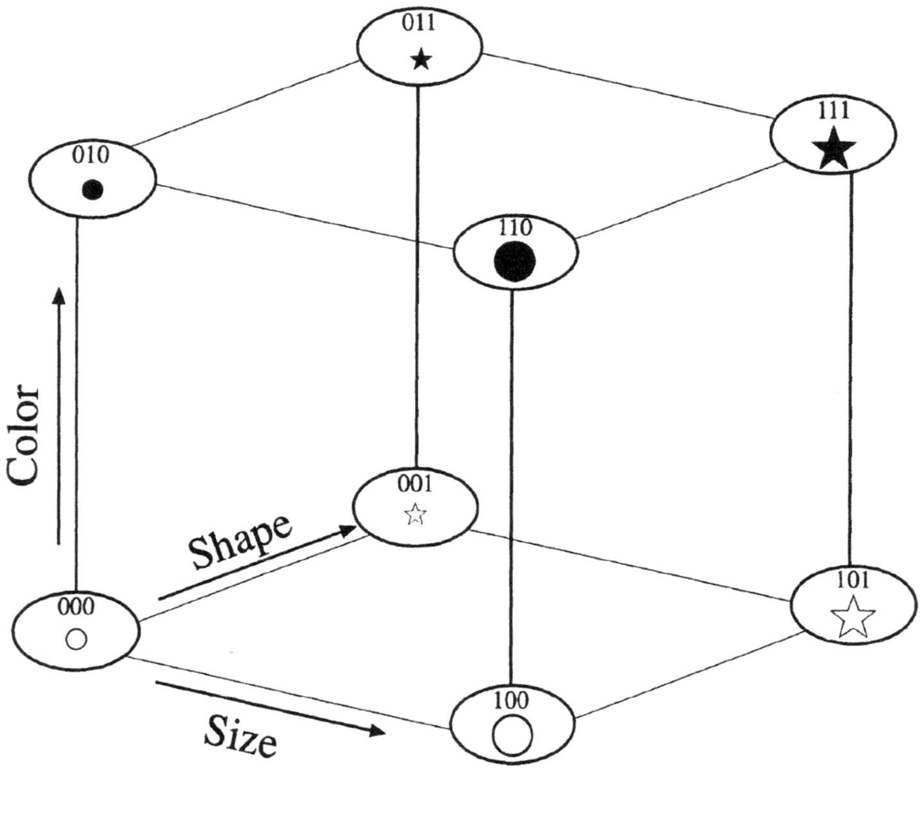

Figure 7.12: Three-dimensional space illustrating eight visual objects defined at the corners as binary digital codes of 0 and 1.

What is Where

Maps have been created to represent the exemplars in the linearly separable categories (Figure 7.15) and non-linearly separable categories (Figure 7.16) designed by

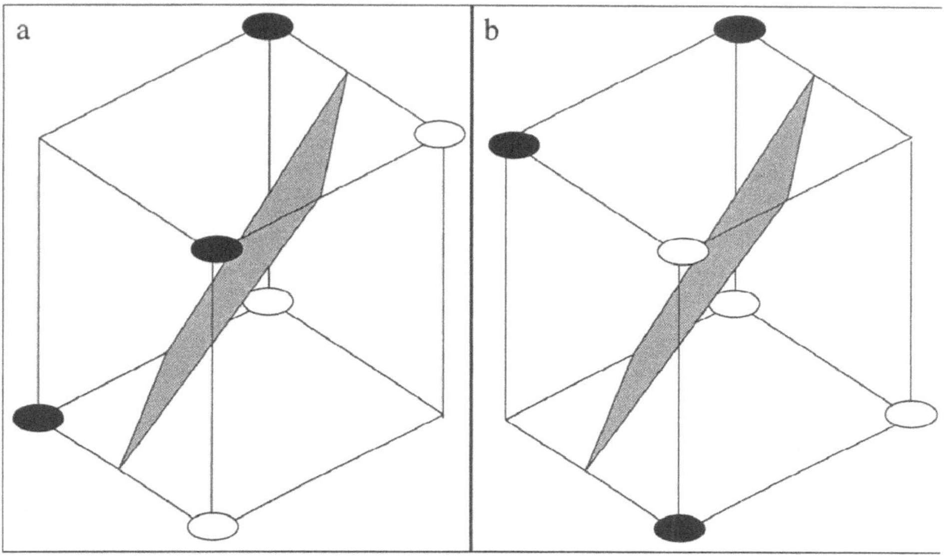

Figure 7.13: Collections of objects used to defined two linearly separable categories (a) and two non-linearly separable categories (b).

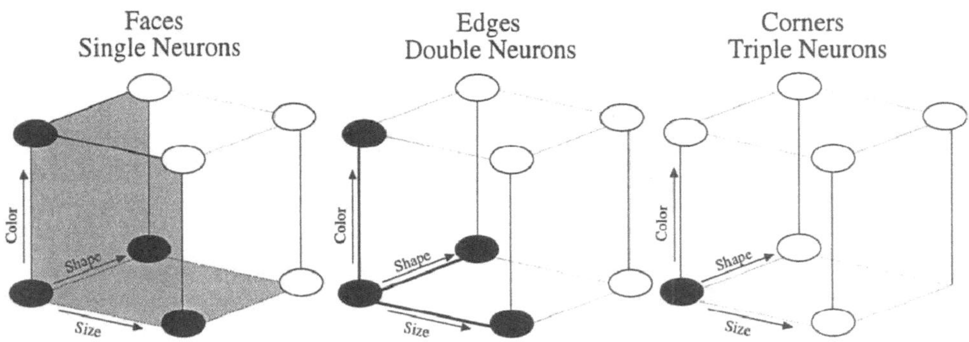

Figure 7.14: Neurons representing dimensions activate faces of the cube, two-way configural neurons activate edges of the cube, and three-way configural neurons activate corners of the cube.

Medin and Schwanenflugel (1981) . The first dimension determines the spatial location of a symbol along the vertical axis as either North (0) or South (1). The second dimension determines the spatial location of a symbol along the horizontal axis as either East (0) or West (1). The third dimension determines the color of the symbol as either Dark (0) or Light (1). The fourth dimension determines the shape of the symbol as either Round (0) or Square (1). The linearly separable categories (Figure 7.15) have a Category A with two of the three symbols located in the South and West, two are Light, and two are Squares. Category B has two of the three symbols located in the North and

East, two are Dark, and two are Round. Note that neither of these prototypes are one of the exemplars for either Category A or B.

The non-linearly separable categories (Figure 7.16) have a Category A with two of the three symbols located in the South and West, two are Light and two are Squares. Category B has two of the three symbols located in the North and East, two are Dark, and two are Circles. Note that the prototype appears as one of the exemplars for both categories.

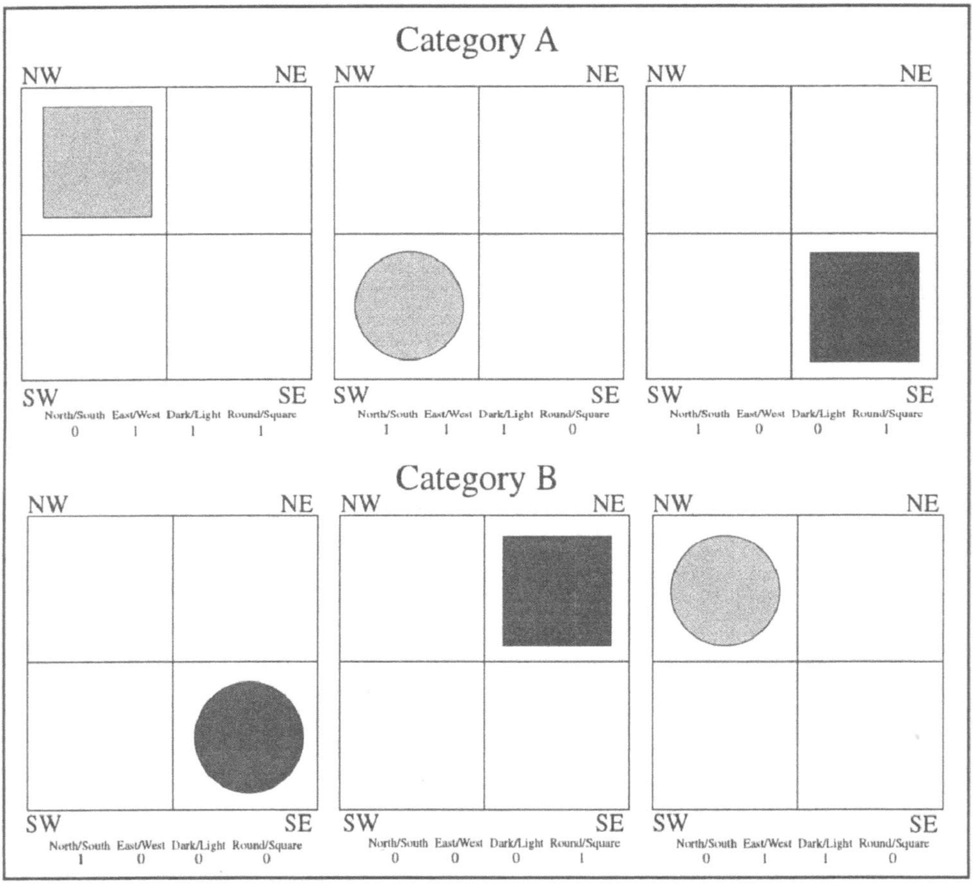

Figure 7.15: Exemplar maps and their binary codes to represent linearly separable Categories A and B.

Learning the Categories

The prediction for a pattern associator neural network is that it should be able to learn the linearly separable categories (Figure 7.15) easier than the non-linearly separable categories (Figure 7.16). A pattern associator network with four input neurons representing each of the dimensions, i.e., Vertical, Horizontal, Color, and Shape, and two output neurons representing Categories A and B was trained for 200 epochs with the linearly separable exemplars (Figure 7.17). The training produced the expected final

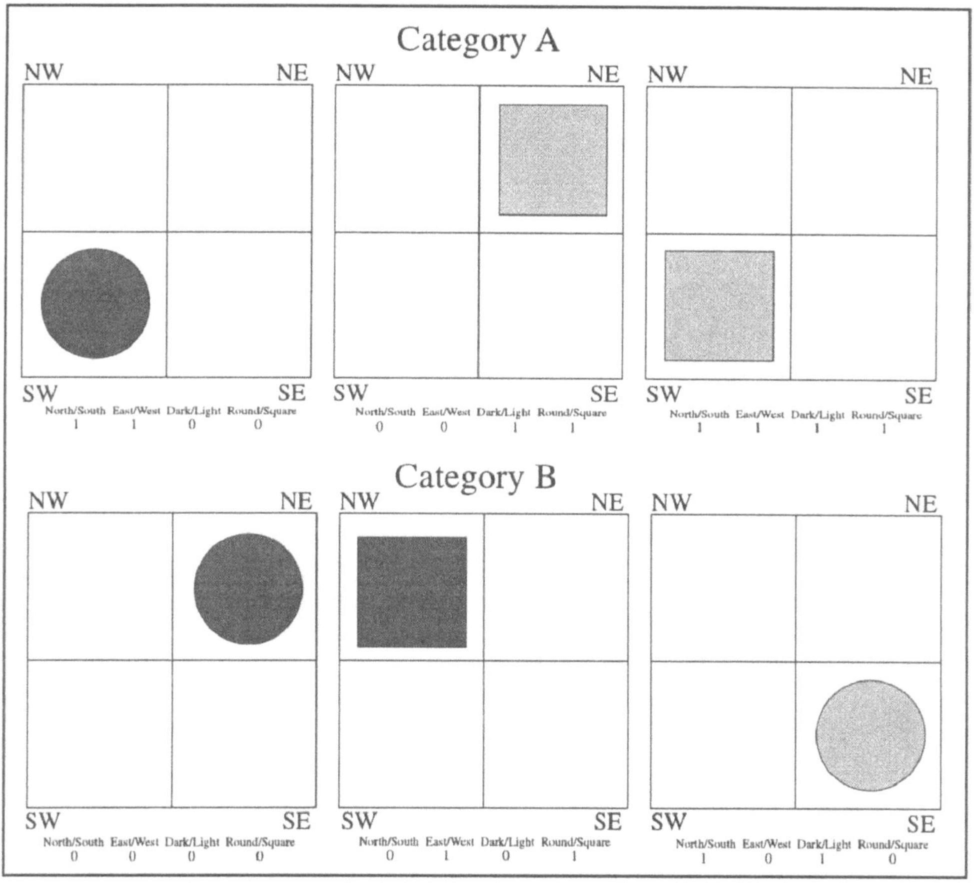

Figure 7.16: Exemplar maps and their binary codes to represent non-linearly separable Categories A and B.

weight pattern with positive connections from the input neurons to Category A's output neuron and negative weights from the input neurons to Category B's output neuron. This is similar to the pattern produced for the family resemblance of face symbols discussed earlier. The weights for the pattern associator neural network were then reset to 0 and trained for 200 epochs using the exemplars for non-linearly separable categories (Figure 7.18). A similar pattern emerged for the weights with positive connections between the four input neurons and Category A's output neuron and negative connections between the four input neurons and Category B's output neuron. The major

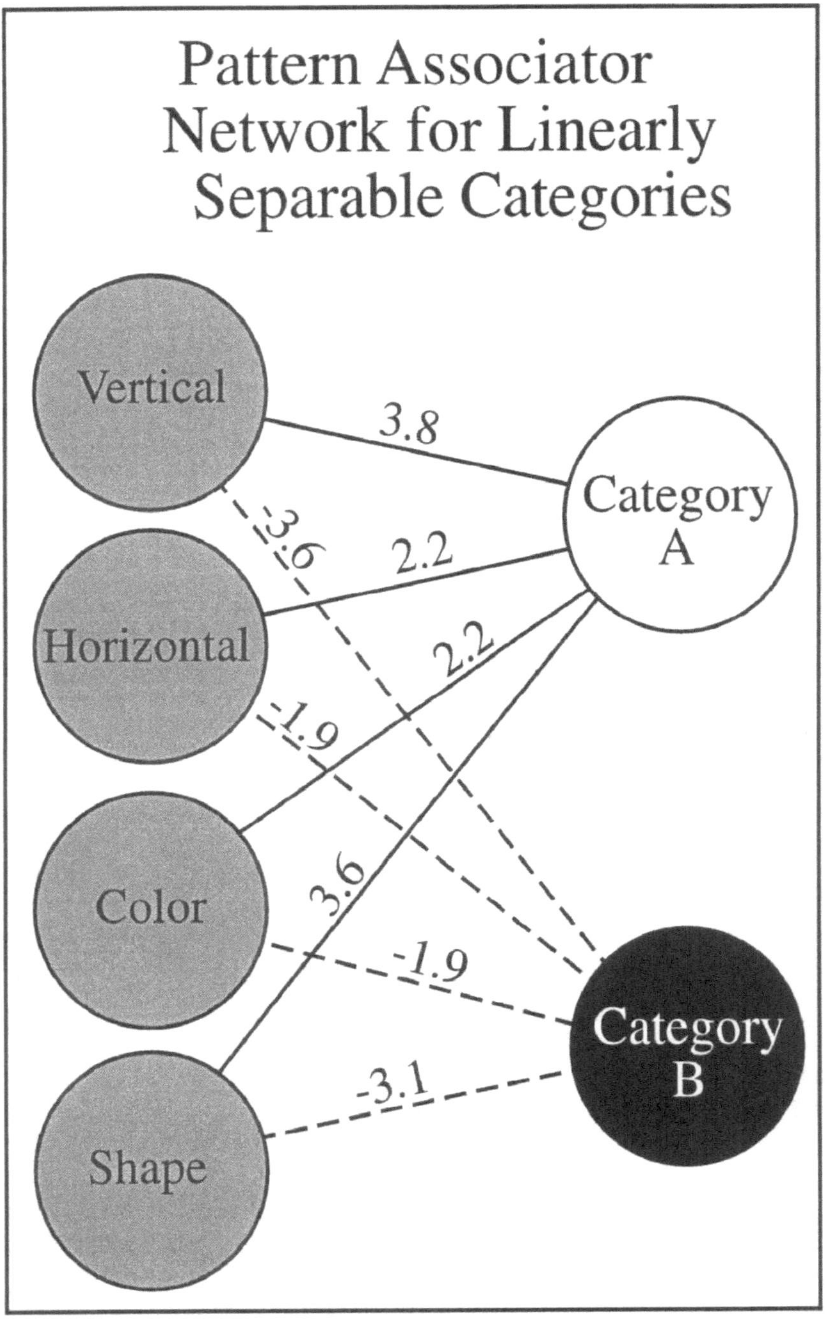

Figure 7.17: Pattern associator model with input neurons for vertical location, horizontal location, color and shape and output neurons for Category A and Category B. Final weights for the linearly separable categories are shown for each connection.

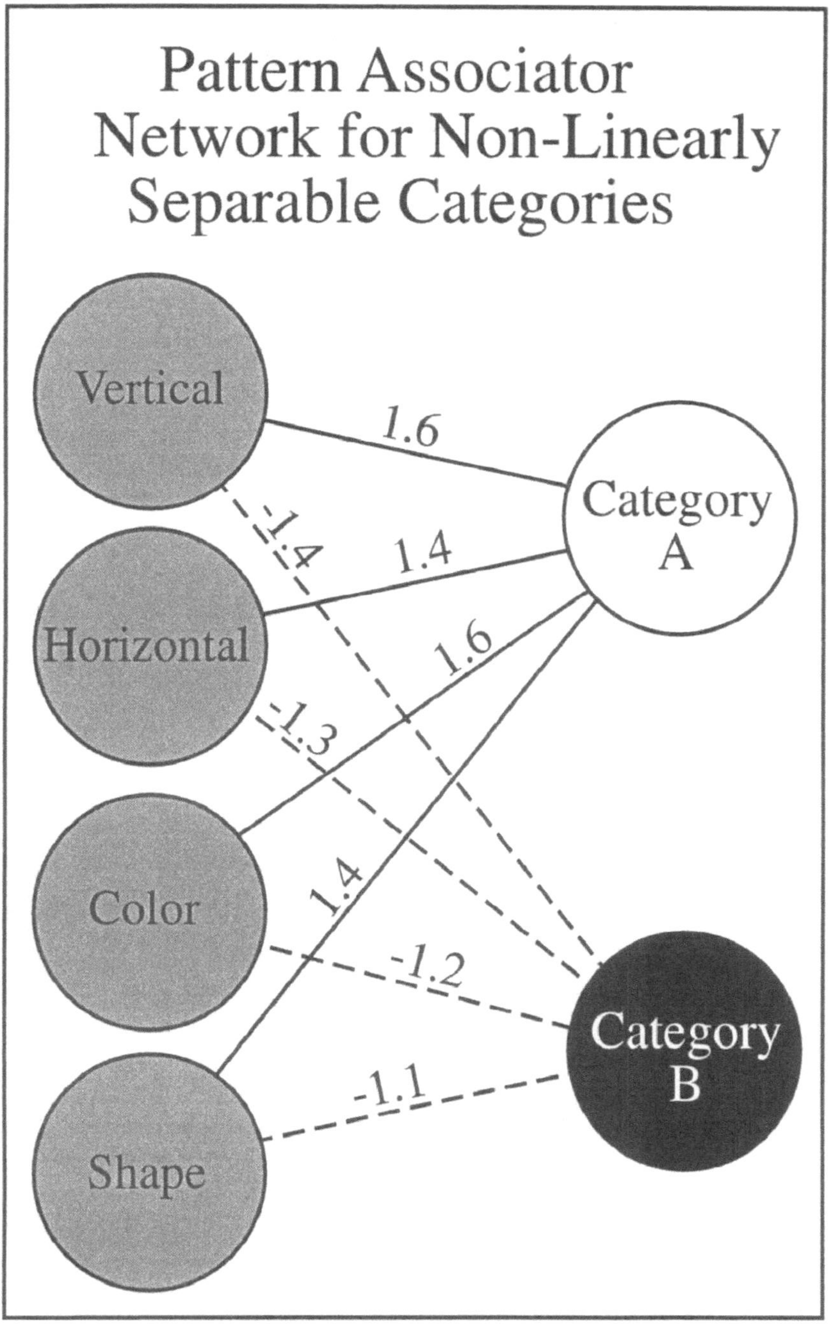

Figure 7.18: Pattern associator model with input neurons for vertical location, horizontal location, color and shape and output neurons for Category A and Category B. Final weights for the non-linearly separable categories are shown for each connection.

difference between the two sets of weight was that uniformly smaller weights were produced by the pattern associator network for the non-linearly separable categories. In addition to this difference, error plots for both learning experiences suggested the pattern associator model was able to consistently reduce error over the 200 epochs for the linearly separable categories but not for the non-linearly separable categories (Figure 7.19). As expected, the pattern associator model was clearly having a difficult time learning the non-linearly separable categories.

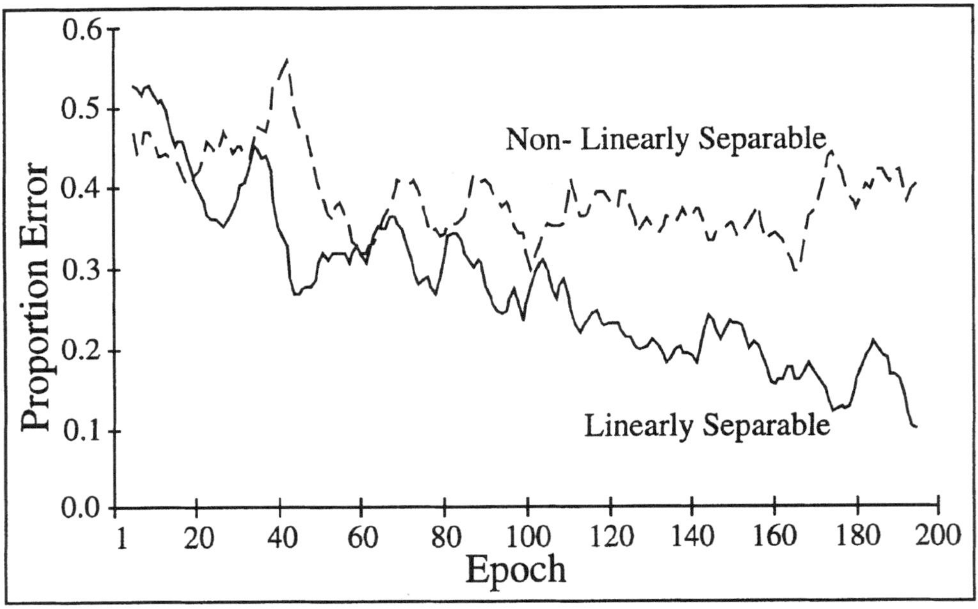

Figure 7.19: Learning curves for the pattern associator neural network model for the linearly separable exemplars (solid line) and the non-linearly separable exemplars (dashed line).

A configural-cue neural network was constructed using the same four input neurons used in the pattern associator model, i.e., Vertical, Horizontal, Color, and Shape, plus six additional configured neurons that combined pairs of dimensions, e.g., Color and Shape or Vertical and Horizontal (Figure 7.20). Input values for the configured neurons were coded as 1 if the two dimensions had the same values. For example, if both Horizontal and Color were coded the same (both 0 or both 1) then the Horizontal/Color configural neuron would be coded as 1. If Horizontal and Color codes were different (0 and 1 or 1 and 0) then the Horizontal/Color would be coded as 0. The network was trained for 200 epochs with the linearly separable exemplars. The final weights were similar to those of the pattern associator network for the four input neurons representing the dimensions. All four neurons had positive connections to the Category A output neuron and negative connections to the Category B output neuron. The pattern was reversed for the configural neurons. All six had negative connections to the Category A output neuron and positive connections to the Category B output neuron (Figure 7.20).

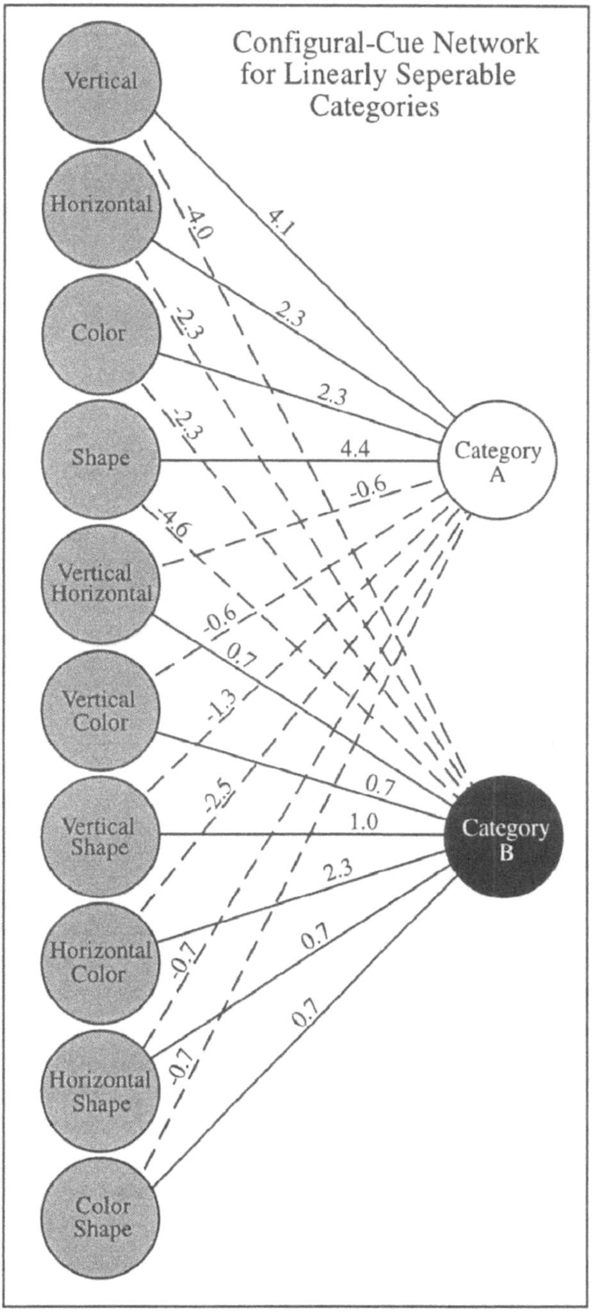

Figure 7.20: Configural-cue neural network with input neurons for vertical location, horizontal location, color and shape dimensions and six configural neurons based on all pairs of dimensions. Output neurons were for Category A and Category B. Final weights for the linearly separable categories are shown for each connection.

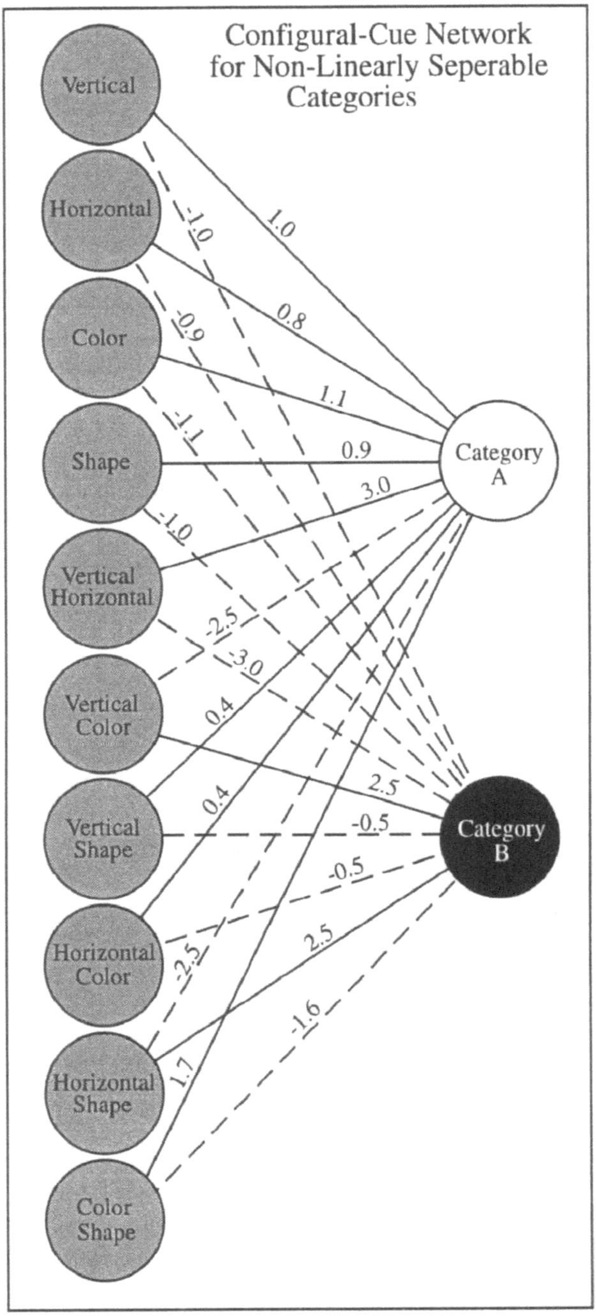

Figure 7.21: Configural-cue neural network with input neurons for vertical location, horizontal location, color and shape dimensions and six configural neurons based on all pairs of dimensions. Output neurons were for Category A and Category B. Final weights for the non-linearly separable categories are shown for each connection.

The weights for the configural-cue network were reset to 0 and it was trained for 200 epochs with the non-linearly separable exemplars (Figure 7.21). The final weights were again similar to those for the pattern associator model for the neurons representing the dimensions. These neurons had positive connections to Category A's output neuron and negative connections to Category B's output neuron. The connections for the configural neurons using the non-linearly separable exemplars (Figure 7.21) were different from those learned for the linearly separable exemplars (Figure 7.20). Each of the six input neurons had a positive connection to one of the output neurons and a negative connection to the other, but, unlike the weights learned for the linearly separable categories, the pattern was not consistent among the configural neurons. This indicated the configural-cue model had responded differently to the two sets of exemplars. As predicted, the error plots indicated it was easier for the configural-cue network to learn the non-linearly separable categories than the linearly separable categories (Figure 7.22). The network made steady progress in reducing the error using the non-linearly separable exemplars, but could not reduce the error for the linearly separable exemplars.

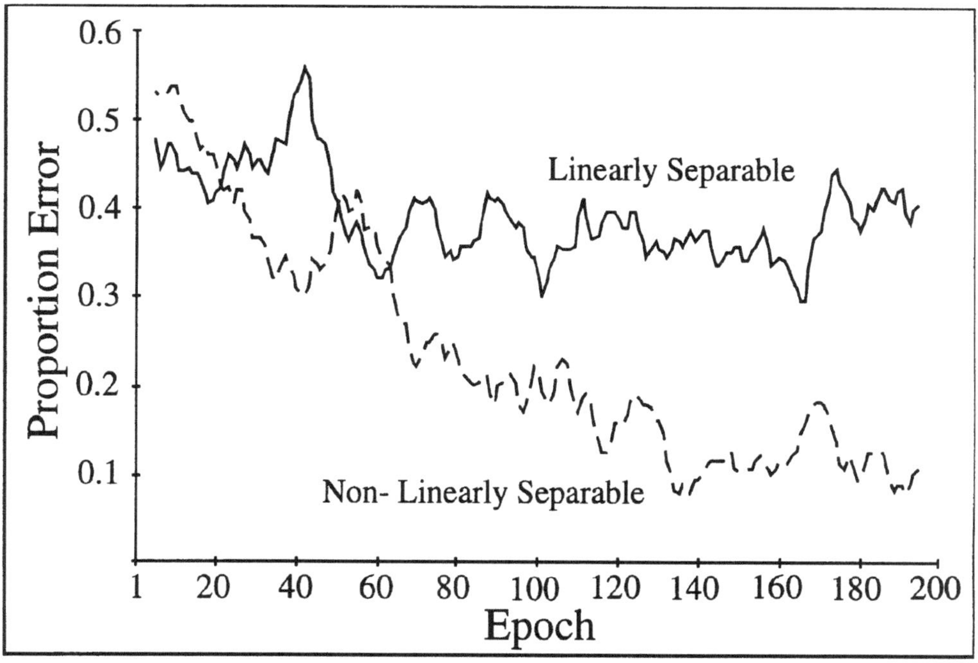

Figure 7.22: Learning curves for the configural-cue neural network model for the linearly separable exemplars (solid line) and the non-linearly separable exemplars (dashed line).

7.3. Using Prototypes

As might be expected, prototypes are thought to be dynamic in that they evolve as new information is provided by new exemplars. Busemeyer and Myung (1988, p. 3) performed an experiment using a "prototype production paradigm." Subjects were shown distributions of dots and told they represented telescopic photography of star patterns located far out in space. Because of atmospheric disturbances none of the photographs were identical even though they were photographs of the same set of stars. "The subjects' task was to identify the true star locations on the basis of the sample of five distorted photos" (Busemeyer and Myung 1988, p. 4). All individual subjects showed strong recency effects in that their estimates of the locations appeared to be most strongly influenced by the most recent photograph they looked at. It was suggested, however, that the magnitude of "the recency effect decreases as the number of exemplars increases" (Busemeyer and Myung 1988, p. 10). This suggests that the impact of the last exemplar on a prototype learned from thousands of exemplars may be relatively minor and that well-learned prototypes, i.e., those based on many examples, may be relatively stable.

Once prototypes are developed we make decisions with them. Examples of their use have been studied as prototype categorization tasks (Busemeyer and Myung 1988; Posner and Keele 1968;1970). Some maps are constructed which provide examples that serve as prototypes in the legend. An unclassed choropleth map might be accompanied by a legend that provides visual information about the range of values on the map (Tobler 1973). The areas on the map, e.g., counties or states, might be N shades of grey based on the individual data for specific areas. An example of a typical high, medium, and low value might be provided as examples in the legend. These legend examples would function as prototypes for categorizing the individual areas on the map. Some cartographers have argued that unclassed maps may help map readers accurately determine individual values for areas, but they may make pattern recognition, e.g., identifying regions of high or low values, more difficult (Dobson 1973;1980). Another researcher reached a different conclusion (Muller 1979,1980). He had subjects classify areas on unclassed maps into high, medium, and low classes. "These results demonstrate that visual regionalization produces a quantized map that is at least as good as the three class generalizations objectively derived from conventional systems" (Muller 1979, p. 248). It is interesting that Muller's test maps did not have legends. It was up to the map readers to determine a prototype for high, medium, and low categories. They apparently were able to generate their own prototypes with little difficulty.

The map reader, using the visual information in the legend and a specific area, classifies the area as being an example of the category represented by the prototype for high, medium, or low. Although the study discussed above gives a general indication that map readers are able to make good classification decisions using prototypes, the cognitive process being used has not been considered. The decision rules used by map readers to make prototype categorization judgments are even more complicated with multidimensional stimuli.

Ashby and Gott (1988, p. 33) have developed a methodology "that allows direct and detailed observation of the decision process in a categorization task." This methodology can be applied to unidimensional problems such as categorizing the shade of grey for an area using prototypes in the legend. It can also be used with multidimensional categorization problems (Ashby and Perrin 1988). When a map symbol is defined by features on two (or more) dimensions, the comparison process becomes more complex. Consider the two dimensions illustrated in Figure 7.23.

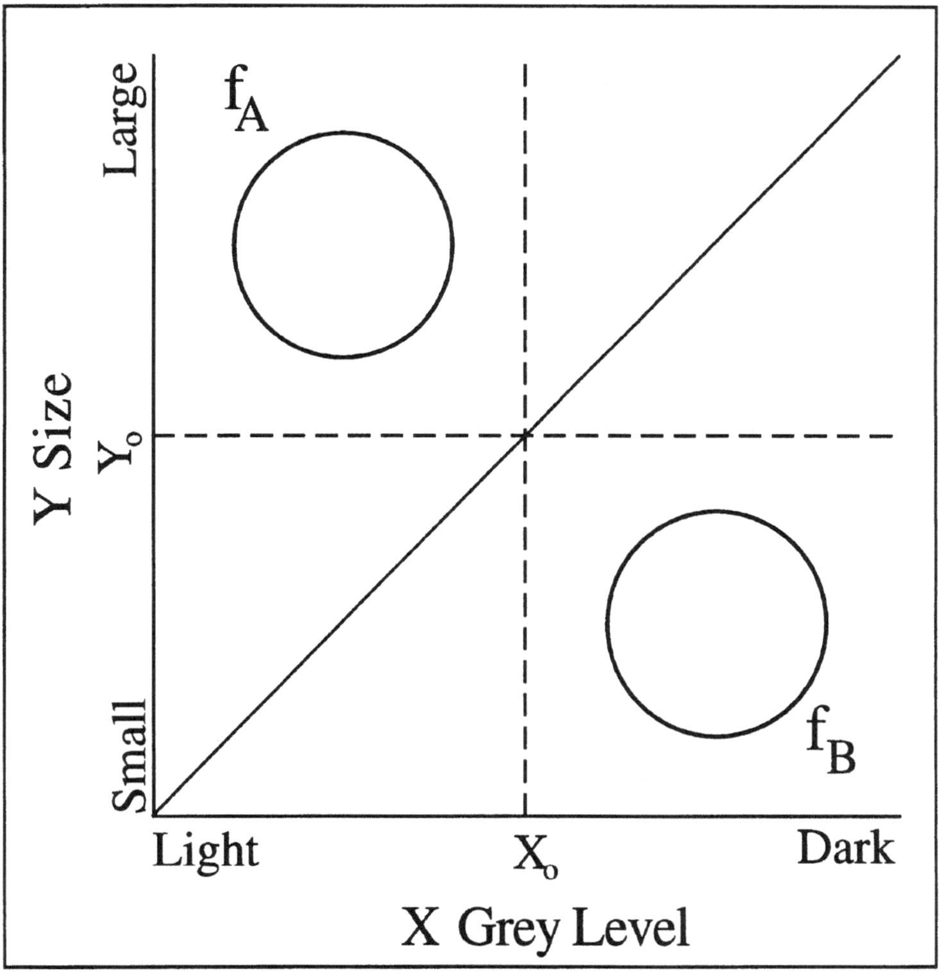

Figure 7.23: Two-dimensional space used to define map symbols. Prototypes A and B are locations at the center of probability density function f_A and f_B. Dashed lines represent the independent decision model and the solid line the information integration model. After Ashby and Gott (1988).

Dimension X and Y represent two dimensions that are used to define a map symbol. If X is the grey level of the symbol, it would vary from light to dark. If Y is the size of the symbol, it would vary from small to large. At the center of each circle is a location that would define specific features for prototype symbols. Using the dashed lines as a reference, prototype A, in the upper left quadrant, would be a large light grey symbol compared to prototype B, in the lower right quadrant, which would be a small dark grey symbol. These particular dimensions (size and grey level) should be separable and could be attended to individually. The two circles represent contours of equal probability for density functions representing the probability that a stimulus at point X,Y in the space will be considered an example of the prototype.

Since the two dimensions are separable one could attend to only one of the dimensions and make the decision. In Figure 7.23 large or light symbols are very likely to be examples of prototype A or small or dark symbols are very likely to be example of prototype B. Suppose that the probability density functions for prototypes A and B are used to generate symbols (defined by X,Y coordinates) to be judged by subjects as being an example of prototype A or B. Someone using an "independent decision model" would consider the size and grey level of the symbol independently to make their decisions (Ashby and Gott 1988, p. 37). The rule for making decisions with this model is represented by the dashed lines in Figure 7.23 Given an example of a symbol defined by an X,Y coordinate for size and grey level, answer A if it is large or light (upper left quadrant) or B if it is small or dark (lower right quadrant). Symbols in the upper right or lower left quadrant would require guessing since there is conflicting information using the independent decision model. Someone using the independent decision model would always answer A for an example in the upper left quadrant, always answer B for examples in the lower right quadrant and have a mixture of A and B answers for examples in the other two quadrants. Such a pattern would be a clear indication that the information on the two dimensions was being attended to separately and that an independent decision model was being used.

If subjects choose to integrate the information on the dimensions before making decisions, then other decision models are possible. Someone doing the same task described above using an "information integration model" would combine the size and grey level information first and then use a different rule to make the decision (Ashby and Gott 1988, p. 37). A simple rule to use with integrated information would be "the prototype that is most similar is the one that is nearest" in the space (Figure 7.23) (Asby and Gott 1988, p. 38). This minimum distance rule is represented by the solid line in Figure 7.23. With integrated information there is no guessing except for examples with X,Y coordinates exactly on the solid line. Using the minimum distance rule any example above the line in Figure 7.23 should be answered A and below the line B. Information in the upper right and lower left quadrants is no longer conflicting information because it is integrated. Note that the slope and the intercept of the decision boundary (solid line in Figure 7.23) are fixed by the mean values on the two dimensions. For specific dimensions used to create symbols there may be some response bias depending on how the information is integrated. A "general linear classifier" can be computed with an intercept and slope that describes the specific rule being used to make decisions with particular combinations of dimensions (Ashby and Gott 1988, p. 38).

If integral dimensions, e.g., the hue and brightness of a color, are being used to define symbols, then it would not be possible to attend to the information for the two dimensions independently. The independent decision model (dashed lines in Figure 7.23) could not be used with integral dimensions. You would expect an integrated decision model to be used with integral dimensions. It would still be of interest, however, since "there are many different ways of integrating information" (Ashby and Gott 1988, p. 42). People who construct maps should be interested in these models for two reasons. First, when multidimensional symbols are being used, the dimensions should be selected so that the map reader will efficiently be able to associate symbols on the map with prototypes in the legend. If map readers use independent decision models to compare map symbols to prototypes, they will be required to guess part of the time. This will result in more error and should be avoided. One way to avoid this problem is to always select dimensions that are integral when producing symbols. If separable dimensions are used, map readers may still be able to integrate the information and use an integrated decision model to make decisions. This, however, should not be taken for granted. Combinations of dimensions that produce acceptable results can be determined by experimental studies. Second, the rules being used by map readers to integrate information and make decisions can be computed. The results reported by Ashby and Gott (1988) were very encouraging for map reading. They provide strong evidence that people use deterministic decision rules when categorizing examples as being similar to prototypes. They found that subjects using integrated information had decision rules that were extremely accurate. Results also indicated that subjects did not use independent decision models even when the experimental design encouraged them to do so. Experimental studies done in a map-reading context can evaluate the map reader's ability to use integrated feature information to make decisions.

Lloyd (1992) considered the decision rules for dimensions commonly use to create map symbols and the consistency of these rules among map readers. Subjects were shown two prototypes that were defined in a two-dimensional space similar to Figure 7.23. Subjects learned the prototypes by being given exemplars based on randomly selected locations in the space. During half of the practice trials the subjects made decision while looking at the prototypes and during the other half of the practice trials the subjects made decision from memory. Experimental trials were all done without the prototypes being visible. When a trial presented a symbol for evaluation the subject simply decided if it was more like Prototype A or Prototype B. The decision and the coordinate locations of the exemplars were recorded for analysis. Six different pairs of prototypes were generated for consideration using dimensions such as length, color, size, and orientation combinations. Three of the sets combined dimensions that could use a common set of resources. For example, in one group Prototype A was a rectangle that was taller than it was wide and Prototype B was another rectangle that was wider than it was tall. A regression was performed for each subject's data using a binary dependent variable (Prototype A = 0 and Prototype B = 1) for the decisions and the coordinate locations of the exemplars as independent variables. The standardized regression coefficients were plotted for each subject to assess how the dimensions were weighted to make decisions. Subjects considering rectangles as exemplars were able to use an information integration model with the two length dimensions and appeared to weight the two dimensions about the same (Figure 7.24a). This is also illustrated by plotting the line provided a linear separation of the space with Prototype A on one side and Prototype B on the other side. The separate line for each subject was estimated by plotting locations predicted by the subject's regression equation to be 0.5 or half way between 0 and 1. These lines for the subjects that considered rectangles generally

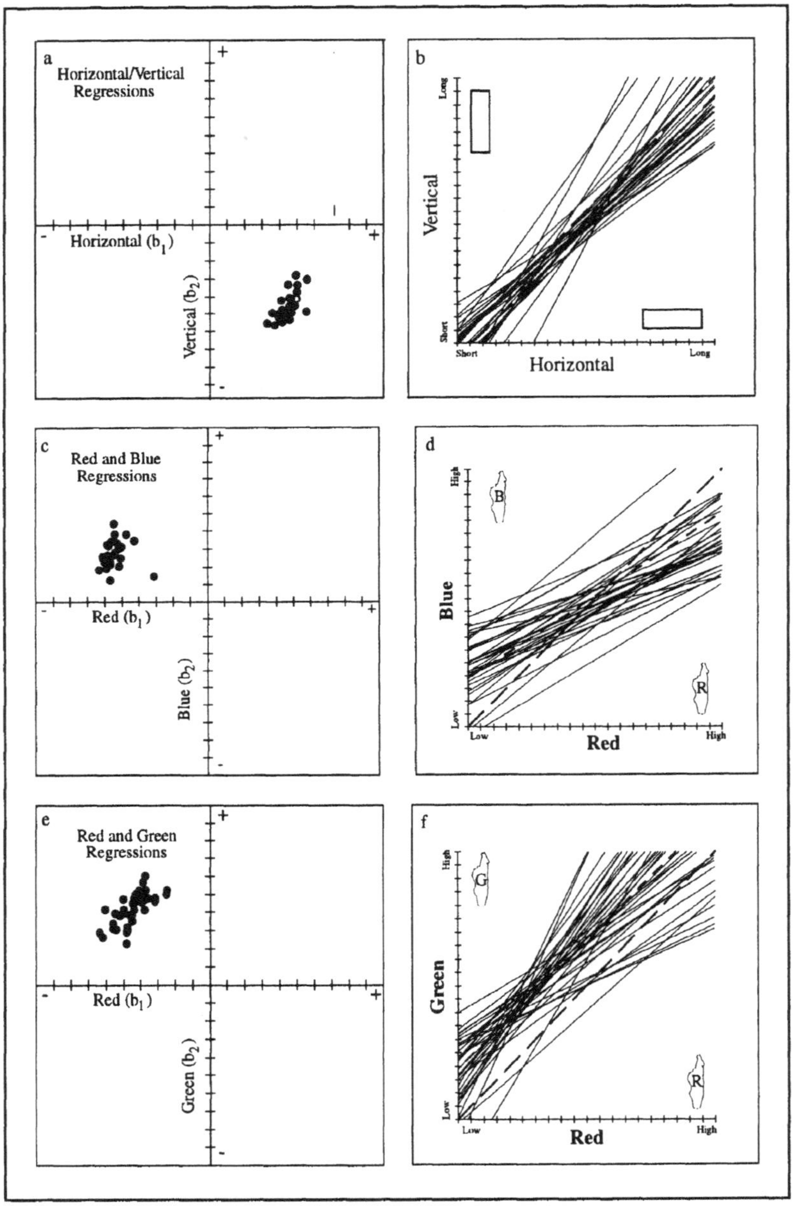

Figure 7.24: Prototypes were defined using horizontal and vertical length of rectangles (a and b), red and blue color of the Island of Madagascar (c and d), and red and green color of the Island of Madagascar (e and f). Dots are for individual subjects and located to indicate the relative importance of the two variables for categorical decisions. Lines represent the partition of the space into parts where Category A and B are selected. Dashed lines represent the ideal and average partitions.

extended from the origin at an angle near to 45^{0} and split the space evenly (Figure 7.24b).

The result for the rectangle prototypes would be ideal for symbols used on maps. Subjects consistently used the same rule to interpret the meaning of the exemplars and were able to integrate the information from the two dimensions. Two other sets of exemplars presented a map of the Island of Madagascar with a color filling the polygon that was some combination of two color guns on the monitor displaying the map. All maps were some combination of the two colors. For example, one set had Prototype A as mostly Red and a little Blue and Prototype B as mostly Blue and a little Red. Subjects were reasonably consistent in the way they responded, but tended to weight the red dimension slightly stronger than the blue dimension (Figures 7.24c and 7.24d). Red was also weighted slightly more when Prototype A was mostly Red and a little Green and Prototype B was mostly Green and a little Red (Figures 7.24e and 7.24f). In the three cases represented in Figure 7.24 subjects were using an information integration model that combined the two dimensions and consistently produced similar results.

Another three sets of prototypes were compared that used dimensions that were not as related. For example, one set of exemplars showed maps of the Island of Madagascar and varied its size from small to large and orientation from vertical to horizontal (Figure 7.25a and 7.25b). This produced a greater amount of variation among the subjects. Some of the subjects appeared to be able to integrate the two dimensions while others put more weight on size and others put more weight on orientation. Subjects were even less successful at integrating the two dimensions when prototypes were defined using brightness as dark to light and size as small to large using circles (Figure 7.25c and 7.25d). Subjects were using an independent decision model and appeared to base their decision mostly on either brightness or size. The most difficult pair of dimensions to integrate were clearly color from blue to green and orientation from vertical to horizontal (Figure 7.25e and 7.25f). No individual subject appeared to be able to integrate these two dimensions. Subjects used color to make their decisions or orientation, but none were able to combine the two dimensions to use an integrated information model. These results suggested that people can integrate the information in multidimensional map symbols better if the dimensions are using similar resources. If the subsystems in the brain that are processing the two dimensions are more connected and better able to share resources, this may make it easier for the integration of information to take place (Kosslyn and Koenig 1992).

7.4. When is What Where?

In addition to the frequency effect, some studies have suggested the sequence in which we encounter exemplars of a category has a significant impact on our learning the category. It has been argued that more influence is exerted by exemplars encountered early and late in a sequence than for those encountered in between (Busemeyer and

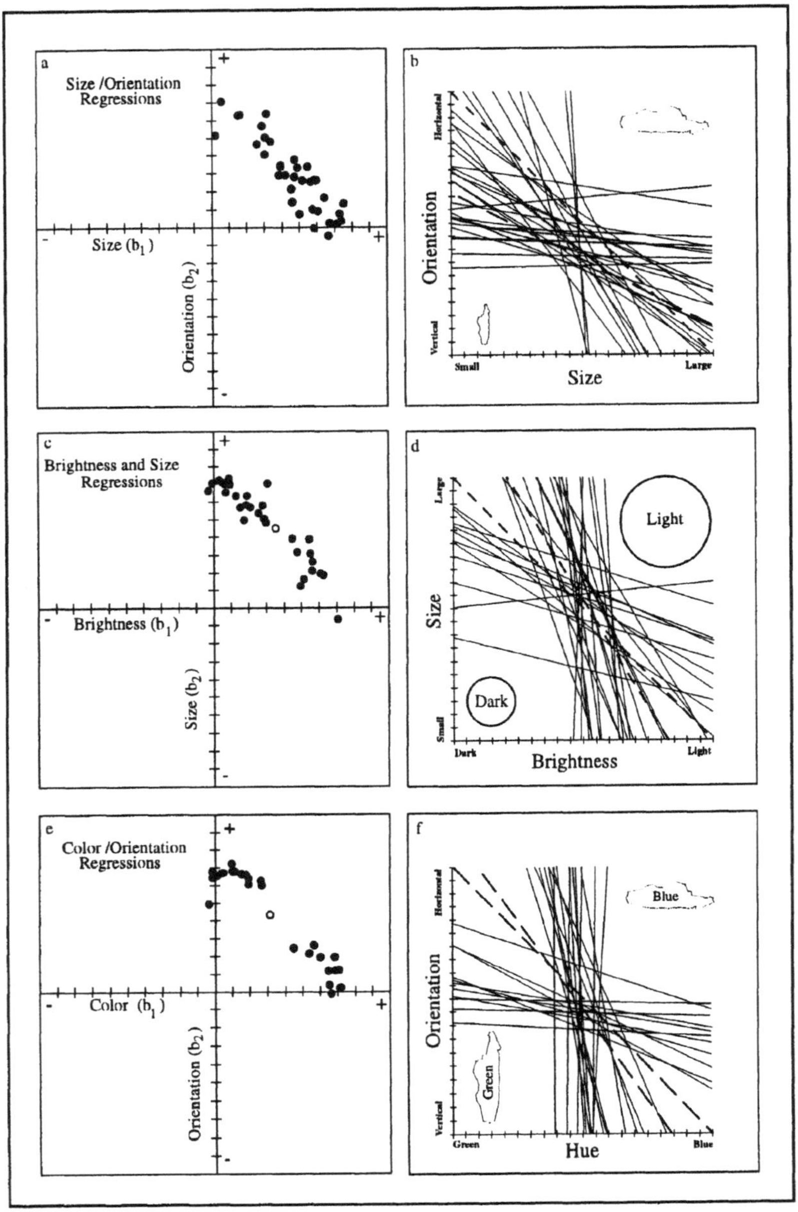

Figure 7.25: Prototypes were defined using size and orientation of the Island of Madagascar (a and b), brightness and size of circles (c and d), and color and orientation of the Island of Madagascar (e and f). Dots are for individual subjects and located to indicate the relative importance of the two variables for categorical decisions. Lines represent the partition of the space into parts where Category A and B are selected. Dashed lines represent the ideal and average partitions.

Myung 1988; Myung and Busemeyer 1992). This is referred to as the primacy/recency effect and means that learning processes are frequently affected by the temporal order of our experiences. Neural network models that learn to classifying input patterns into output categories or learn prototypes from a sequence of input pattern are known to be influenced by the order of the input patterns (McClelland and Rumelhart 1986, 1989; Rumelhart and McClelland 1986). Patterns ordered in a particular sequence may cause the network to learn quickly or slowly. Different orders may produce final solutions with more or less overall error. The ordering of events also has implications for processing information from maps. Sequencing has been investigated by cartographers through the presentation of mapped information with animation techniques. The goal was to enhance a map reader's ability to process information on maps efficiently (MacEachren 1992a; Egbert and Slocum 1992; Monmonier 1992; Patton and Cammack 1996; Slocum and Egbert 1993).

7.4.1. EARLY MAPS OF NORTH AMERICA

Lloyd (1994) conducted a study with human subjects that had them learn a sequence of exemplar maps that represented an Island off the eastern coast of North America. After the subjects learn the prototype, they then rate the typicality of exemplar maps. The maps were categorized by the European country that was supposed to have made the maps. All the maps in the study had a common background that was the same for all categories of maps. The maps all had a green land area on the western edge of the map and grid lines in the ocean to facilitate judgments of relative location (Figure 7.26). Also held constant within each map category were the Island's nonspatial properties of shape and color (Kosslyn and Koenig 1992). Each category of maps had a unique color and shape for the Island.

Each exemplar map varied in its spatial characteristics. The Island could be coded in any of 9 spatial locations defined by 3 vertical positions 3 horizontal positions. The Island could be coded in three different sizes and orientations. This allowed for 3^4 or 81 possible maps to be encoded.

7.4.2. USING A LEARNED PROTOTYPE

During the learning stage of the experiment subjects paid particular attention to the locations of the Islands on maps and to the Islands' sizes and orientations as they view 81 maps for a particular country. Attending to the exemplar maps to determine how that country tended to display the Island allowed the subjects to form a prototype for the category named for a particular country. The experiment was designed to build in a particular sequencing bias for each country's maps so that the influence of the primacy/recency effect could be investigated. First the 81 maps for a country were put in a random order (Figure 7.27), then 27 or 1/3 of the original maps were replaced by maps that established a bias in terms of location, size and orientation for each country. The English category was biased with extra maps showing the Island in northern and western locations. The Spanish category was biased with extra maps showing the Island with smaller sizes and horizontal orientations. The French category was biased with extra maps showing the Island in southern and eastern locations. The Dutch category was biased with extra maps showing the Island with larger sizes and vertical orientations. The biases were inserted to be part of the sequence of 81 maps in four different ways (Figure 7.23). To test for a primacy effect the English category was biased in the

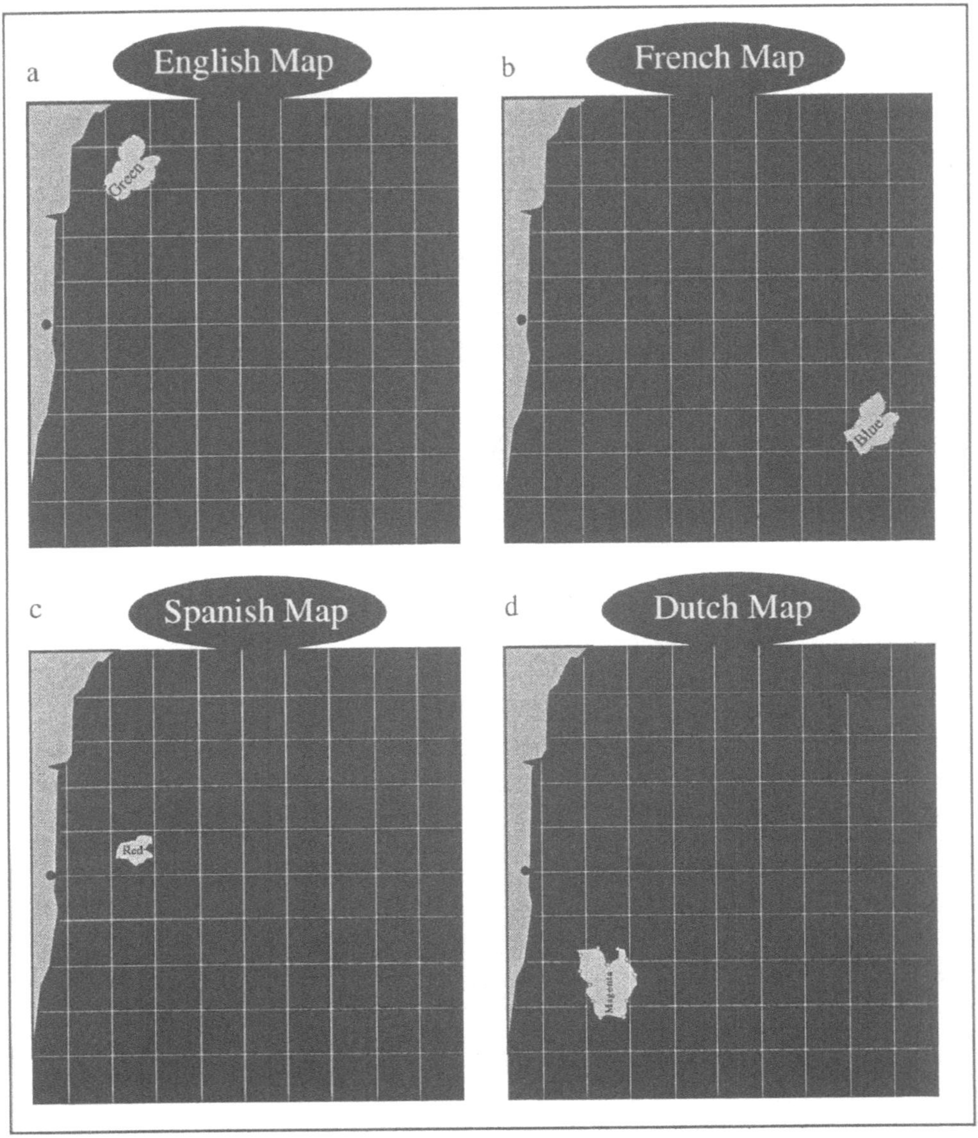

Figure 7.26: An English Map (a) with the medium oblique island in the northwest, a French Map (b) with a medium oblique island in the southeast, a Spanish Map (c) with a small horizontal island in the middlewest, and a Dutch Map (d) with a large vertical island in the Southwest.

first third of the sequence of 81 maps. To test for a recency effect the French category was biased in the last third of the sequence of 81 maps. As control groups the Spanish category was biased in the middle third of the sequence of 81 maps and the Dutch category was biased by replacing every third map in the series.

Subjects used the prototype they had learned by viewing the exemplars for a Country category in the evaluation stage of the experiment by indicating how typical each of the 81 original maps was for the category. These typicality ratings were

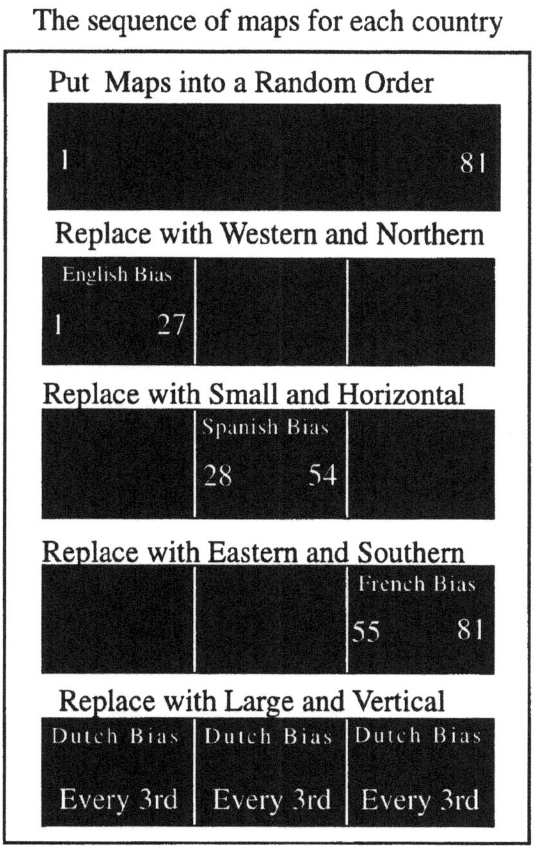

Figure 7.27: The 81 maps were first put into a random order and then biased in different way for each country category.

analyzed to determine if the biases for the categories had been learned and if the sequencing of the exemplars had affected learning. For the categories that had the bias clustered as the first, middle or last third of the sequence the subjects appeared to learn the bias. The primacy effect appeared to have the strongest influence. The subjects learning the English category was evaluated as having the strongest prototype. The Dutch subjects, who viewed an unclustered bias distributed throughout the sequence of 81 exemplars, appeared to have difficulty learning the prototype. These results indicated that spatial properties can be encoded as prototypes for maps and that the order in which exemplars are experienced can significantly influence people's ability to encode spatial prototypes.

7.5 Conclusions

This chapter has considered prototypes as structures encoded in memory to represent the essence of a category. These structures represent the family resemblance that defines the characteristics of categories. Examples were discussed that modeled the learning of prototypes with neural networks using map symbols and maps. The next chapter considers the cognitive processes used to determine similarity.

CHAPTER 8

SIMILARITY

8.1. Introduction

This chapter focuses attention on the similarity of maps, but this concern is only a part of geographers' larger interest in the similarity and differences of places.

> *When, however, we study more complex integrations in geography, we find a much smaller number of essentially similar specimens. As in many other sciences, we attempt to overcome this difficulty by recognizing categories or types within which differences in cases appear less marked than the similarities. But in thus classifying objects, or phenomena, each of which involves a host of independent or semi-independent elements, we do not have specimens that are similar in all essential elements. They are similar only in terms of the particular categories we have chosen, and may differ notably in other respects which research may demonstrate to be less important.*
> *Hartshorne (1959, p. 150)*

Hartshorne's quotation expressed the common problem geographers face when trying to determine the similarity of places. The same issues are important when map readers visually compare two maps and determine their similarity. Goldstone (1994, p. 3) has presented three general arguments for studying similarity. He argued that similarity plays an important role in cognitive theories, provides a tool for studying internal structures, and processes that are difficult to study directly, and occupies an important middle ground between perception and higher level knowledge. The development of spatial theory related to search, memory structures, categorization and many other phenomena are connected to judgments of similarity.

Maps with quantitative symbols at N fixed locations can be represented geometrically as normalized vectors in an N dimensional space. The cosine of the angle between pairs of vectors would then represent a measure of the similarity, i.e., the statistical correlation for two maps (Johnston 1978). Since many distribution patterns for pairs of maps can result in the same statistical correlation coefficient and features other than pattern may affect visual similarity, geometric models may not represent visual similarity well.

Maps also can be thought of as distributions of features like colors, shapes, textures, sizes, and spatial frequencies. The apparent visual similarity of some maps may be described better by models based on matching features than those based on statistical correlation (Tversky 1977). This chapter argues that the visual similarity of two maps is dependent on the matching of their features. This may be at least partially true for quantitative maps that can be compared by computing statistical correlation coefficients. It is particularly true for map that represent nominal classes of information, e.g., land use patterns, because it may not be appropriate to compute standard statistical correlations for such maps.

Maps represent environments for people to view and successful map design requires two types of knowledge. First, knowledge about the map construction process is obviously important along with the skills that come with experiences producing maps. Second, and less obvious, is the knowledge map designers can learn and use about how maps are processed by map readers. Previous Chapters have discussed some of the processes used by map readers such as searching for a target (Chapter 5), classifying objects (Chapter 6), and identifying objects (Chapter 7). The current chapter is concerned with how the similarity of two objects (maps) is determined. It will consider how map readers learn the processes used in map reading. Cartographers with knowledge of the cognitive processes used by map readers can use this knowledge to improve their designs (Lloyd, Rostkowska-Covington, and Steinke 1996). Statistical classifications and training processes for neural networks are frequently divided into two types (Jensen 1986; McClelland and Rumelhart 1989). Supervised processes require and expert to assist in the classification or training (Figure 8.1). A human with knowledge of the environment being studied identifies pixels in a remote sensing scene that represent particular landcover classes and these pixels are used to profile classes for the statistical classifier. Some neural network architectures require an expert to provide correct values for the output neurons. Error is then measured as differences between the values produced for the output neurons by the neural network and those provided by the expert. The goal is to have the predictions of the statistical classifier or neural network reflect the *truth* provided by the human expert. Other statistical classifiers and neural networks perform similar tasks without the aid of an expert. Since no external *truth* is available these processes look for consistent patterns in the input data. These ideas also apply to the exploring of environments and reading maps. Unsupervised processes use bottom-up information from the environment (or map) being considered while supervised processes us both bottom-up information and the top-down information provided by the expert.

Nigrin (1993) identified three distinct types of learning processes that Lloyd, Rostkowska-Covington, and Steinke (1996) have applied to map reading. First, *supervised learning* occurs when a student is taught how to read maps by a teacher. It is possible that the same person could be the map designer and the educator (Figure 8.2). A map designer might write a text accompanying published maps on how they should be read or possibly even teach students in a course on how to read the maps. Individual map readers whose learning processes have been directly or even indirectly supervised by the designer of the maps are probably rare. Map designers usually construct their maps with little knowledge of who will be reading the maps or how the maps will be processed.

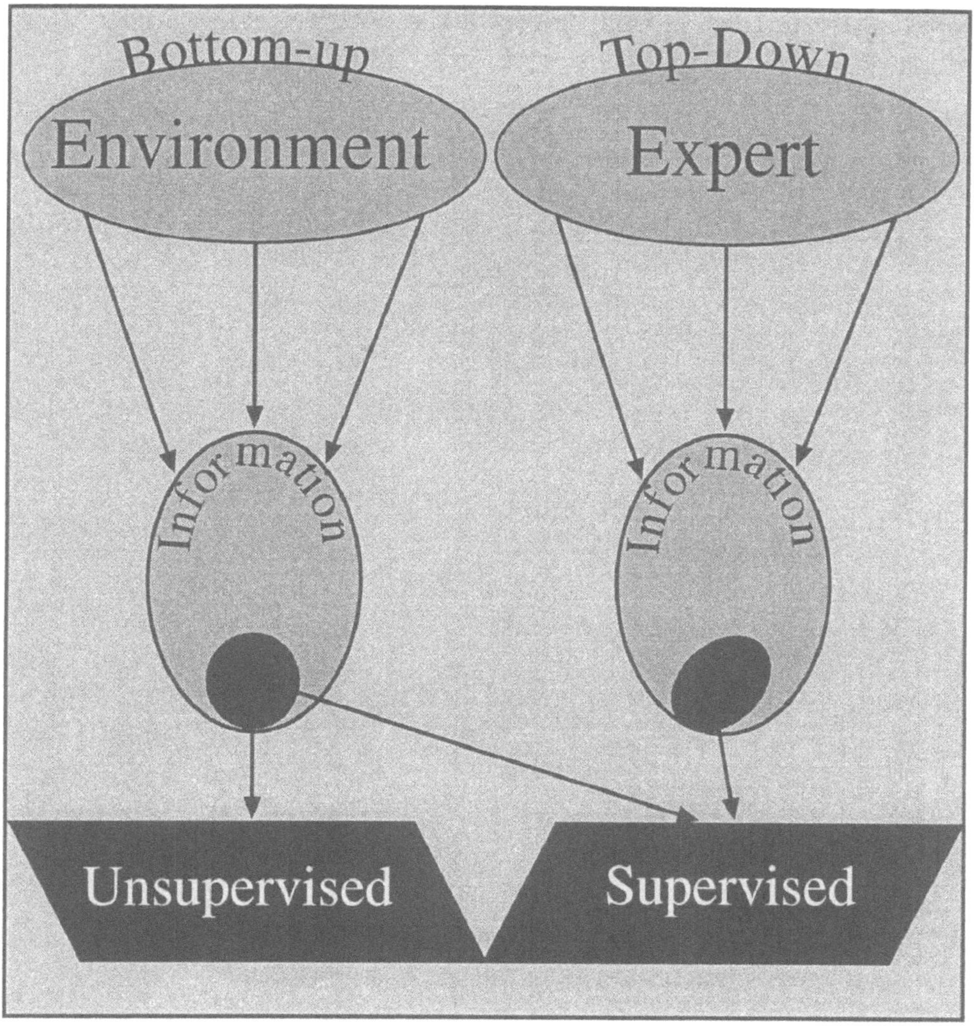

Figure 8.1: Supervised processes combine bottom-up information from the environment and top-down information from an expert while unsupervised processes only use bottom-up information.

It is probably true that most map readers bring relatively little formal training to their task of processing spatial information. Map readers do bring to the task common sense and a number of generalized cognitive processes developed for processing any visual information. Unlike *supervised learning*, the second and third types of learning experiences have no direct supervision by an expert (Nigrin 1993). With the second type, *unsupervised learning*, map readers acquire knowledge without knowing if the processing has produced a correct result. With *unsupervised learning* there is neither an expert to tell map readers they are making proper interpretations nor does this type of learning have feedback from the consequences of the map readers' task. *Unsupervised learning* acquires information from maps, but no decisions are made that could provide feedback on the success of the learning experience. Examples of *unsupervised learning* are the many casual experiences we have with maps that are not motivated by imminent decisions.

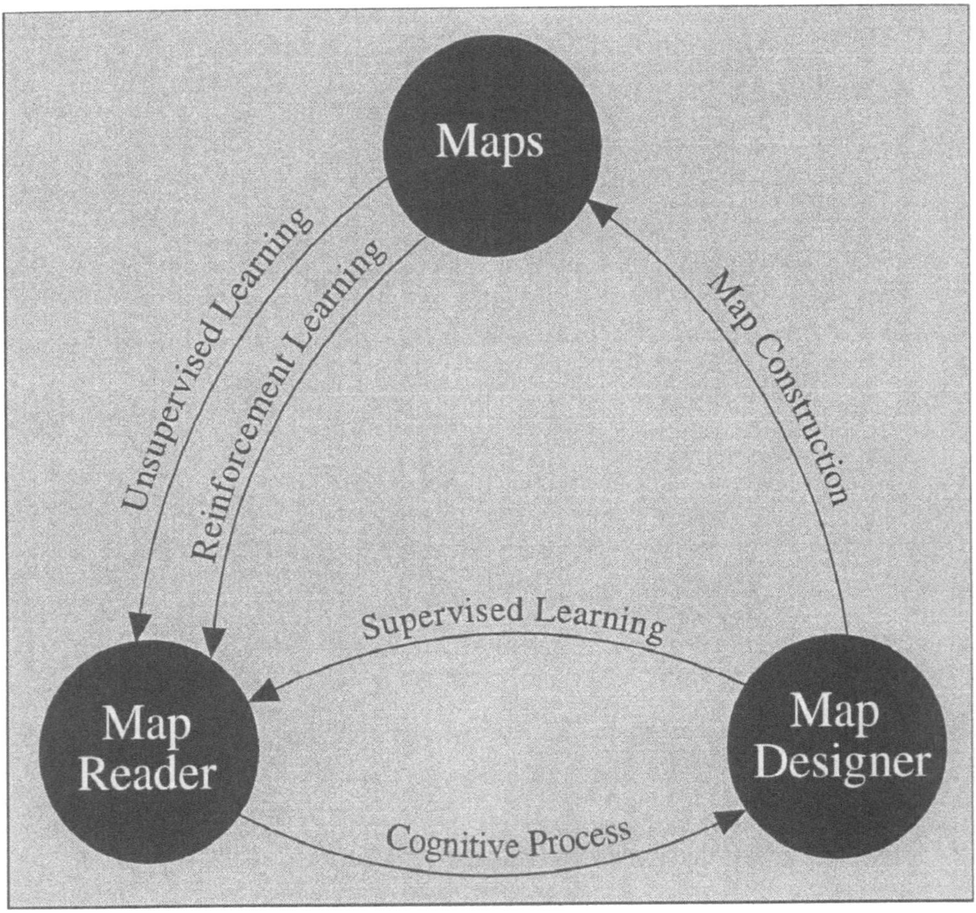

Figure 8.2: Supervised, unsupervised, and reinforced learning processes used in used in map reading. After Lloyd, Rostkowska-Covington, and Steinke (1996).

The third type of learning, *reinforcement learning*, also does not have direct supervision by an expert but does involve indirect feedback from the consequences of decisions made by the map reader (Nigrin 1993). A person encoding information from a map to learn the locations of some landmarks in an environment is a familiar example. Finding the landmarks in the environment would indicate a successful map-reading process and not finding the landmarks would indicate a failure. The learning process used is positively reinforced by success, and negatively reinforced by failure. *Reinforcement learning* is clearly not supervised by an expert, but based on the trials and errors of an individual learning by making mistakes. A failed process eliminates a possible strategy but does not provide the key for successful learning. After a number of experiences that resulted in failure, the map reader may eventually find a successful solution.

An expert does not assist for either *unsupervised learning* or *reinforcement learning*, but both may be at least partially successful, if not optimal, solutions for learning. The basic difference between the two is that *unsupervised learning* may produce very good or very bad results without the map reader being aware of the outcome. It may take persistence and a little luck for *reinforcement learning* eventually to achieve optimal success. Given that most individuals who read maps are not using skills acquired through *supervised learning*, an important goal for map designers should be to understand the cognitive processes being used to process spatial information. This is because, for the most part, these processes are a product of *unsupervised* and *reinforcement learning* and may not conform to theoretical expectations implicit in map construction and *supervised learning* processes.

8.2. Comparing Maps

Consider doing a common map reading task. You have two maps lying side by side and you are to compare them and ultimately state how similar they are on a scale of 1 to 10. Now describe the process you used to decide the appropriate rating. If you are like most people you will not describe a comparison process involved with numerical processing. You will not talk about mental equations or computing the sum of squared differences. Visual comparisons of pairs of maps appear to be rather intuitive. They are very much like processes used to compare any pair of everyday objects in the environment. These processes are also likely to be the product of *unsupervised learning* and *reinforcement learning.*

8.2.1. THE SIMILARITY OF CHOROPLETH MAPS

It is well known that the visual comparisons of maps do not produce the same results as formal statistical comparisons of maps. An early study by McCarty and Salisbury (1961) made comparisons between the visual similarity and statistical correlations of maps and did not find a strong match between the two measures. They suggested other factors affecting differences in the visual appearance of maps, such as blackness and complexity, may explain the weak correspondence between visual similarity and statistical correlation. For choropleth maps, constructed by filling polygons, e.g., states or counties, with shades of grey proportional to some numeric value for the polygon, blackness refers to the overall light to dark tone of a map. For example, two maps look different on this visual dimension if one map has darker polygons (Figure 8.3a) and the other map has lighter polygons (Figure 8.3b). These maps, however, have a strong positive linear association as measured by a correlation coefficient ($r = 1.0$). Maps c and d are inversely correlated ($r = -0.67$), but are the same on the blackness dimension with each having two polygons with the same shade of grey (Figure 8.3).

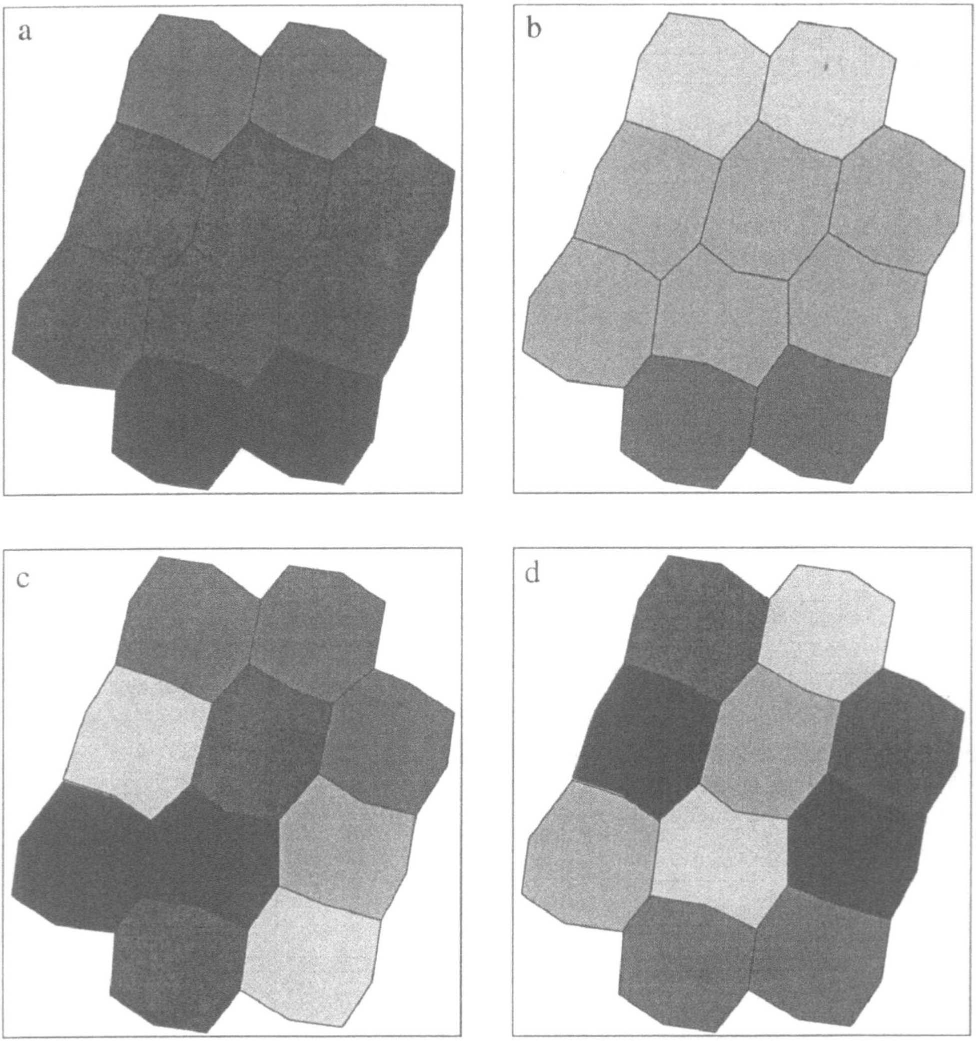

Figure 8.3: The *dark* map (a) is visually different from the *light* map (b) on the blackness dimension but have a strong positive correlation. Map c and map d are identical on the blackness dimensions with equal numbers of each shade of grey.

Complexity refers to the texture produced by the arrangement of the shades of grey used to fill the polygons. Simple textures have neighbors that are the same shade of grey. The simple map illustrated in Figure 8.4a has 26 of 31 boundaries between polygons with the same shade of grey on either side of the boundary. The complex map illustrated in Figure 8.4b has all of its 31 boundaries with different shades of grey on either side of the boundaries. The first map has an organized appearance with one color boundary across the center of the map and two regions. The second map has a random appearance with its many color boundaries and no distinct regions. Although the two maps are at the opposite end of the complexity dimension, they are not strongly

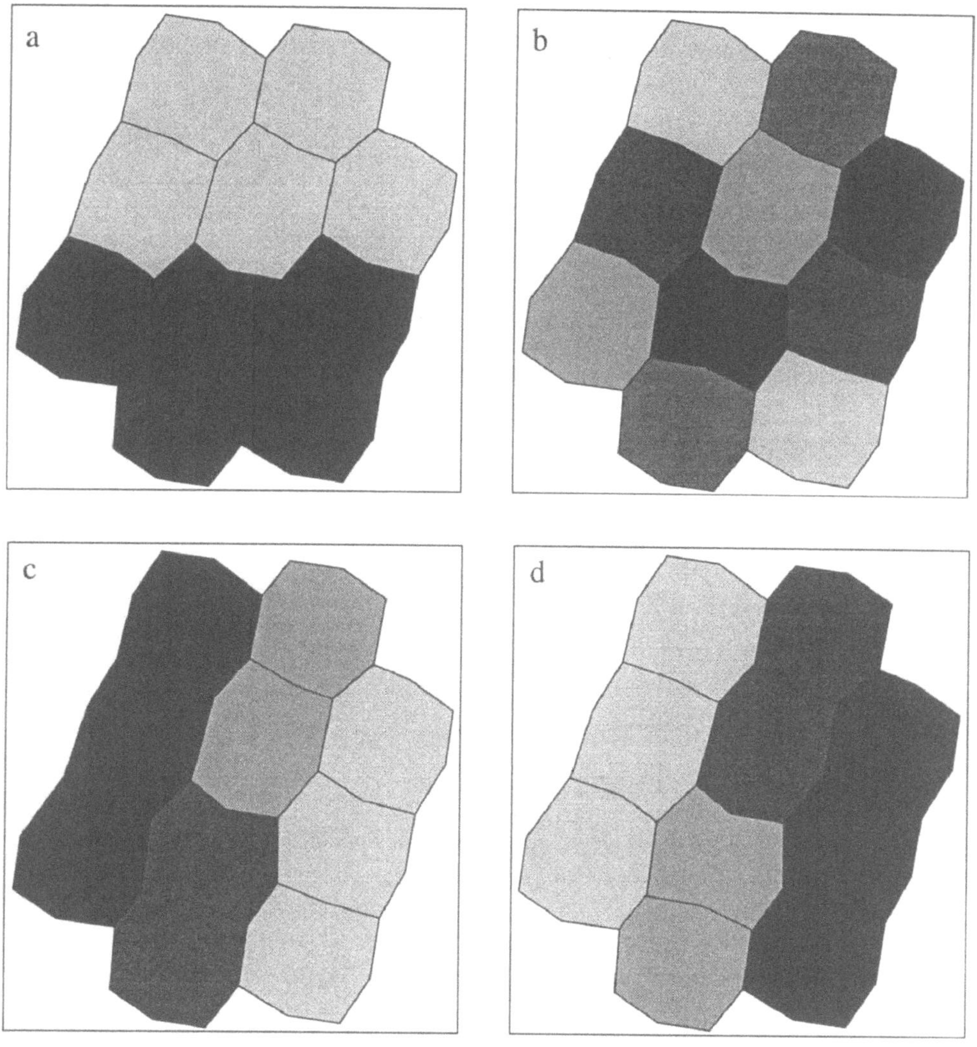

Figure 8.4: The *simple* map (a) is visually different from the *complex* map (b) on the complexity dimension. Two maps (c and d) have different spatial patterns, but are similar on the complexity dimension.

correlated (r=0.0). The maps illustrated in Figures 8.4c and 8.4d are the same on the complexity dimension but have a strong negative correlation (r = -1.0). Correlation coefficients measure the linear association of the values for the polygons on two maps. The coefficient is strongly positive (near +1.0) if dark and light grey values tend to match for individual polygons on the two maps. The two maps shown in Figures 8.5a and 8.5c have s strong positive correlation (r = 0.9). The scatter plot of the grey values illustrates the strong positive linear relationship. Correlation coefficients are strongly negative (near -1.0) if dark tends to match light and light match dark on the two maps. The two maps shown in Figures 8.6a and 8.6c have a strong negative correlation

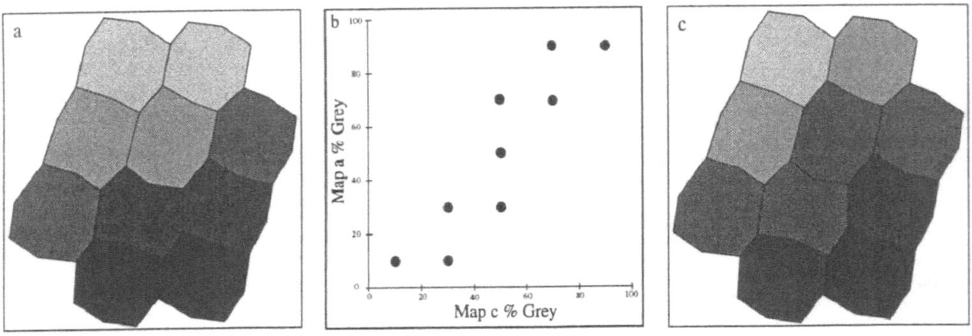

Figure 8.5: Two maps with a strong positive statistical correlation.

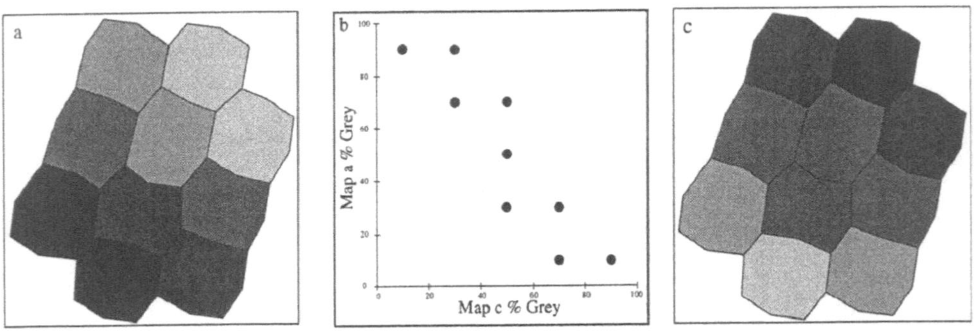

Figure 8.6: Two maps with a strong negative statistical correlation.

(r= -0.9). The scatter plot of the grey values illustrates the strong negative relationship. Correlation coefficients indicate no linear relationship (near 0.0) if there is no consistent match between light and dark polygons on the two maps. The two maps in Figures 8.7a and 8.7c have patterns that are not correlated. The scatter plot of the grey values illustrates no covariations between the values on the two maps.

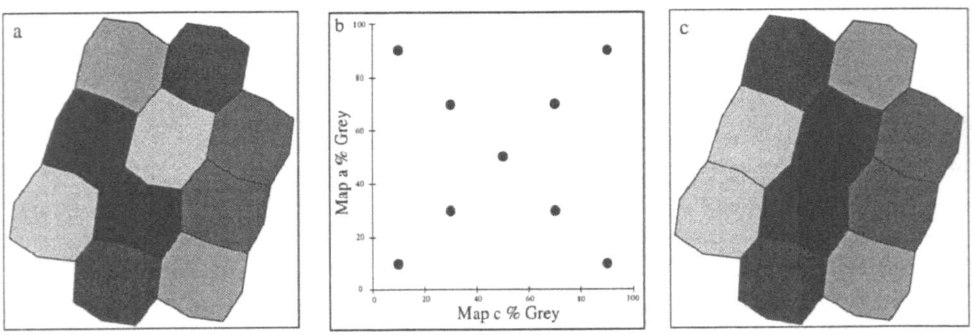

Figure 8.7: Two maps with a weak statistical correlation.

8.2.2. THE VISUAL SIMILARITY OF MAPS AND MAP IMAGES

A number other early studies have supported the notion that the visual similarity of maps is partially related to statistical correlation, but also significantly affected by blackness and complexity (Lloyd and Steinke 1976; Muller 1976; Muehrcke 1973; Olson 1970).

Lloyd and Steinke (1977) treated the statistical correlation as the signal that human subjects should be seeing when visually comparing maps and blackness as noise that can interfere with the signal. They showed map readers pairs of choropleth maps of lower Michigan that represented data from the United States census. The subjects were asked to indicate the visual similarity for pairs of maps. Subjects were divided into three groups and each group viewed maps that were constructed using a different method. An *equal interval* method classed the raw data by dividing the range of the data into five equal segments. A *minimum deviation* method classed the data so that the within class error was minimized (Jenks and Caspall 1971). These were common methods that did not control differences in blackness among the maps. The third group viewed maps that were produced using an *equal area* method for classification. This method weighted the data by the area of the polygon before producing classes with equal frequencies. This produced a set of maps with exactly the same amount of blackness on each map. Maps for this group did not vary in blackness and caused differences for this visual dimension (the noise) to be eliminated.

The final analysis was an individual difference multidimensional scaling using grey values for the three sets of maps, the raw data, blackness for the equal interval and minimum deviation maps, and the three sets of cognitive estimations of the visual similarity of the maps as "subjects" of the analysis. The two-dimensional space produced by the analysis represented the signal on the horizontal axis and the noise on the vertical axis (Figure 8.8). The alignment of the cognitive estimates for the equal area subjects with the horizontal dimension indicated the signal was being received by these map readers. The alignment of the cognitive estimates for the *equal interval* and *minimum deviation* subjects with the vertical dimension indicated differences in blackness among the maps (the noise) had affected these subjects' estimates of visual similarity. These results suggested that the visual similarity of maps can be expressed in a way that more closely approximates statistical correlation if other effects such as blackness can be controlled.

Two additional studies done by Steinke and Lloyd (1981 and 1983b) used the same maps of lower Michigan to measure the relative importance of correlation, blackness, and complexity to explain the similarity of maps. The first study had map readers judge the similarity of pairs of maps while looking at the maps. An analysis was performed on each subject's data using the subject's estimates of the similarity of pairs of maps as the dependent variable and objective measures of correlation, blackness, and complexity for the maps as independent variables. Each subject was plotted in a

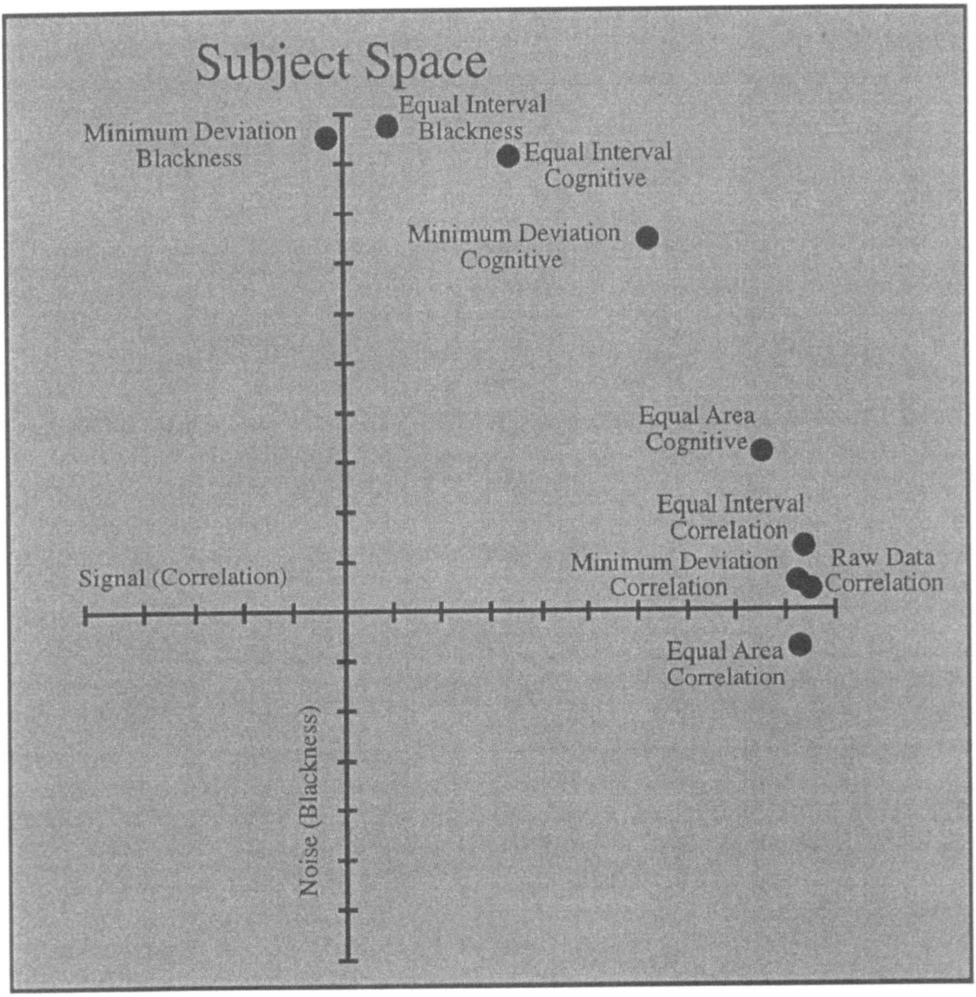

Figure 8.8: A two-dimensional space representing correlation on the horizontal axis and blackness on the vertical axis. The alignment of the similarity judgments of the *equal area* cognitive group with the horizontal axis indicated their judgments were similar to statistical correlations. The alignment of the *minimum deviation* and *equal interval* groups with the vertical axis indicated their judgments were similar to blackness differences among the maps. After Lloyd and Steinke (1977).

triangular space based on the relative importance of correlation, blackness, and complexity to the subject's judgments of visual similarity (Figure 8.9). Each axis along the triangle measured the relative importance of one of the variables within a range from 0 (not important) to 100 (extremely important). The position of a subject in the space

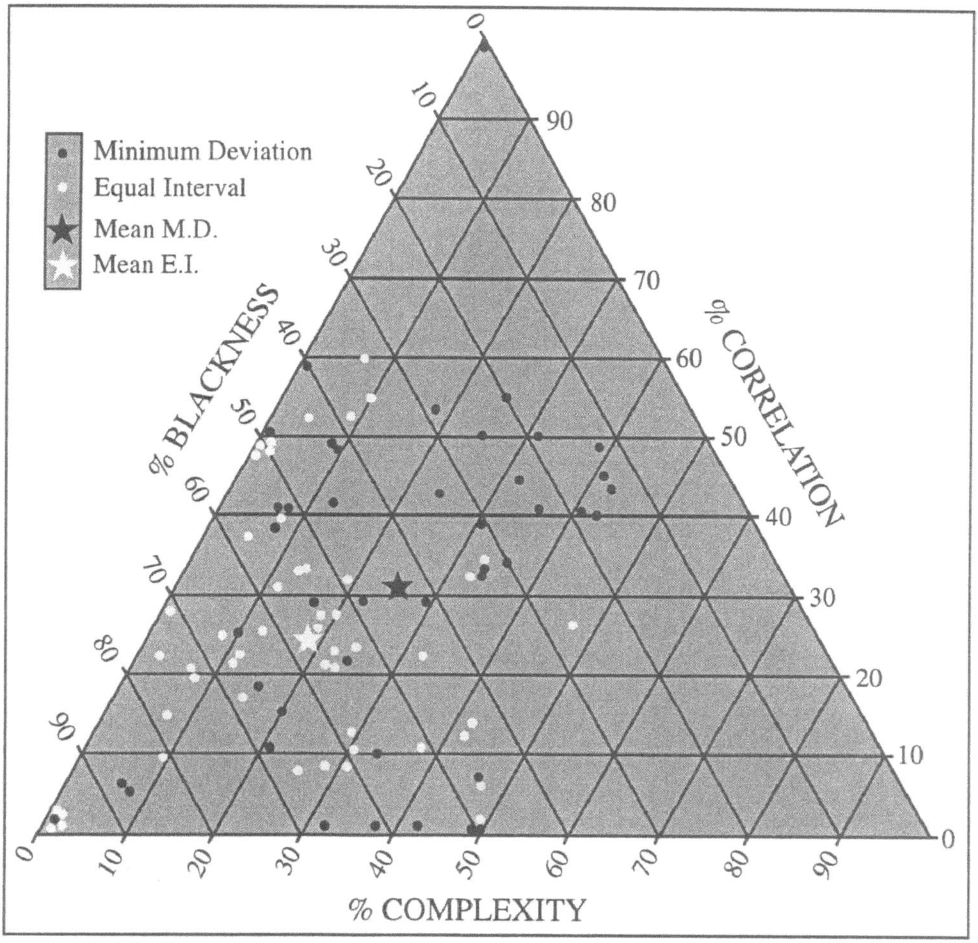

Figure 8.9: The triangular space represents correlation, blackness, and complexity dimensions. Dots mark locations of individual subjects in the space that were members of three experimental groups. The subject's coordinate location on the three axes indicates how that subjects integrated the three dimensions when making similarity judgments. After Steinke and Lloyd (1981).

indicated how that subject traded off correlation (the spatial pattern), blackness (the tone), and complexity (the texture) when judging visual similarity. Although the distribution of the subjects viewing *equal interval* and *minimum deviation* maps showed some individual differences, the average subject tended to be influenced most by blackness. Complexity did not appear to strongly influence many subjects.

The second study repeated the same experiment with one important change (Steinke and Lloyd 1983b). Subjects first studied the *minimum deviation* maps to remember the patterns and then judged the visual similarity without the maps being present. The goal was to investigate whether processing map images in human memory was similar to processing maps that were being viewed. An analysis was performed again on each individual subject's data to determine the relative importance of correlation, blackness, and complexity. Subjects who judged the visual similarity of the *minimum deviation* maps from the earlier experiment (Steinke and Lloyd 1981) and subjects from the memory experiment were both plotted in a triangular space to compare how they integrated correlaton, blackness and complexity information (Figure 8.10).

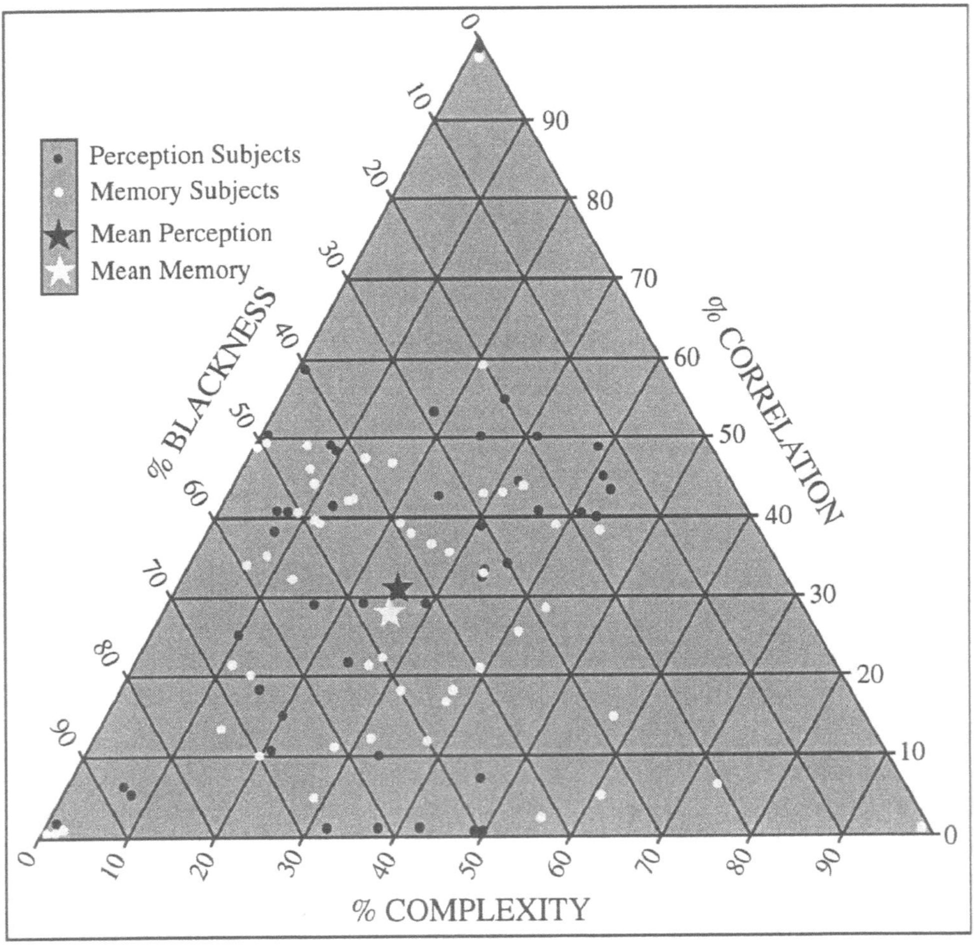

Figure 8.10: The triangular space represents correlation, blackness, and complexity dimensions. Black dots mark the locations of subjects who judged the similarity of maps they were viewing and white dots mark the locations of subjects who judged the similarity of map images in their memories. After Steinke and Lloyd (1983).

Results indicated the distributions of the two groups (perception and memory) were very similar. The means on the three dimensions for the perception group suggest blackness was most important (43 percent), followed by correlation (32 percent) and then complexity (25 percent). The memory subjects had mean values that also indicated blackness was most important (45 percent), followed by correlation (30 percent) and then complexity (25 percent). The means were found not to be statistically different on any of the dimensions. This indicated that subjects judging the similarity of map images from memory were integrating the correlation, blackness, and complexity information very much like the perception subjects who had done the same task while viewing the maps. The results supported the notion that images can be functionally equivalent to the objects they represent (Kosslyn 1980). The results also suggested that factors unrelated to the spatial patterns represented by the data can have a large impact on visual similarity when map readers compare choropleth maps. Some maps appeared to be more or less similar than would be expected based just on the correlation of their data.

8.3. Theoretical Similarity of Maps

Visual similarity cannot be computed with standard correlation coefficients for reference maps that display distributions of nominal features. Pairs of reference maps still, however, provide a definite visual impression of similarity when they are considered side by side. The maps being compared might look more similar if they have the same types of features symbolized, e.g., both might have a lake or a mountain range. This would be true to some degree even if the lakes or mountain ranges were different sizes and located in different relative locations on their respective maps. The maps being compared would look less similar if one map has a unique feature not shared by the other map, e.g., a volcano or a river delta.

Any two maps that you might compare will have some information on them that is shared and some that is unique. A simple intuitive argument is that the visual similarity of maps is a function of both the common and the distinctive information found on the maps by a map reader. Amos Tversky (1977) first proposed this theoretical notion as a general model for making comparisons. He argued that similarity is determined by matching the features of the objects being compared and called his idea the *contrast model.*

8.3.1. THE CONTRAST MODEL

A series of experiments that considered how features are used to make similarity judgments have been performed by Amos Tversky and others (Ben-Shakhar and Gati 1987; Gati and Tversky 1982,1984,1987; Sattath and Tversky 1987; Tversky 1977). These researchers have argued that geometric models of similarity, i.e., those that compute similarity as the distance between objects in a space of N dimensions, make dimensional and metric assumptions that may not be true for all situations. An important distinction is made between quantitative dimensions and qualitative dimensions. One must be able to compute accurately the distance between points for geometric models to be appropriate. This may not be possible for information measured on qualitative dimensions that indicate only the presence or absence of a feature, e.g., has an airport or does not have an airport. Although features like color may be considered as values on a quantitative dimension, they may also function as qualitative features when making visual comparisons. Two choropleth maps that both have legend classes defined as shades of green might be considered to be more similarity than two showing different colors. One map with legend classes defined as shades of a single color might be considered different from another that uses a variety of different colors. Their unique features contribute to their visual dissimilarity. Tversky (1977) also argued that the metric assumptions of:

Minimality:	$d(\mathbf{a},\mathbf{b}) >= d(\mathbf{a},\mathbf{a}) = 0$,	(8.1)
Symmetry:	$d(\mathbf{a},\mathbf{b}) = d(\mathbf{b},\mathbf{a})$, and	(8.2)
Triangular Inequality:	$d(\mathbf{a},\mathbf{b}) + d(\mathbf{b},\mathbf{c}) >= d(\mathbf{a},\mathbf{c})$	(8.3)

must be true for distance computations to be valid.

The minimality assumption states that all locations on a given dimension are separated from location *a* by some distance and that two objects having the same location on the dimension will be separated by no distance. Minimality violations in map reading occur when there is not sufficient contrast between symbols or colors. For example, county X on a choropleth map might be classified as belonging to class A and assigned the appropriate color. More people, however, might remember county X as belonging to class B than to class A because the visual contrast between the two classes was not sufficient. Tversky provided two very geographic examples of symmetry violations in his initial discussion of the contrast model. "Similarity judgments can be regarded as extensions of similarity statements, that is, statements of the form 'a is like b'.... We say 'an ellipse is like a circle,' not 'a circle is like an ellipse,' and we say 'North Korea is like Red China,' rather than 'Red China is like North Korea'" (Tversky 1977, p. 328). He also provided a geographic example for a triangular inequality violation that applied to 1977 political alignments:

> *However, the triangle inequality implies that if a is quite similar to b, and b is quite similar to c, than a and c cannot be very dissimilar from each other. The following example (based on William James) casts some doubts on the psychological validity of this assumption. Consider the similarity between countries: Jamaica is similar to Cuba (because of geographical proximity); Cuba is similar to Russia (because of their political affinity); but Jamaica and Russia are not similar at all.*

It has been shown that distinctive features are extremely important for identification tasks such as searching for a target on a map (Brennan and Lloyd 1993; Lloyd 1988). "The detection of a distinctive feature establishes a difference between stimuli, regardless of the number of common features" (Gati and Tversky 1984, p. 342). Common features are more important for classification tasks like determining if a linear map symbol represents a road or a river. The similarity of objects, however, is determined by both their common and distinctive features. People trade off the visual impact of common and distinctive features and "the relative weight varies with the nature of the task" (Gati and Tversky 1984, p. 342). Methods have been developed and tested that can be used to quantitatively measure the relative importance of a specific feature as either a common or distinctive feature when judging visual similarity. Gati and Tversky (1984) made a distinction between characteristics of objects (maps) which they defined as either additive attributes or substitutive attributes. Additive attributes define the presence or absence of a particular feature and substitutive attributes define the presence of exactly one element from a possible set. For example, let **b** represent some common background for a map display and say it is a basemap of the state of Pennsylvania in the United States. Also define two substitutive attributes, **p** and **q,** that define the style of a particular map. Although it may seem unusual to compare maps produced in different styles, some spatial data may be appropriately displayed in various styles of maps. It is important to know how style affects the communication process because map readers may need to compare maps produced in different styles. A discussion of the effectiveness of different map styles for communicating general spatial patterns has been provided by MacEachren (1982). Lets say style **p** is a choropleth map and style **q** is an isopleth map. Also define two additive attributes **x** and **y** that represent elements that might be present or absent from the display. Let attribute **x** be a title and attribute **y** be a legend for the map. The following displays could now be created and presented in pairs for a map reader to evaluate. A **bp** display is a choropleth map of Pennsylvania with no legend and no title (Figure 8.11). A **bq** display is an isopleth map of Pennsylvania with no legend and no title (Figure 8.12).

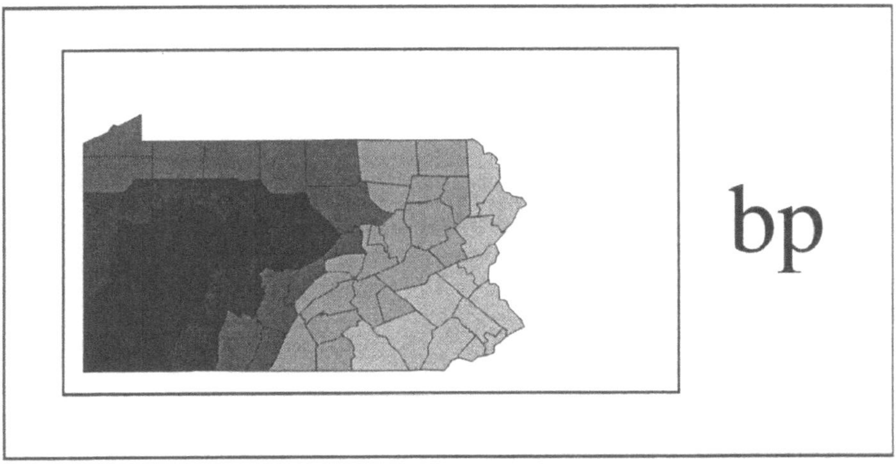

Figure 8.11: A representation of a Pennsylvania (b) choropleth map (p).

A **bpx** display is a choropleth map of Pennsylvania with a title and with no legend (Figure 8.13). The Steelers in the title refers to a football team affiliated with the city of Pittsburgh in Western Pennsylvania. A **bqx** display is an isopleth map of Pennsylvania with a title and with no legend (Figure 8.14).

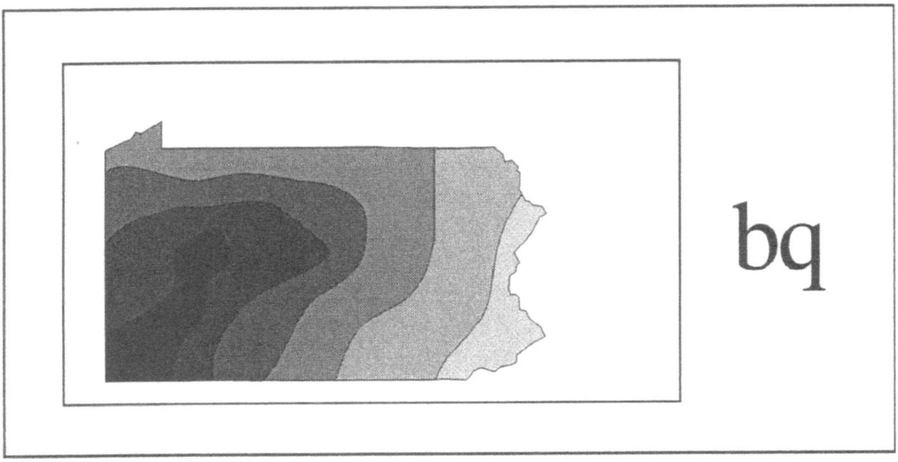

Figure 8.12: A representation of a Pennsylvania (**b**) isopleth map (**q**).

A **bpy** display is a choropleth map of Pennsylvania with a legend and no title (Figure 8.15). A **bqy** display is an isopleth map of Pennsylvania with a legend and no title (Figure 8.16). To assess the effect of additive component **x** (title) as a common feature, denoted C(**x**), map readers would be asked to compare displays **bp** and **bq** (Figure 8.17) and rate their similarity and to do the same for displays **bpx** and **bqx** (Figure 8.18). The difference between these similarities is an estimate of C(**x**). In equation form:

$$C(\mathbf{x}) = S(\mathbf{bpx},\mathbf{bqx}) - S(\mathbf{bp},\mathbf{bq}) \quad (8.4)$$

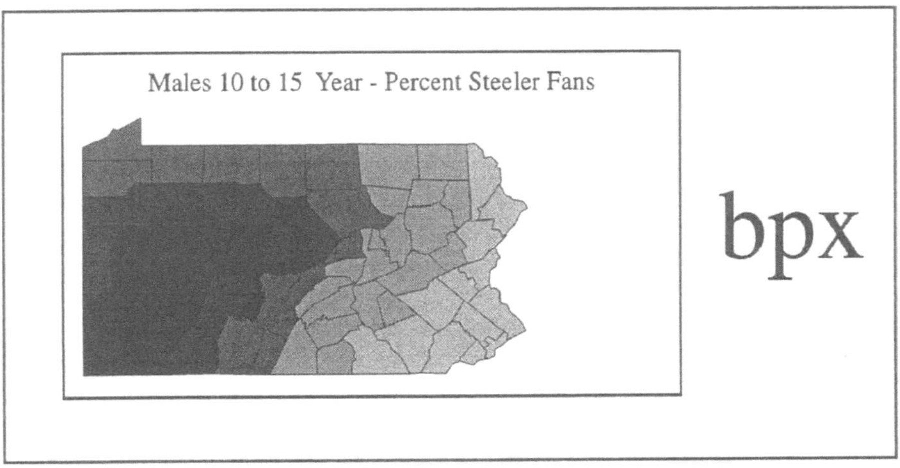

Figure 8.13: A representation of a Pennsylvania (**b**) choropleth (**p**) map with a title (**x**).

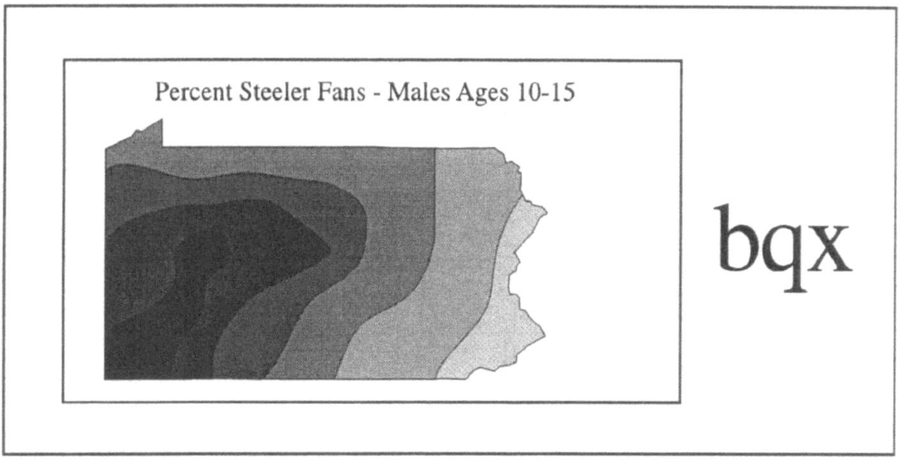

Figure 8.14: A representation of a Pennsylvania (**b**) isopleth (**q**) map with a title (**x**).

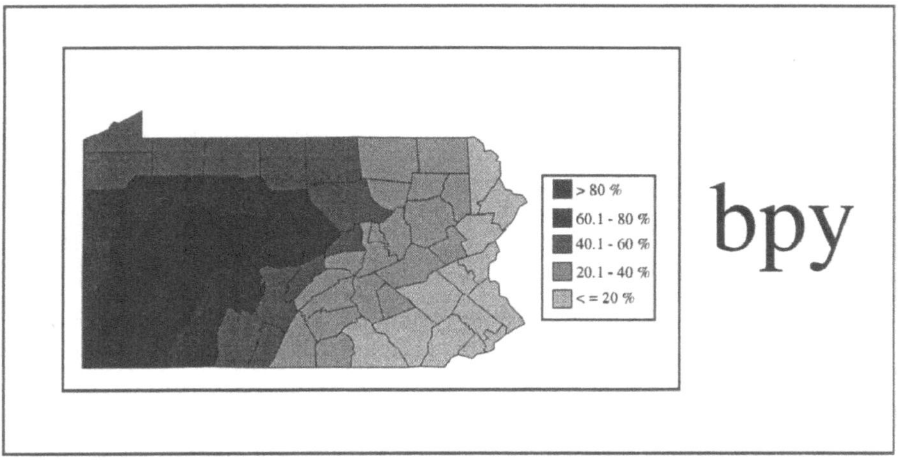

Figure 8.15: A representation of a Pennsylvania (**b**) choropleth (**p**) map with a legend (**y**).

To assess the effect of additive component **x** (title) as a distinctive feature, denoted D(**x**), map readers would be asked to compare displays **bp** and **bpy** (Figure 8.19) and rate their similarity and to do the same for displays **bpx** and **bpy** (Figure 8.20). The difference between these similarities is an estimate of D(**x**). In equation form:

$$D(\mathbf{x}) = S(\mathbf{bp},\mathbf{bpy}) - S(\mathbf{bpx},\mathbf{bpy}) \quad (8.5)$$

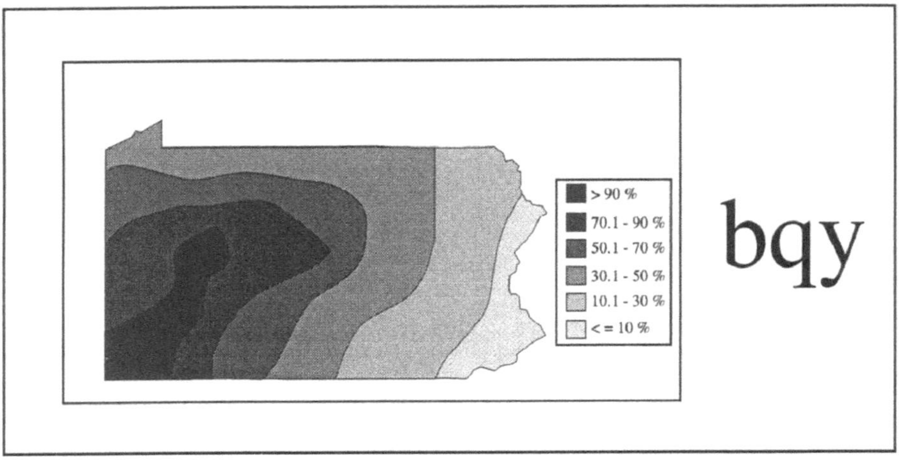

Figure 8.16: A representation of a Pennsylvania (**b**) isopleth (**q**) map with a legend (**y**).

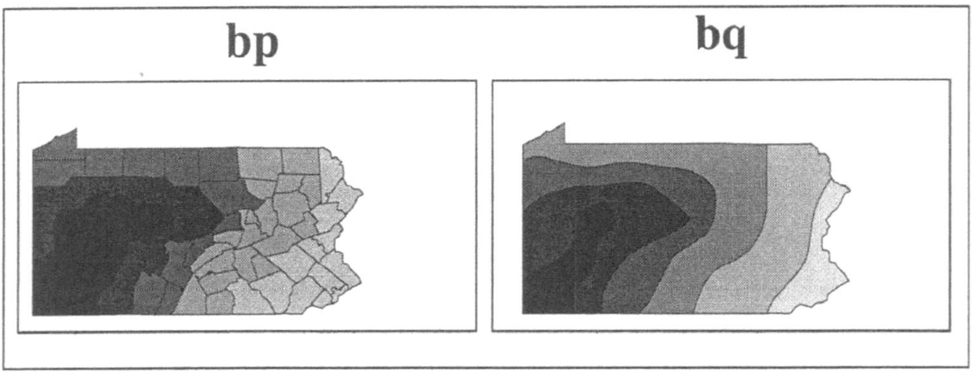

Figure 8.17: Comparison of Pennsylvania (**b**) choropleth (**p**) and isopleth (**q**) maps to determine their similarity s(**bp**, **bq**) as part of the calculation of C(**x**).

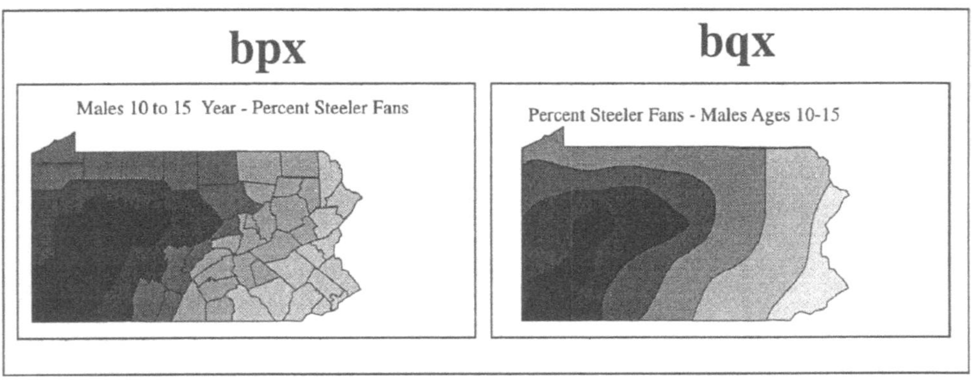

Figure 8.18: Comparison of Pennsylvania (**b**) choropleth (**p**) and isopleth (**q**) maps with a title (**x**) to determine their similarity s(**bpx**, **bqx**) as part of the calculation of C(**x**).

Figure 8.19: Comparison of Pennsylvania (**b**) choropleth (**p**) maps without and with a legend (**y**) to determine their similarity s(**bp**, **bpy**) as part of the calculation of D(**x**).

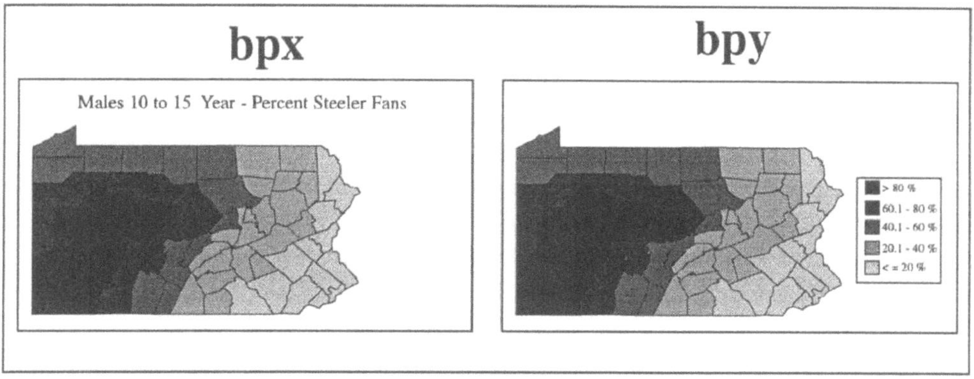

Figure 8.20: Comparison of Pennsylvania (**b**) choropleth (**p**) maps with a title (x) to one with a legend (**y**) to determine their similarity s(**bpx**, **bpy**) as part of the calculation of D(**x**).

The effect of **x** as a common feature relative to its effect as a distinctive feature is defined as W(**x**) and computed as follows:

$$W(\mathbf{x}) = C(\mathbf{x})/[C(\mathbf{x}) + D(\mathbf{x})] \quad (8.6)$$

The value of W(**x**) ranges from 0, when C(**x**) = 0, to 1 when D(**x**) = 0, and equals 0.5 when C(**x**) = D(**x**). The value of W(**x**) can be used to determine the relative strength of feature **x** as a common and distinctive feature.

This basic methodology has been thoroughly tested on both verbal and pictorial stimuli. Descriptions of people or meals served as verbal stimuli and schematic faces or sketches of landscapes served as pictorial stimuli. When verbal descriptions were used, common features were found to be more important for judging similarity than distinctive features. When pictorial displays were used, distinctive features were more important for judging similarity than common features. Verbal descriptions of pictorial displays were compared with the results for pictorial displays and appeared to be evaluated like other verbal stimuli, i.e., common features were more important than distinctive features.

In another experiment Gati and Tversky (1987) initially had subjects rate the similarity of both verbal and pictorial stimuli. The verbal description and pictorial displays were presented again after removing one common and one distinctive feature. "The analysis of both correct recall and misclassified recall suggest that components that were encoded as common are relatively more salient in verbal than in pictorial stimuli, whereas components that were encoded as distinctive exhibit the opposite pattern" (Gati and Tversky 1987, p. 99). These experiments suggested judgments of the similarity of maps should be dependent on both their common and distinctive features, but that distinctive features should dominate. If one were to read verbal descriptions of two maps, however, one would expect similarity judgments to depend on both common and distinctive features, but common features should dominate.

8.3.2. DEPICTIONS AND DESCRIPTIONS OF MAPS

An experiment that contrasted the similarity comparisons of subjects who viewed depictions of simple reference maps with similarity comparisons of subjects that read verbal descriptions of the same maps has directly illustrated Tversky's (1977) contrast model as it related to cartographic maps (Covington 1992; Lloyd, Rostkowska-Covington, and Steinke 1996). Simple reference maps with qualitative information were constructed to represent the type of symbols found on reference maps and to construct stimuli for a cognitive experiment. A total of 114 maps was constructed by organizing symbols like those illustrated in Figure 8.21 in different ways.

Particular pairs of maps represented a configuration that was needed to measure a value for the C(**x**) (Equation 8.4) or D(**x**) (Equation 8.5) indices. On a particular pair of maps the basemaps were constructed with different symbols for **p** or **q**, e.g., water features or transportation features and different symbols were used to represent **x** or **y** features, e.g., industrial area or forest. For example, one map (**p**) could show a water feature and another map (**q**) could show a highway feature (Figure 8.22). Adding an industrial feature (**x**) to both base maps could produce two more maps (**px** and **qx**) (Figure 8.22). Similarity judgments with these four maps could be used to compute a value for the effect of the industrial symbol as a common feature or C(**x**) with Equation 8.4. Organizing the features in different ways produced other maps. For example, a forest feature (**y**) could be added to the basemap with water features (**p**) to produce another map (**py**) and a residential feature (**x**) could be added to the same basemap (**p**) to produce yet another map (**px**) (Figure 8.23). These maps would be used to compute the effect of urban as a distinctive feature D(**x**) (Equation 8.5).

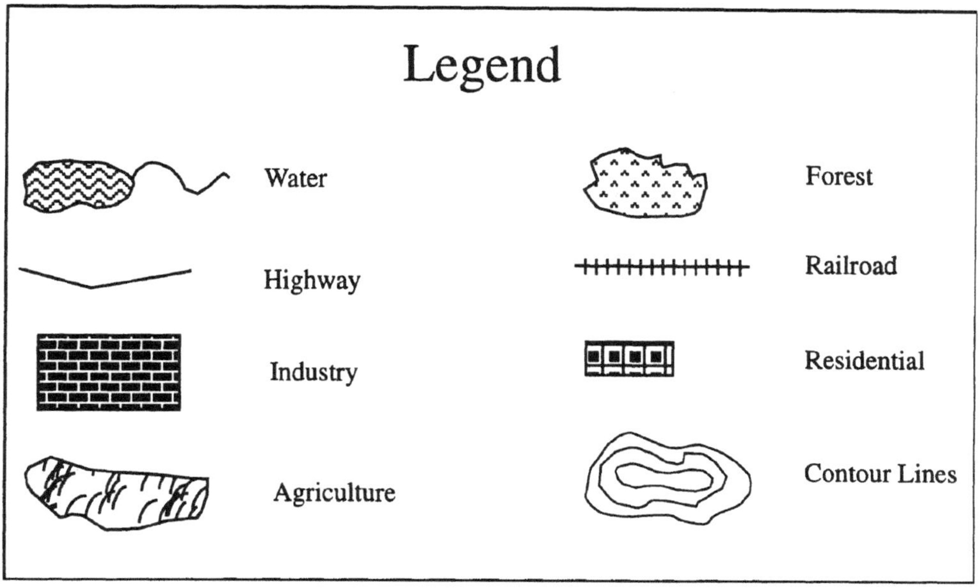

Figure 8.21: Combinations of symbols like these were used to construct pairs of maps. After Lloyd, Rostkowska-Covington, and Steinke (1996).

A verbal description was also created that corresponded to each map depiction. These described, in simple terms, what someone would see if they were looking at a particular map. In addition to the simple cases, where one common or distinctive feature was added to create a new version of the map or map description, three other types of maps were created. One type added a common feature, but put it in a different location. Another type added two common features and a third added two distinctive features. The experiment divided subjects into two groups. The map group viewed pairs of maps on a monitor and estimated similarity by marking a linear scale from 0 (not very similar) to 100 (very similar). Subjects in the verbal group were presented the verbal descriptions and responded in a similar fashion. This created 10 different types of comparisons (Table 8.1). The time it took to respond with an answer (reaction time) for each trial was recorded as well as the similarity estimates. The mean similarity judgments and reaction times were computed for the eight different comparisons and plotted on a graph to represent the 10 types of comparisons (Table 8.1 and Figure 8.24). Each comparison, i.e., M_1through M_5 and V_1 through V_5, is represented on the graph as a triangle. The horizontal leg of a triangle indicates change in similarity caused to adding a common or distinctive feature (**x**). The vertical leg of a triangle indicates change in reaction time caused by adding a common or distinctive feature (**x**). The overall change is represented by a vector that is also the hypotenuse of a triangle. The direction of this change vector relative to the vertical axis (up means reaction time increased and down means it decreased) and horizontal axis (left means similarity decreased and right means it increased) indicates the impact caused by the addition of feature **x**.

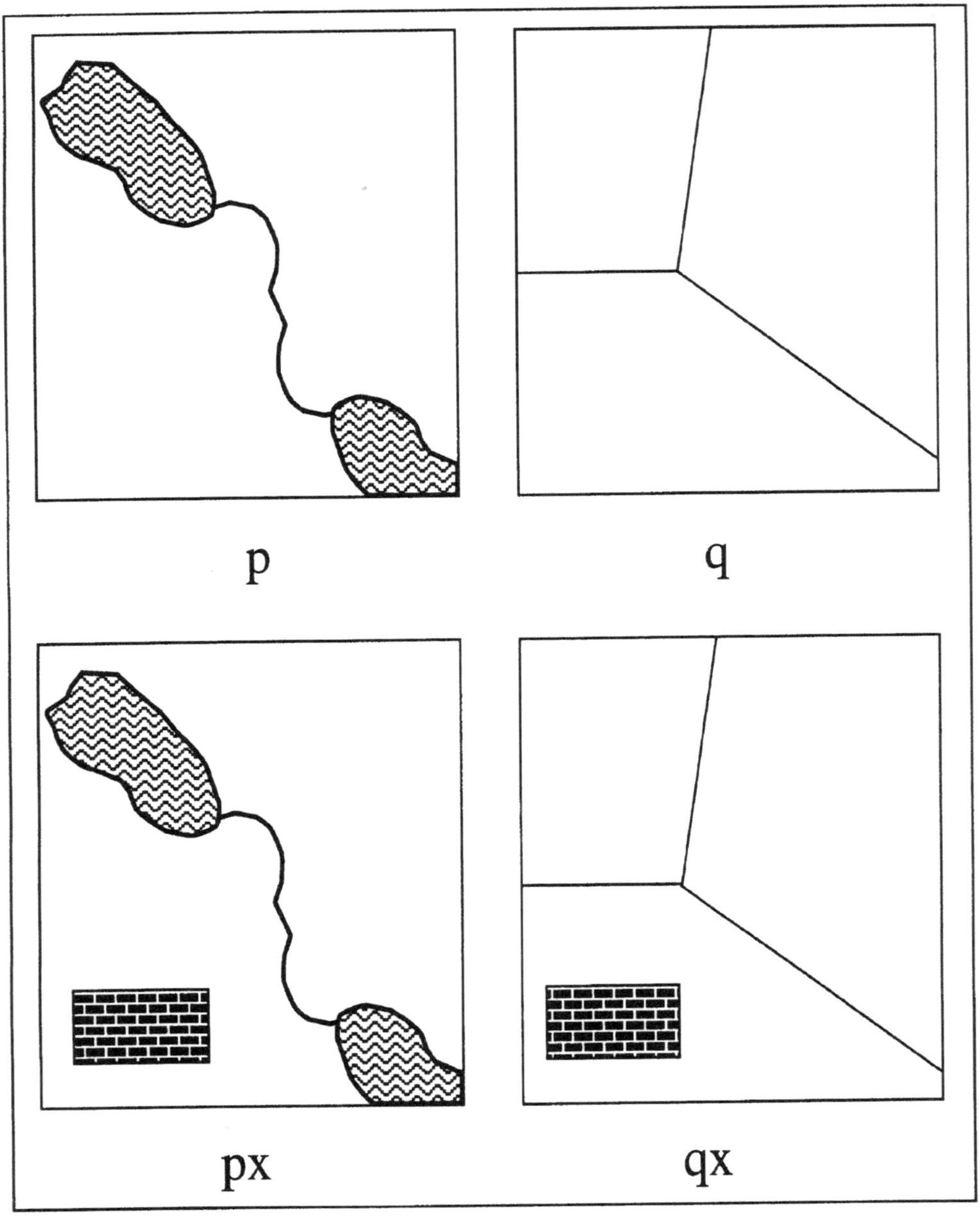

Figure 8.22: An example of pairs of maps used to compute C(**x**) with Equation 8.4 to provide an estimate of the effect of adding a single common feature on subjects' judgments of map similarity. After Lloyd, Rostkowska-Covington, and Steinke (1996).

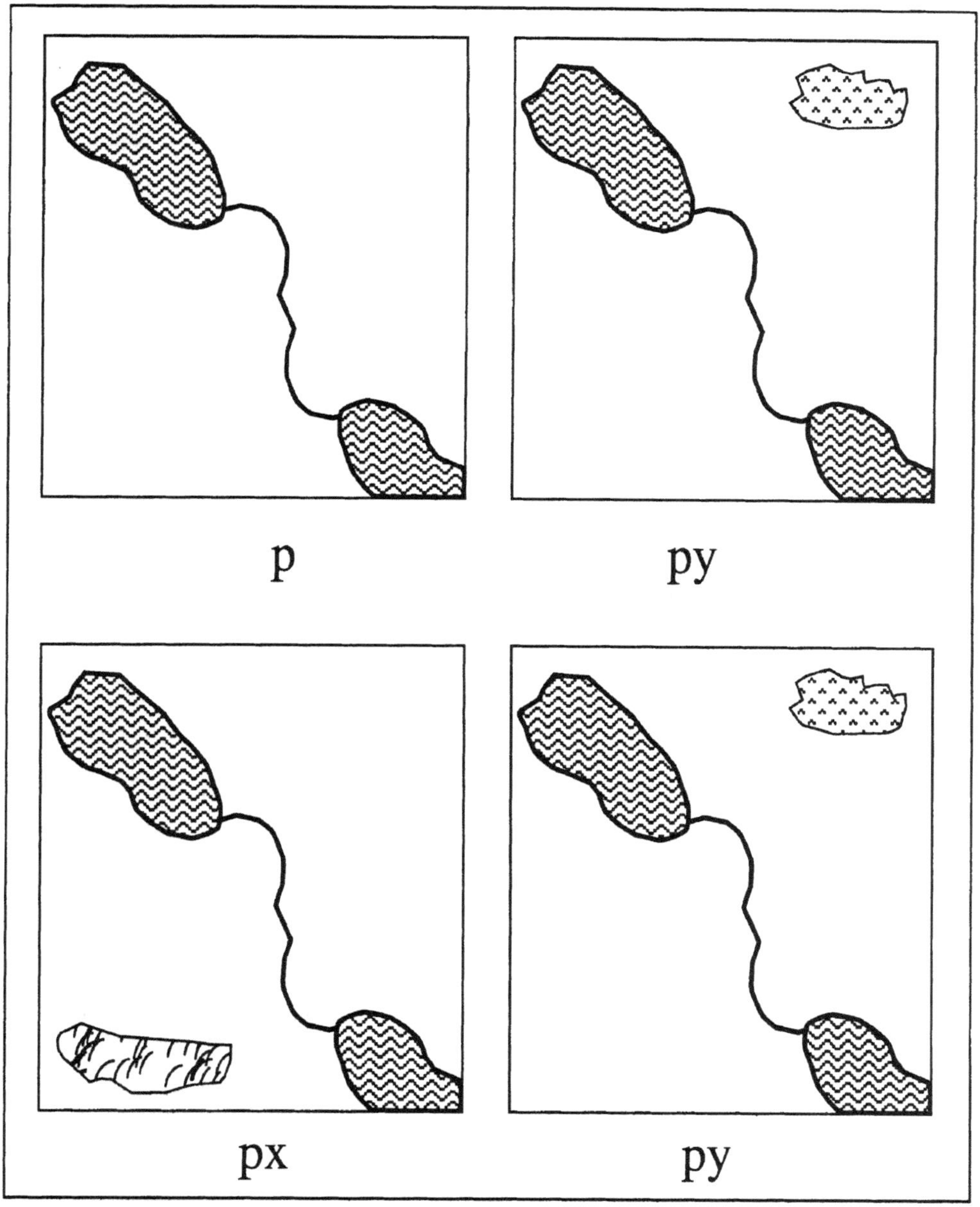

Figure 8.23: An example of pairs of maps used to compute D(**x**) for Equation 8.5 to provide an estimate of the effect of adding a single distinctive feature on subjects' judgments of map similarity. After Lloyd, Rostkowska-Covington, and Steinke (1996).

Table 8.1: The types of similarity comparisons created for the experiment to determine the effect of **x** as a first or second common or distinctive feature (Lloyd, Rostkowska-Covington, and Steinke 1996).

	Map Depiction	Verbal Description
Common Feature	1 Feature (**p**, **q**) and (**px**, **qx**) 2 Features (**py**, **qy** and (**pxy**, **qxy**) 1 Feature (at different location) (**p**, **q**) and (**px**, **qx**$_a$)	1 Feature (**p**, **q**) and (**px**, **qx**) 2 Features (**py**, **qy** and (**pxy**, **qxy**) 1 Feature (at different location) (**p**, **q**) and (**px**, **qx**$_a$)
Distinctive Feature	1 Feature (**p**, **py**) and (**px**, **py**) 2 Features (**p**, **py**) and (**p**, **pxy**)	1 Feature (**p**, **py**) and (**px**, **py**) 2 Features (**p**, **py**) and (**p**, **pxy**)

The theoretical arguments offered by Tversky's (1977) contrast model could be evaluated for reference map comparisons with these aggregate results. For example, similarity should increase when a common feature is added to either a map or a verbal description of a map. The prediction of the contrast model is that all six indices that measure adding a common **x** should increase the similarity of the map or map description (Figure 8.24). These predictions were generally found to be true when one or two common features were added at the same location on maps (M_1 and M_2) and on map descriptions (V_1 and V_2). Similarity was increased only a small amount when a single common feature was added at a different location on a map (M_3) and actually decreased similarity when added at a different location on a map description (V_3) (Figure 8.24). It would appear that maps are not like other objects. The locations of features were apparently being considered by subjects as they assessed the visual similarity of maps. This effect was even stronger for map descriptions. This was probably because

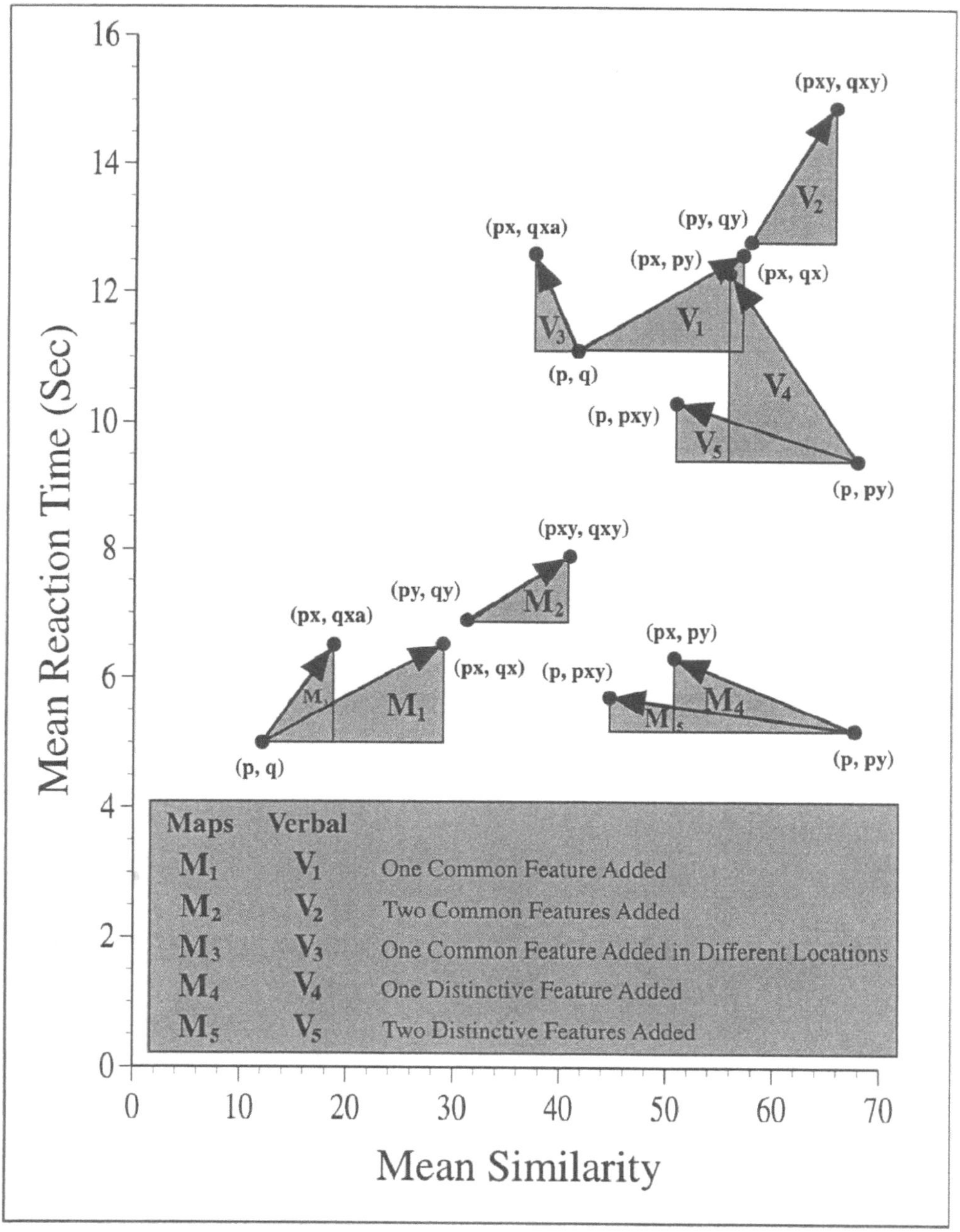

Figure 8.24: Change vectors for the five types of map comparisons and five types of verbal description comparisons. Vectors pointing up indicate an increase in reaction time, those pointing left indicate a decrease in similarity, and those pointing right indicate an increase in similarity. After Lloyd, Rostkowska-Covington, and Steinke (1996).

the statements that described the map made the differences in locations more distinctive. Since the location of a feature on a map is frequently a major part of the information being communicated to the map reader, this result is not surprising. Maps may be special objects for this reason and accurate measurements of their similarity will have to account for this effect.

The contrast model also predicts that all four indices that measure adding a distinctive **x** should decrease the similarity of the maps or map descriptions. This prediction was support for both the map comparisons (M_4 and M_5) and the verbal descriptions (V_4 and V_5). Another effect that was always true related to the addition of a second common or distinctive feature. The second feature always had less impact on similarity than the first feature. This can be seen by considering the lengths of the horizontal legs of triangles M_1 versus M_2, M_4 versus M_5, V_1 versus V_2, and V_4 versus V_5. Adding more features to increase or decrease visual similarity appeared to have diminishing returns. Reaction times also were as expected. Verbal descriptions always took longer to process than maps. This is expected because the maps could be parallel processed while the verbal descriptions required a serial process (Figure 8.24). The W(**x**) index was also computed for individual subjects for the five common and five distinctive cases (Table 8.1 and Equation 8.6). The contrast model predicts that W(**x**) should be less than 0.5 for visual map comparisons and greater than 0.5 for verbal description comparisons (Gati and Tversky 1984). This indicates that subjects should pay more attention to distinctive features for map comparisons and pay more attention to common features for verbal description comparisons. The results reported by Lloyd, Rostkowska-Covington, and Steinke (1996) supported this assertion with the average map comparison having a W(**x**) of 0.448 and the average verbal description comparison having a W(**x**) of 0.555.

8.3.3. THE EFFECT OF TOP-DOWN INFORMATION

A second experiment reported by Lloyd, Rostkowska-Covington, and Steinke (1996) related the contrast model to a different type of spatial display. In this experiment subjects compared pairs of visual displays that represented qualitative maps of landuse. All the maps were grid-like displays organized in a window of 3 X 3 cells. They were supposed to represent randomly picked subsets from a larger raster display of landuse classified from remotely sensed data. The nine cells in a window were either a red, blue, green, or yellow color to represent four different landuses. Maps were actually artificially generated to represent eight levels of similarity using a simple equation that has been used for a variety of problems (DeSarbo, Johnson, Manrai, Manrai, and Edwards 1992; Goldstone, Medin, and Gentner 1991; Tversky 1977):

$$\textit{Similarity}(A, B) = \alpha * f(A \cap B) - \beta * f(A - B) - \gamma * f(B - A) \qquad (8.7)$$

The similarity of map A and map B is a function of three separate components. The first component ($A \cap B$) measures the features that Map A and B share. The second component (A - B) measures the features that Map A has that are absent from Map B.

The third component (B - A) measures the features that Map B has that are absent from Map A. The coefficients α, β, and γ weight the relative importance of the three components. The equation uses only common features to measure similarity if α is set to 1 and both β and γ are set to 0. The equation uses only distinctive features to measure similarity if α is set to 0 and both β and γ are set to 1. The equation can also be used to measure dissimilarity by setting α to a negative value and both β and γ to positive values. Since the contrast model considers similarity to be a function of both common and distinctive features, all coefficients were set to non-zero values. The number of cells on the two maps that matched in color were counted to determine the first component for Equation 8.7. The second component was determined by counting the unique colors on Map A and the third component was determined by counting the unique colors on Map B. The two maps in Figure 8.25 have 12 common features, Map A has 3 distinct features, and Map B has 3 distinct features. Ten pairs of maps were

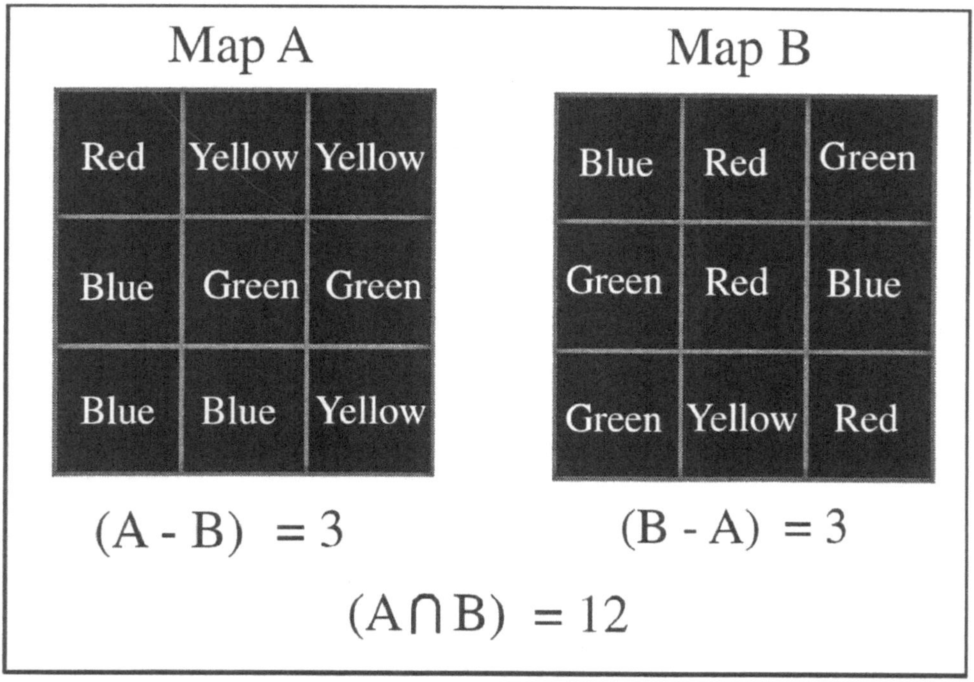

Figure 8.25: An example of a pair of maps and their common (A B) and distinct (A -B) and (B - A) features.

generated for each of the eight (low to high) similarity categories. The 80 pairs of map were viewed and rated on their similarity on a 0 (not very similar) to 100 (very similar) scale by two groups of subjects. The two groups of subjects considered the identical set of 80 maps, but were given different instructions about the origins of the maps. One group was told the maps were all from the same location, but for two different times. The other group was told the maps were from two different places for the same time. These two different premises were created to determine if top-down information in a map reader's memory affected similarity judgments. Since the bottom-up information coming from the maps was identical, any differences in the two groups judgments might be caused by what they believed to be true about the maps. Results indicated a significant relationship between the similarity index computed from Equation 8.7 and the subjects estimates of similarity. This suggested that subjects were using both common and distinctive features to make their judgments. The results also indicated that pairs of maps believed to be for the same place, but for different times, were rated significantly more similar than pairs of maps believed to be for different places, but for the same time. This was particularly true for pairs that were considered to be in the highest categories by the Tversky similarity equation (Equation 8.7). Subjects who believed the maps were for the same place, but for different times, also had significantly faster reactions times. This was an indication that these subjects made their estimates with more confidence. It was concluded that the top-down knowledge that the maps were for the same place provided a subtle message that the maps ought to be similar. It was easier for subjects to see similar maps if they expected to see similar maps.

8.3.4. THE EFFECT OF LOCATION

The contrast model bases similarity on the degree to which two objects have common and distinctive features. A feature is considered common if it exists on both objects. The location of the feature is not considered important. A feature is considered distinctive if one objects has the feature and the other does not. The assessment of similarity involves listing features for two objects and comparing the overlap between the objects. A match (common feature) increases similarity and a mismatch (unique feature) decreases similarity. Goldstone, Medin, and Gentner (1991) have argued that attribute features and relational features operate differently to express the similarity of two objects. They used the term attribute to mean any single component or property of an object. Relations were defined as the description of a connection between two or more attributes. This definition of an attribute feature is very similar to Tversky's (1977) use of the term. It would include both additive attribute (the presence or absence of a particular feature) and substitutive attributes(the presence of exactly one element from a possible set). For example, an attribute feature on the map in Figure 8.26a is a *Railroad* symbol *or a rectangular Station* symbol. The *Railroad* symbol could exist on the map or not and the *rectangular Station* might be a *circular Station*. Relational attributes can be expressed as conceptual propositions that relate one feature to others to express some connection between the features. For example, The *Farm* on map *a* (Figure 8.26) *is north of the Railroad* and the *Village is south of the Railroad*. The relation between the *Farm* and the *Railroad* is an attribute of the map and changing this relationship could effect how similar the map is to other maps.

The distinction between attribute features and relational features becomes complicated by the nature of maps. The choice of a particular symbol to express the location of something in space is constrained somewhat by cartographic conventions, but a range of acceptable choices usually exist (Robinson, Morrison, Muehrcke,

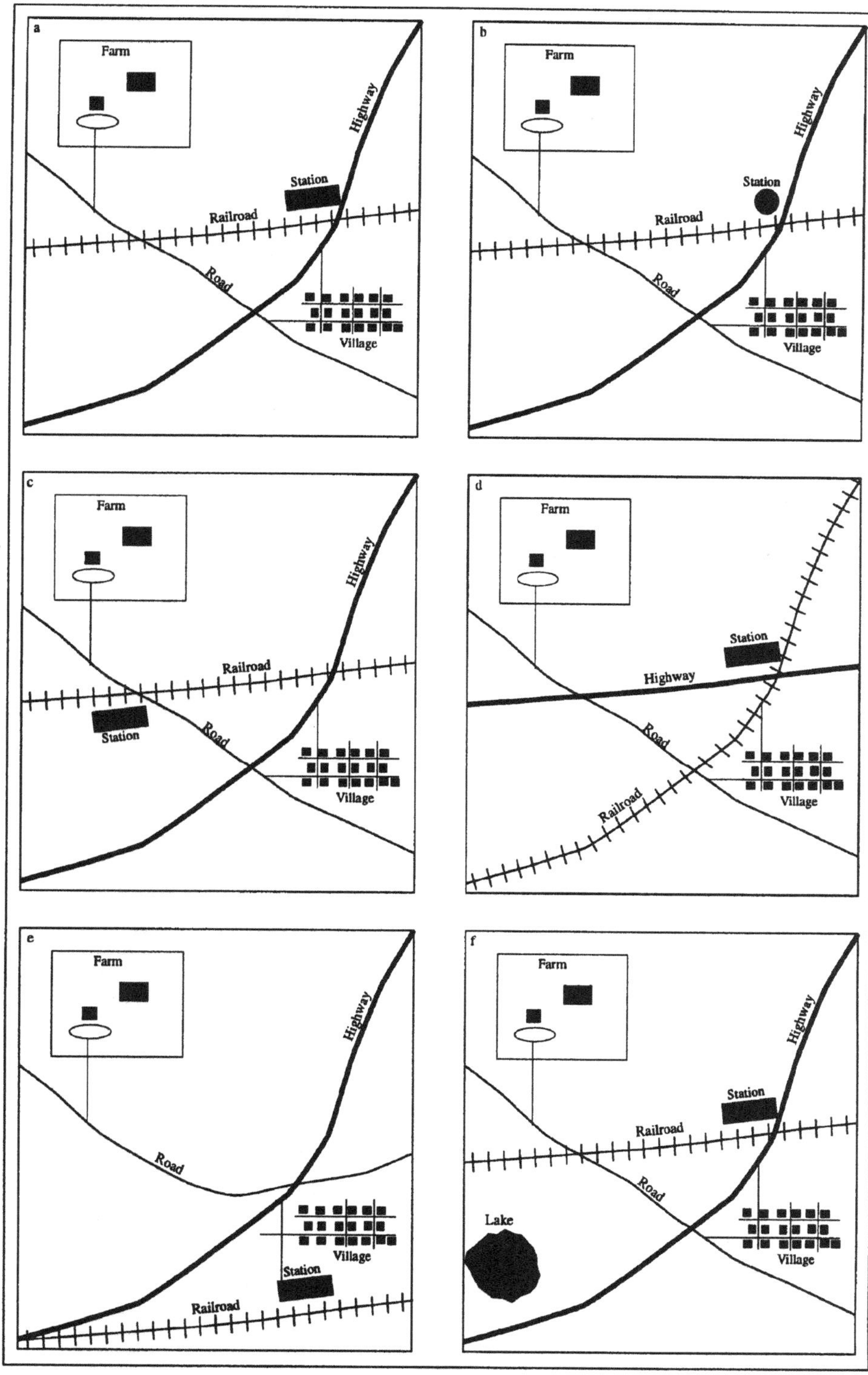

Figure 8.26: Maps with attribute, relational, and functional features.

Kimmerling, and Guptill 1995). The *Station* symbols used on the maps *a* and *b* represented in Figures 8.26 might both be reasonable choices made by two different cartographers. The shape of the two symbols is different, but they have equivalent functions and locations. The function of a feature might be a third distinction that is particularly important for judging the similarity of maps. Since maps are artificial objects constructed to represent an environment, a common function for a map symbol might be as important a consideration when judging similarity as a distinctive shape. If shape is considered a substitutive attribute for the maps in Figure 8.26, then a comparison of maps *a* and *b* would treat the station as a distinctive attribute using the contrast model. This would make the two maps less similar. Relational and functional attributes for the *Station* symbol are common features for the two maps and would increase similarity. If *what* and *where* are the two most important messages the symbol provides the map reader, then maps a and b (Figure 8.26) are equivalent.

A comparison of map a with map *c* (Figure 8.26) illustrates the importance of location. The *Station* symbol has shape and function as common features on both maps, but relational attributes are distinctive features. The *Station* is located north of the *Railroad* on map *a* and south of the *Railroad* of map *c*. The *Station* is located east of the *Road* on map *a* and west of the *Road* on map *c*. The *Station* is north of the *Highway* on both maps. The contrast model would consider maps a and c to be equivalent because they have all common features and no distinctive features. This is only true because relational features are not considered. Since maps tell us *what* is *where,* the relational information should be more important for judging the similarity of maps than it would be for judging the similarity of other common objects.

Comparing map *a* with map *d* (Figure 8.26) illustrates a case where both relational and functional attributes become distinctive features. Maps *a* and *d* have all common and no distinctive features according to the contrast model. Although map *d* shows the location of the *Highway* and *Railroad* reversed as linear features compared to map *a*, both maps have a *Railroad* and *Highway* features and they are symbolized in the same ways. Although reversed in their function, both maps have lines that follow the same paths across the map. Relational features change only because the functions of the lines have changed, illustrating again that functional and locational attributes are important for judging the similarity of maps.

Although map *a* and *e* (Figure 8.26) have a common set of matched features, it is obviously important to a map reader using the maps to navigate to the *Station* that the *Railroad* and the *Station* is south of the *Village* on map *e* and north of the *Village* on map *a*. Top-down knowledge that requires you to project yourself into the map with a purpose could clearly change how you weight the importance of attribute, relational, and functional features. Although a new additive feature, such as the *Lake* in map *f* (Figure 8.26) clearly should decrease the similarity of maps *a* and *f*, how much the similarity is affected by this type of change compared to changes in relational and functional attributes is an important research problems in map reading.

Goldstone (1994, p. 5) argued that similarity assessments share deep similarities with analogical reasoning. He defined the term *correspondence* as a connection or association between parts from two scenes. Maps might be considered visual scenes and the map symbols parts of the scenes. Analogies are constructed by identifying a correspondence between elements that are consistent in two situations (Gentner 1983; Holyoak and Thagard 1989). For maps, similarity would be dependent on the degree to which the correspondence of the symbols on the maps is globally consistent. Goldstone (1994) defines two types of matches between parts of a scene. A **MIP** is defined as a match in place. For this type of match, a pair of symbols on two maps would match with respect to both their attribute features and locations. For example, the *Farm* on all the maps represented in Figure 8.26 match in place. A **MOP** is defined as a match out of place. For this type of match, a pair of symbols on two maps would match with respect to their attribute features, but not with respect to their locations. For example, the *Station* symbols on maps *a* and *c* (Figure 8.26) match out of place. Goldstone (1994) argued that both **MIPS** and **MOPS** increase similarity between scenes, but that **MIPS** increase similarity more than **MOPS**. This argument, that both attribute features and location are important, was the bases of Goldstone's (1994) similarity as interactive activation and mapping (SIAM) model. This model has some similarity to McClelland and Rumelhart's (1989) interactive activation competition neural network in that objects and features are organized in a network of nodes that excite or inhibit each other based on their correspondence relationships. The model has three types of nodes and connections: feature-to-feature, object-to-object, and role-to-role. The term *role* here is similar to the term *function* in the above discussion. Map symbols on two maps may have different features (shapes), but have the same function (to mark the location of the *Station* in Figure 8.26). The connections between nodes are based on consistency rather than being learned. If feature or role nodes in the network are consistent on two maps, they would send excitatory activation to each other. If the nodes were not consistent, they would inhibit each other. All nodes in the model take on activation values between 0 and 1. At the beginning of the process , node activations are set to 0.5. When the model is cycled, the node representing the similarity between map *a* and map *b* would increase or decrease depending on the similarity between the two maps expressed in the connections.

8.4. Conclusions

This chapter has considered what makes objects like maps appear to be similar or different. Tversky's (1977) contrast model argued that similarity is a matter of feature matching. Common features make objects, or descriptions of objects, appear more similar and distinctive features make them appear less similar. The similarity of maps may depend on the matching of common and distinctive features, on the spatial relationships among the features, and on the function of the features. The next chapter demonstrates the use of neural networks for modeling geographic problems.

CHAPTER 9

NEURAL NETWORK APPLICATIONS

9.1. Introduction

This chapter provides examples of research problems that can be modeled with neural networks. Four classes of problems of general interests to geographers are identified and real-world examples are presented and discussed. The examples illustrate different types of data that can be used with neural networks and different approaches that can be taken to train networks.

9.2. Types of Problems

The examples represent two important dimensions that define the type of the problem being addressed. The first dimension relates to the nature of the information being learned by a neural network and the second relates to the nature of the learning process being used to train the network (Figure 9.1).

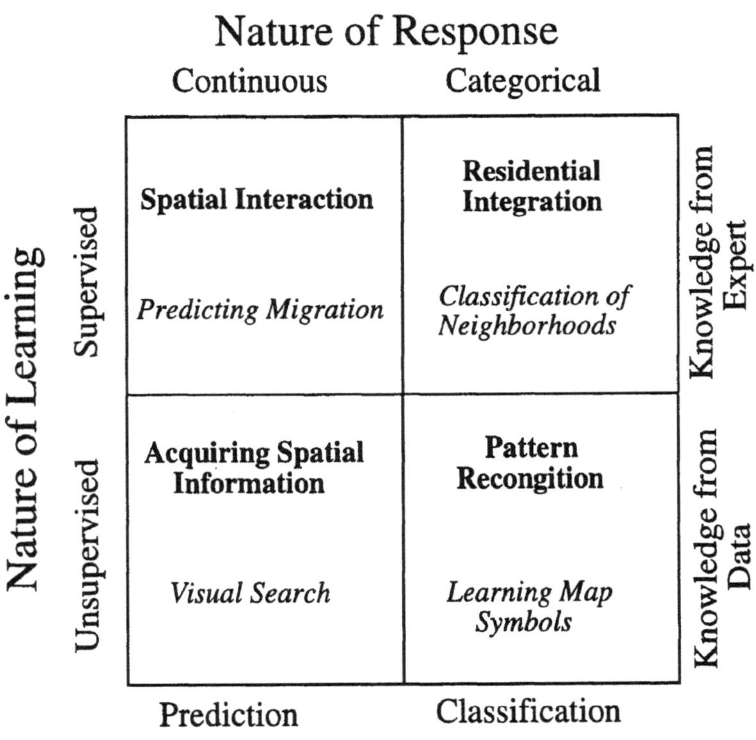

Figure 9.1: Classification of examples by the nature of the output and training.

9.2.1. NATURE OF THE INFORMATION

The first dimension defines the nature of the neural network's output. The activations of neurons in an output layer of a neural network may be used for prediction or classification. If the goal is prediction, then the inputs to the network will cause the output neuron(s) to take on some level of activation along a continuous scale. For example, the final activation of a neuron could be produced by an output function such as the logistic function that varies between 0.0 and 1.0 (Figure 9.2).

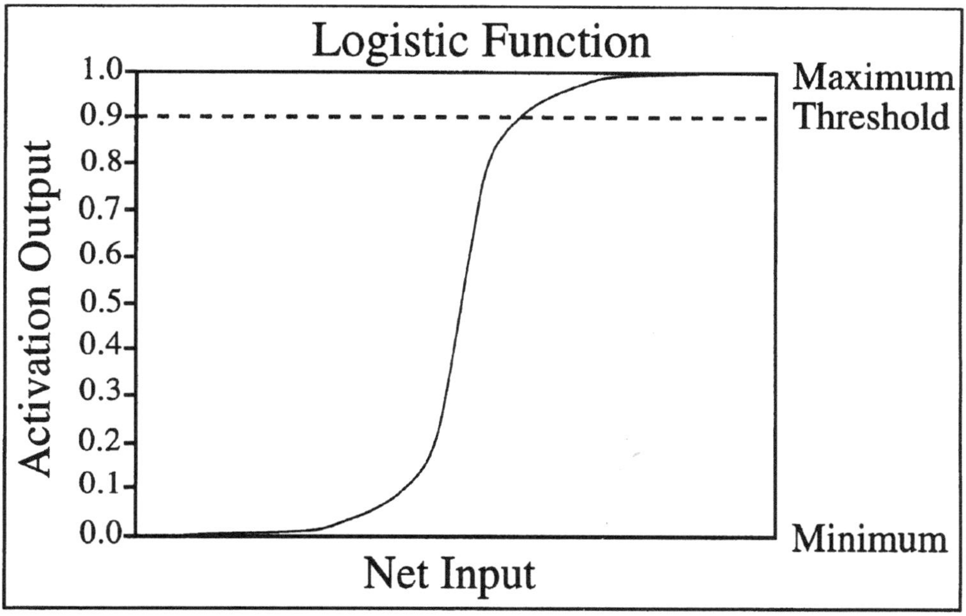

Figure 9.2: The logistic function is a non-linear function that produces activation outputs between 0.0 and 1.0.

An alternate goal might be to use the inputs to the neural network to activate 1 of N classes. If the goal is classification, one of the output neurons should be excited or turned on and the other output neurons should be inhibited or turned off. The network's function in this case is to learn to associate input patterns with classes of outputs. The final activation of an output neuron in a network being used for classification could be determine by a winner-takes-all rule or threshold rule. A winner-takes-all rule turns on the output neuron with the highest activation and turns off its competitors. A threshold rule turns on any neurons that have reached a threshold activation and turn off those with activations less than the threshold.

9.2.2. NATURE OF THE TRAINING

The second dimension of the examples considers the nature of the learning process used to train the network (Figure 9.1). Supervised training requires that some *truth* information be provided by an *expert* to correct the network and, thus, allow it to learn from its mistakes. These networks learn by comparing the current activations of the network's output neurons to target outputs provided by the expert. This is why the learning is said to be supervised. Differences between the current output and target output can be treated as error and used to adjust the weights connecting neurons in the network. For networks being used to make predictions, the target output could be the observed values of some dependent variable. In this case, the network is expected to learn how to translate the values provided to input neurons into the expected activation for the output neuron(s). The simplest case mimics a regression model. For example, there might be two input neurons and a single output neuron. The input neurons function like independent variables and the output neuron functions like a dependent variable. The basic difference is that a regression procedure models the relationships between the independent variables and dependent variable by using all the observations simultaneously to compute a least-squares-fit equation. An analogous neural network learns the appropriate connections between its neurons by presenting the values for each individual observation to the input neurons and producing an activation for an output neuron. The difference between the current activation and the target activation is then used to adjust the network's weights. This process is repeated with all the additional observations in the training set to complete an epoch or round of training. The network is given additional epochs of training until predictions match target expectations or until further training does not improve the network.

Supervised learning can also be used for classification problems. For example, information from an earth satellite's sensor system is represented as digital activations in bands of the electromagnetic spectrum for pixels (picture elements) that represent an area on the earth surface (Civco 1993; Heermann and Khazenie 1992; Ritter and Hepner 1990). This information and neural networks are used by remote sensing scientists to model land use on the earth. The digital information from the bands is presented to input neurons. The network produces activations for output neurons that represent nominal land use classes. This result is compared to the known land use for the location on the earth represented by the pixel determined by an expert. Information for training pixels is repeatedly presented to the network until the network is able to activate the correct land use when presented the information from the remotely sensed information. Once the network is successfully trained it can be presented information for new pixels not in the training set to determine the appropriate land use class for the area represented by the pixel.

9.3. The Gravity Model

The first example represents the upper left category of Figure 9.1. The model predicts the values for a continuous variable and learns through a supervised training

Table 9.1: Data used to compute the gravity model using regression and a neural network. Data for all variables are scaled so that the highest value equals 0.9 and the lowest value equals 0.1.

Destination State	Population	Distance	Migration
Alabama	0.2269	0.1796	0.2030
Arizona	0.1587	0.7064	0.1235
Arkansas	0.1649	0.2981	0.1201
California	0.9000	0.8547	0.3530
Colorado	0.1764	0.5751	0.1373
Connecticut	0.2101	0.3038	0.1830
Delaware	0.1088	0.2170	0.1170
Florida	0.3633	0.1868	0.5254
Georgia	0.2736	0.1075	0.6212
Idaho	0.1155	0.7679	0.1082
Illinois	0.5396	0.2811	0.1767
Indiana	0.3276	0.2415	0.1407
Iowa	0.2016	0.3830	0.1076
Kansas	0.1781	0.4396	0.1283
Kentucky	0.2177	0.1755	0.1536
Louisiana	0.2349	0.3121	0.1556
	0.1269	0.4208	0.1100
Maryland	0.2464	0.1925	0.2252
Massachusetts	0.3184	0.3344	0.1704
Michigan	0.4483	0.3075	0.1725
Minnesota	0.2416	0.4755	0.1147
Mississippi	0.1769	0.2442	0.1439
Missouri	0.2772	0.3189	0.1463
Montana	0.1148	0.7226	0.1041
Nebraska	0.1469	0.4811	0.1089
Nevada	0.1064	0.8019	0.1086
New Hampshire	0.1166	0.3547	0.1076
New Jersey	0.3787	0.2509	0.2410
New Mexico	0.1279	0.5857	0.1186
New York	0.8300	0.3113	0.3975
North Carolina	0.2937	0.1000	0.9000
North Dakota	0.1117	0.5717	0.1054
Ohio	0.5208	0.2132	0.1967
Oklahoma	0.1908	0.4019	0.1296
Oregon	0.1717	0.8811	0.1153
Pennsylvania	0.5673	0.2415	0.2391
Rhode Island	0.1251	0.3226	0.1249
South Dakota	0.1136	0.5245	0.1000
Tennessee	0.2465	0.1679	0.2341
Texas	0.1353	0.4619	0.1249
Utah	0.1296	0.6943	0.1118
Vermont	0.1046	0.3547	0.1036
Virginia	0.2760	0.1491	0.3984
Washington	0.2255	0.9000	0.1416
West Virginia	0.1576	0.1660	0.1179
Wisconsin	0.2666	0.3717	0.1170
Wyoming	0.1000	0.6377	0.1023

process. Modeling spatial interaction is a familiar problem in geography and the gravity model is a familiar solution to this problem. Since gravity models are often computed using multiple regression, the results for a regression model are also discussed and compared with results produced by a neural network model. Both models used the same data listed in Table 9.1. The migration column is from the 1970 United States

Census for the number of migrants leaving South Carolina and moving to the destination state listed in the first column. The values have been scaled so that the highest is 0.9 and the lowest is 0.1. The values in the population column are scaled in the same way and are based on total population for the each state from the same census. The values in the distance column are based on the distance from the center of South Carolina to the center of each destination state and also have been scaled to range between 0.9 and 0.1. Scaling each variable treats them as equally important inputs to the neural network and makes them compatible with a logistical output function that produces values between 0.0 and 1.0.

The general notion behind a gravity model is that spatial interaction, in this case migration, should increase as population size increases and decrease as distance increases. For this specific example, one should hypothesize that the number of migrants leaving South Carolina and taking up residence in each of the 47 destination states should increase with the population size of the destination state and decrease with the distance between South Carolina and the destination state. The general characteristics of these relationships for the variables in Table 9.1 are illustrated in Figure 9.3 as scatter

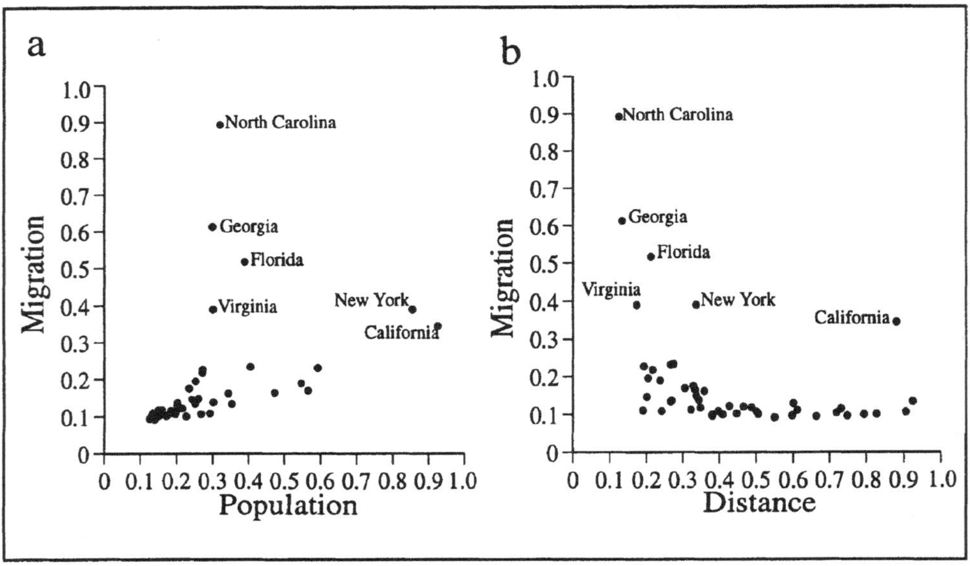

Figure 9.3: Scatter plot of variables population and migration from Table 9.1 (a) and scatter plot of variables distance and migration from Table 9.1 (b).

plots. The scatter plot of population and migration indicates a general positive relationship, but not one that is clearly linear (Figure 9.3a). The states with the smallest populations are not receiving many migrants, but, as population increases, two distinct patterns are apparent. There is a general trend that increases gradually with the largest states (New York and California) having the largest number of migrants at the termination of the trend. The other trend extends vertically from the center of the distribution and includes neighboring states. South Carolina's closest neighbors, Georgia and North Carolina, are shown as receiving the greatest number of migrants.

The scatter plot of migration and distance (Figure 9.3b) indicates a general negative relationship, but it also is not linear. The largest numbers of migrants go to neighboring states with the numbers decreasing rapidly in a typical distance decay pattern. The two largest states (New York and California) show up as deviations from the general pattern and receive more migrants that expected for their distances from South Carolina.

9.3.1 THE REGRESSION MODEL

Using regression to model migration treats population and distance as independent variables and migration as the dependent variable. Three parameters, an intercept, *a*, and two regression coefficients, b_1 and b_2 , are computed using the data. This allows the general equation shown in Figure 9.4a to be used to make predictions.

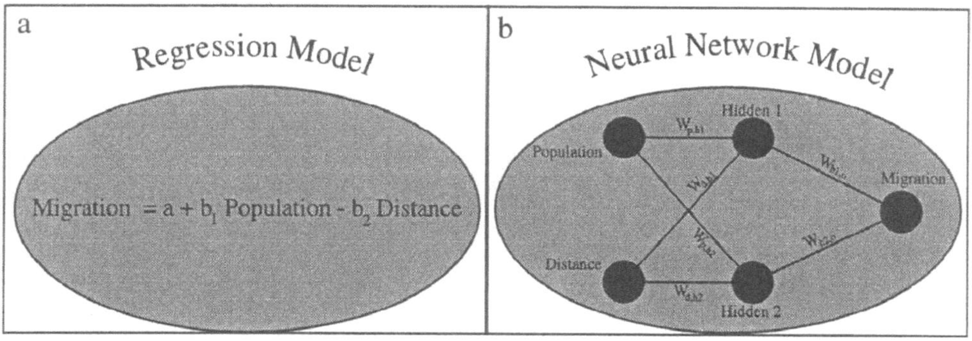

Figure 9.4: Representations of the gravity as a linear regression equation (a) and as a network of connected neurons (b).

The predictions made by the model define a plane. For this migration problem the prediction surface is illustrated in Figure 9.5a. The *a* parameter is where the plane intercepts the vertical (migration) axis. The two regression coefficients are the slopes of the plane along the population axis (b_1) and distance axis (b_2). The tilt of the plane indicates the expected positive change of migration as population increases and the expected negative change of migration as distance increases. As one might expect, a regression model (Figure 9.3a) that predicts migration as a linear function of population and distance cannot explain much of the non-linear variation present in the data (Figure 9.1). A plot of the regression model's predictions versus the actual migration values (Figure 9.6a) indicates the model over predicts many of the states that receive a moderate amount of migrants and severely under predicts neighboring states (Florida, Georgia, and North Carolina) that receive a large number of migrants.

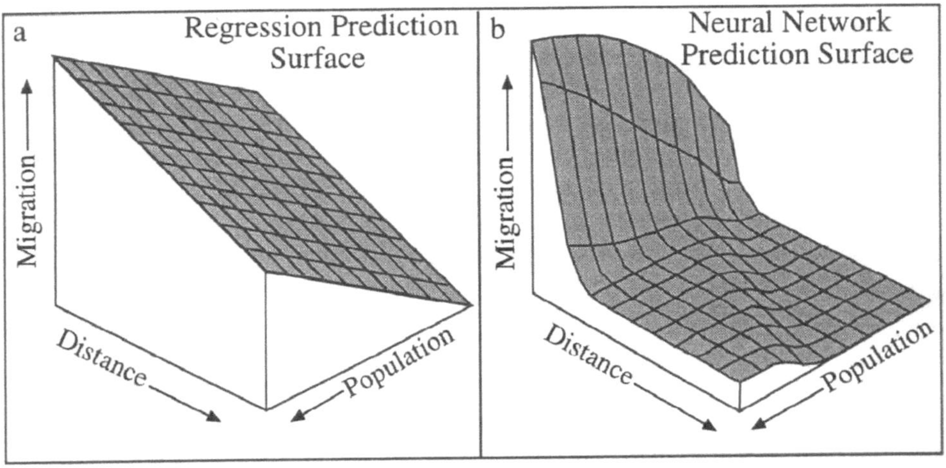

Figure 9.5: Surfaces representing the predicted migration for the linear regression model (a) and neural network model (b).

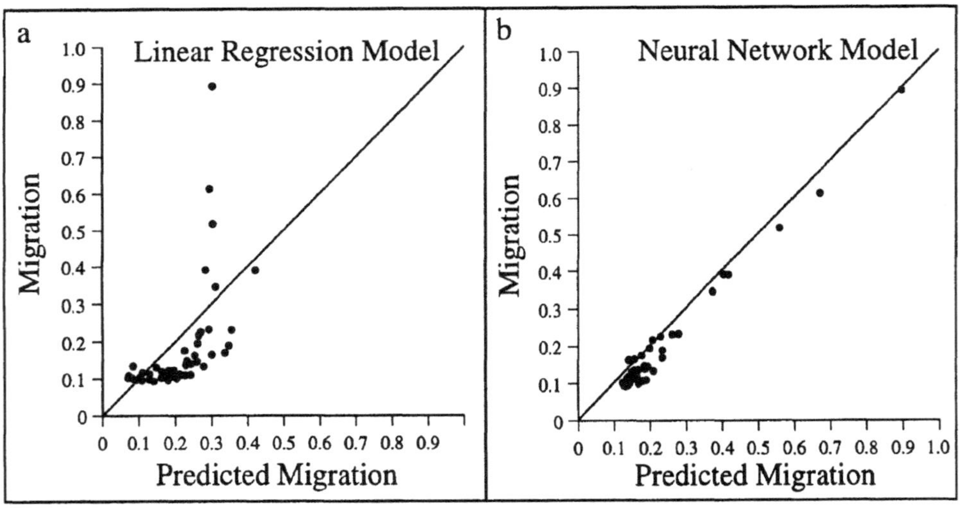

Figure 9.6: Predicted migrations (horizontal axis) versus actual migration (vertical axis) for linear regression model (a) and neural network model (b).

9.3.2. THE BACKPROPAGATION NEURAL NETWORK

A backpropagation network was created that defined two input neurons, representing population and distance, two hidden neurons, and one output neuron, representing migration (Figure 9.5b). The six weights connecting pairs of neurons were initially set to some small value by a random process. Each row of the data (Table 9.1) represents an input value for the population and distance neuron and a target for the output neuron. The network is trained by randomly presenting exemplar patterns to the input neurons. These values are, multiplied by the current weights and summed by the hidden units to create the input to each hidden unit (McClelland and Rumelhart 1989).

If *i* equal a sender neuron (input neuron) and *j* equals a receiver neuron (hidden neuron):

$$net_j = \Sigma w_{i,j} I_i \qquad (9.1)$$

These net_j values are then processed by a logistic function to create an output for each hidden unit (Figure 9.2). If a_j = the activation of receiver neuron j:

$$a_j = \frac{1}{1 + e^{-net_j}} \qquad (9.2)$$

These outputs (a_j) for the hidden units are multiplied by the weights connecting the hidden neurons with the output neuron and summed to create its input. If *j* equals a sender unit (hidden neuron) and *o* equals the receiver unit (output unit):

$$net_o = \Sigma w_{j,o} a_j \qquad (9.3)$$

The outputs units input is then processed by a logistic function to create its activation (a_o).

$$a_o = \frac{1}{1 + e^{-net_o}} \qquad (9.4)$$

This activation is the network's prediction of the migration for the exemplar state given that state's inputs and the current state of the weights in the network. The activation produced by the network for an exemplar is then compared with the target activation (actual migration) to determine the size of the error. Given that the weights were initially set by a random process the prediction is initially not likely to be accurate. For example, the first value for the training set (Alabama) has a value of 0.2269 for population and 0.1796 for distance. These would be passed through the network to produce an activation that would be compare with 0.2030, the target value for migration to Alabama from South Carolina.

After each forward pass of information through the network the error computed as the difference between actual and target activation for the output neuron is propagated back through the network and used to adjust the weights connecting pairs of neurons. If the error coefficient for the connection between a hidden (sender) neuron and the output (receiver) neuron for a given input pattern is defined as δ_o :

$$\delta_o = (t-a_o)a_o\,(1-a_o) \qquad (9.5)$$

The predicted migration from Equation 9.4 is subtracted from the actual migration (t) and multiplied by the derivative of the Equation 9.4:

$$\frac{da_o}{d\mathrm{net}_o} = a_o\,(1-a_o) \qquad (9.6)$$

The delta rule is then used to make the weight adjustment for weights connecting a hidden neuron to the output neuron:

$$\Delta w_{j,o} = \beta\,\delta_o\,a_o \qquad (9.7)$$

The change in the weights is proportional to the learning constant (β) times the error coefficient from Equation 9.5 (δ_o) times the current activation from Equation 9.4 (a_o).

Adjustments to the weights connecting the input neurons and the hidden neurons are slightly different since there is no specific target associated with the hidden units. The error for the output is propagated back through the system to adjust these weights. If the error coefficient for the connection between an input (sender) neuron and hidden neuron (receiver) on the forward pass is defined as δ_j:

$$\delta_j = a_j\,(1-a_j)\sum w_{j,o}\delta_o \qquad (9.8)$$

The derivative of the logistic function (Equation 9.2) producing the activation for a hidden unit is multiplied by the error from the output layer weighted by its connections back to the hidden layer. The delta rule can also be applied to compute the adjustments for connection weights between the input and hidden layers:

$$\Delta w_{i,j} = \beta\,\delta_j\,a_j \qquad (9.9)$$

The change in the weights is proportional to the learning constant (β) times the error coefficient from Equation 9.8 (δ_j) times the activation produced for the hidden neuron on the forward pass from Equation 9.2 (a_j).

Since the changes in the weights are proportional to the size of the error, the initial changes are likely to be relatively large and, as the process continues, get smaller until they stabilize. At the end of the learning process the network should be able to make predictions that are as close as possible to the target values. The predictions of the network used for this example (Figure 9.4b) can be represented as a surface that relates the values presented to the input neurons (population and distance) to the output neuron's (migration) activation. This prediction surface (Figure 9.4b) is distinctly non-linear. The prediction surface generally shows migration increasing with population and decreasing with distance, but the interactions between the input and hidden neurons and between the hidden neurons and output neuron have allowed the model to account for the non-linear relationships between the dependent and independent variables (Table 9.1). A plot of the migration predicted by the neural network model versus actual migration (Figure 9.6b) indicates the model's high degree of success. The neighboring states, that were severely under predicted by the regression model (Figure 9.6a) are accurately predicted as are the other states. The proportion of the total migration variance explained by a model (R^2) indicates the relative success of the two models with the regression model explaining 30.1 percent of the variance and the neural network explaining 96.8 percent of the variance. The standard error of estimate ($SE_{Y'}$), or the square root of the unexplained variance, measures the typical deviation between the target values (actual migration) and the prediction surface. The neural network (0.028) clearly can predict migrations more accurately than the regression model (0.130). Regression techniques frequently use logarithmic transformations to make relationships between dependent and independent variables more linear. When the original population, distance, and migration data are transformed by a $\log_e$ function before being scaled, the results (R^2 =0.776 and $SE_{Y'}$ = 0.084) are better than the original regression, but are still not as good as the backpropagation neural network results.

9.4. Residential Integration

The upper right corner of Figure 9.1 represents a problem using a supervised learning strategy for a classification problem. This is a classic pattern association problem where a neural network learns to associate input patterns with categories of outputs. The example problem considers urban neighborhoods represented by census tracts and models the association of their characteristics with their being in a racial class.

9.4.1. URBAN STRUCTURE

Urban neighborhoods can be measured on many different characteristics. These tend to vary along two basic underlying dimensions. Classic urban structure theory suggests that some neighborhood characteristics, those based on economic controls, should be distributed in sectors within a city (Cadwallader 1996). Variables such as income, house value, education, and occupation should be distributed in high, middle, and low socio-economic sectors. Other neighborhood characteristics, those based on life-cycle controls, should be distributed in concentric zones within a city. Variables such as age, family size, or house style, should be distributed in high, middle, and low life-cycle concentric zones. High life-cycle points to larger families, with middle aged parents and dependent children, living in low density suburban locations. Low life-cycle points to smaller families of younger or older adults with fewer children, living in higher density city neighborhoods. A third important underlying dimension that is related to clusters of racial (or ethnic) minorities in urban areas has also been suggested for some cities. This cluster effect is more likely to occur if the majority racial group has substantially superior numbers. Racial groups are more likely to share the space more equally in cities that are occupied by groups with numbers that are closer to balanced. Sharing the geographic space might mean carving out large chunks of space that are dominated by one group or the other, it might mean the two groups living in integrated neighborhoods, or it might mean some mixture of integrated and segregated residential neighborhoods.

9.4.2. CLASSIFICATION

Lloyd (1996a) used a backpropagation neural network to study the integration of neighborhoods in Columbia, South Carolina (McClelland and Rumelhart 1989). The 1990 percentage of black population for Columbia's urbanized census tracts indicated the values ranged from a maximum of 99.6 percent to a minimum of 0.0 percent with the mean equaling 35.1 percent and the standard deviation equaling 31.1 percent. These summary statistics suggested the black population was more than a small minority, but did not indicate how the black and white populations were distributed within the urban area or how these groups related to other variables. Three neighborhood racial classes were defined to provide a clearer impression of the nature of the distribution of the urban population by race. One class collected tracts that were occupied predominately by black population and was defined as tracts having 2/3 or more black residents. Another class gathered tracts that were occupied predominately by white population and was defined as tracts having 1/3 or fewer black residents. The final class accumulated tracts that were integrated or shared by approximately an equal number of black and white residents. This class was defined as tracts with more than 1/3, but less than 2/3, black residents. The nature of the census tracts in each of these three classes provided addition insight into the distribution of the black population within the urban area (Figure 9.7). Most tracts (52) classified as predominately white, but 30 of the 82 tracts considered fell

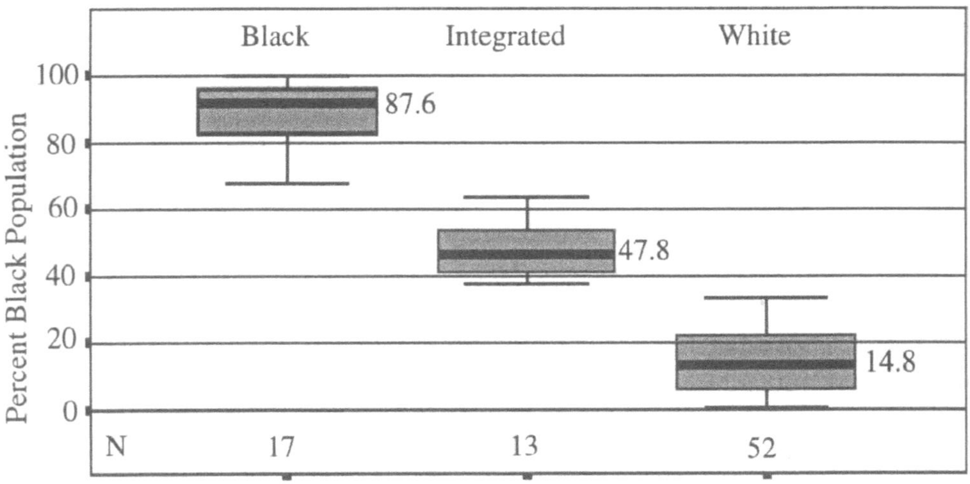

Figure 9.7: Summary statistics for the Black, Integrated, and White classes. After Lloyd (1996a).

into the predominately black (17) or integrated (13) classes (Figure 9.7). The class means indicated the typical tract that was predominately black had a 12.4 percent white minority and the typical tract that was predominately white had a 14.8 percent black minority. The average integrated tract had a population that was approximately equal for the two racial groups (52.2 percent white and 47.8 percent black). The summary statistics for the three classes suggested the black population was not clustered into a small number of completely segregated neighborhoods. Some tracts were occupied by a black majority and a white minority, some were occupied by equal numbers of black and white residents, and some were occupied by a white majority and a black minority. A map of the three racial classes suggested the predominately black and white tracts generally were distributed in sectors within the urban area and that the integrated neighborhoods occupied transition locations between the black and white sectors (Figure 9.8). Although black residents could be found in almost all neighborhoods in the urban area, the spatial distribution of the three classes did not suggest a random distribution. Since economic variables were generally distributed in sectors, it seemed reasonable to hypothesize that the distribution of the classes was based on economic factors.

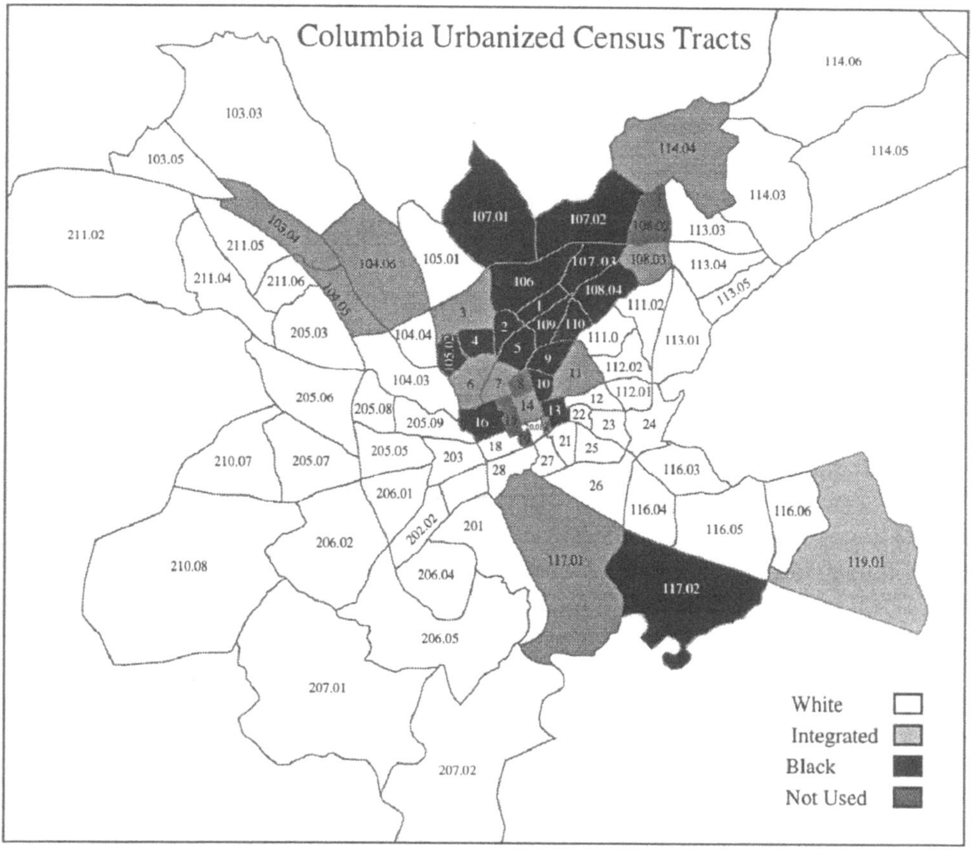

Figure 9.8: Map of Columbia, South Carolina urbanized area with the black, integrated, and white classes indicated. Tracts without numbers were institutional tracts not considered by the analyses. After Lloyd (1996a)

Factorial Ecology

The general nature of the urbanized census tracts can be defined by variables that measure both socio-economic (sector theory) and life-cycle (concentric zone theory) characteristics (Table 9.2). Analyses of the variation among the classes indicated all

Table 9.2: Class means for 11 socio-economic and life-cycle variables for the Black, Integrated, and White classes.

	Black	Integrated	White
% < 18 Years Old[Y]	27.8*	21.9	22.2
% > 65 Years Old[N]	10.4	11.0	11.2
Persons Per Household[Y]	2.9	2.3	2.5
% without High School[Y]	40.4*	27.2*	14.3*
% with Colledge Degree[Y]	18.1*	31.8*	42.3*
% Females in the Labor Force[N]	59.4	63.1	64.4
% Service Occupation[Y]	28.4*	16.6*	10.5*
% Managerial and Professional Occupation[Y]	16.8*	29.0	35.4
% < the Poverty Level[Y]	28.8*	16.8*	9.1*
$ Median House Value[Y]	49,558*	72,385	84,130
$ Median Family Income[Y]	20,895	27,284	42,148*

[Y]Significant differences using a one- way ANOVA
[N]No significant differences using a one-way ANOVA
*Class is significantly different from other classes using Student-Neuman-Keuls test

economic variables had significant differences among the classes. The life-cycle variables generally did not show significant differences among the classes. The exception was that predominantly black census tracts were significantly higher for the percent of the population less than 18 years old. These generalizations can be illustrated by considering the locations of the census tracts in a two-dimensional space created by a factor analysis of the 11 variables. The space represented socio-economic status on the vertical axis and life-cycle status on the horizontal axis (Figure 9.9). The location of

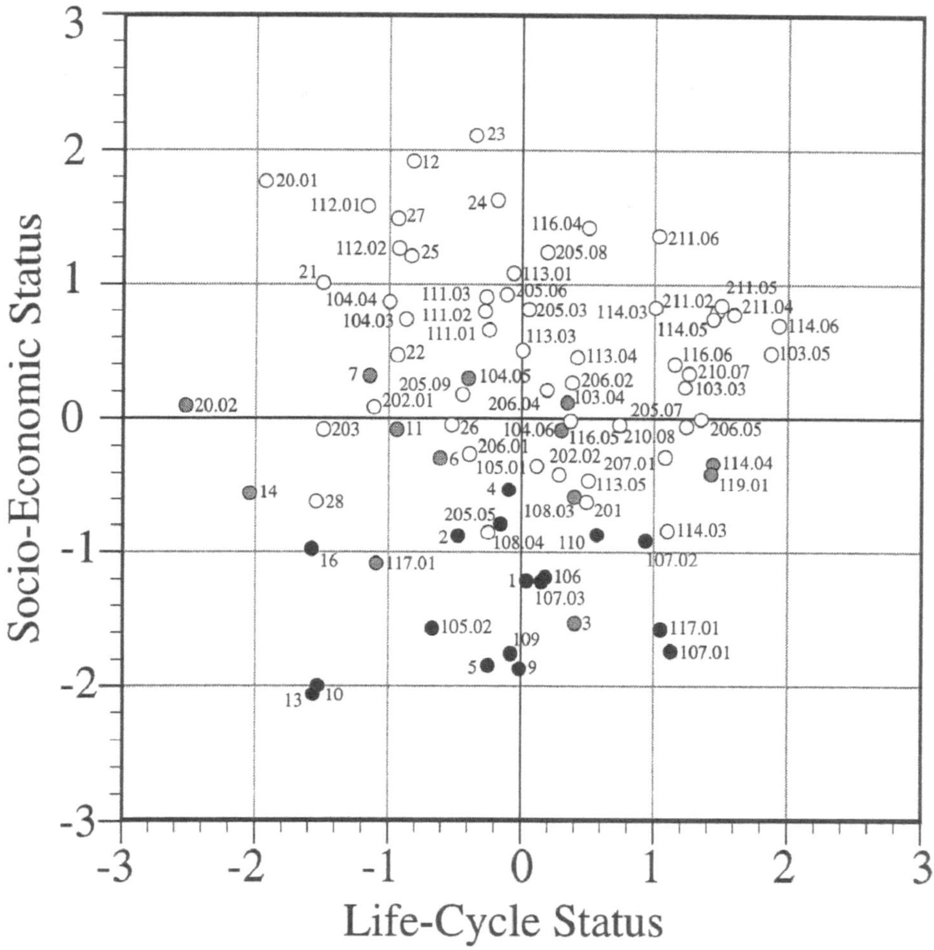

Figure 9.9: A space representing the basic urban structure of Columbia. The horizontal axis, an index measuring life-cycle status, and the vertical dimension, an index measuring socio-economic status, were determined by a principle axis Factor Analysis of the 11 variables listed in Table 9.2. The specific locations of the census tracts in the space were determined by the tracts' factor scores on the dimensions. After Lloyd (1996a).

each census tract on these two basic underlying dimensions revealed its basic character. For example, tract # 13 had the lowest socio-economic states, tract # 23 had the highest socio-economic status, tract # 20.02 had the lowest life-cycle status, and tract # 114.06 had the highest life-cycle status. Tract # 20.01 was very low on the horizontal axis and very high on the vertical axis, indicating it was a wealthy inner-city neighborhood. Extreme cases in the upper right quadrant indicated wealthy suburban neighborhoods like tract #211.06. Extreme locations in the lower right quadrant, like tract #107.01, indicated poor suburban neighborhoods and extreme locations in the lower left quadrant, like tract #10, indicated poor inner-city neighborhoods. The dot indicting each census tract's location in the space was coded to indicate its classification as a predominately black, integrated, or predominantly white census tract (Figure 9.9). The distribution of the dots indicated that most predominantly white neighborhoods were positive on the socio-economic status axis and none were more than one standard deviation below the mean of zero. The distribution also showed that no predominantly black neighborhoods were positive on the socio-economic status axis and that more than half were more than one standard deviation below the mean of zero. Most of the integrated neighborhoods were near the center of the socio-economic axis with four on the positive side and nine on the negative side. Two integrated tracts were more than one standard deviation below the mean on the socio-economic status axis. No particular pattern emerged for any of the three racial classes relative to the life-cycle axis. Black, integrated, and white tracts could be found at all locations along the horizontal axis (Figure 9.9). These patterns suggested that the variation of socio-economic variables could be used to predict the racial classification of a neighborhood much better than the variation of life-cycle variables.

Neural Network Model

Lloyd (1996a) defined a neural network model with input neurons representing the 11 variables from Table 9.2, four hidden neurons, and three output neurons representing the black, integrated, and white racial classes (Figure 9.10). The model learned to turn on the correct output neuron and turn off its competitors when it was presented the 82 census tracts as input patterns. Training was repeated until the root mean squared error reached a threshold limit of 0.05. Weights were adjusted after each pattern was passed forward through the network by propagating the error back through the network as described in section 9.3 of this chapter. Given the complex and indirect nature of the connections between an input and output neuron, it was relatively difficult to trace relationships through the magnitude of the weights in the network. This was because the activation of a receiver neuron was based not only on inputs from the sender neuron of interest, but all the other neurons in the sender layer. The output from a sender neuron did not go to just the receiver neuron of interest, but to all the receiver neurons. After the network was trained, it was possible to monitor the response of one of the output neurons, e.g., the integrated neighborhood class, given a range of values from an input neuron, e.g., median family income. Response surfaces that illustrated the activation of an output neuron given values for two input neurons provided visual representations of the relationships between input patterns and output classifications. It was necessary to illustrate selected cases since it is impossible to show the inputs of 11 inputs and 3 responses in a single illustration.

The Response of the White Class. The activation responses of the three output neurons (racial classes) were represented by the vertical axis for plots related to the white class (Figure 9.11), black class (Figure 9.13), and the integrated class (Figure 9.14). In each case the other two dimensions represented the input values from two selected input

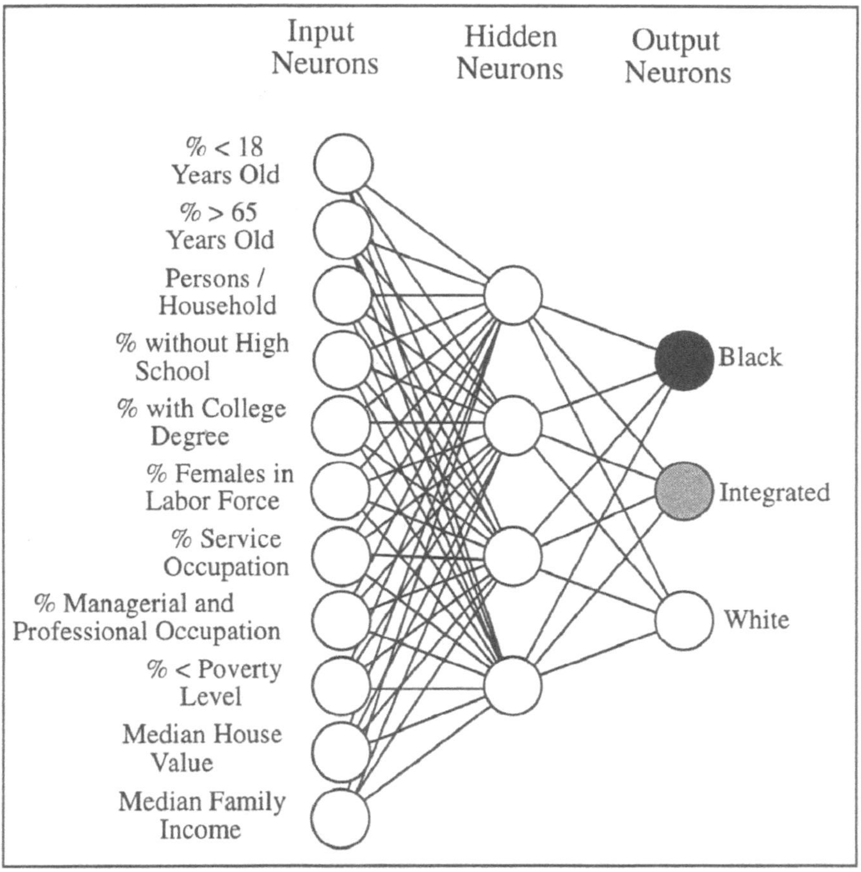

Figure 9.10: The backpropagation neural network used to learn to classify the urban census tracts into the black, integrated, and white classes. After Lloyd (1996a).

neurons. Some surfaces illustrated simple relationships as in Figure 9.11a. Here the **White Class** has a high activation if the **Percent Without High School** neuron is sending a value below a threshold and low if the value is greater than the threshold. The steep vertical cliff in the plot is the visual representation of a simple *threshold rule* (Figure 9.12a):

$$\text{If } X>T_x \text{ then increase Z else decrease Z} \tag{9.10}$$

If the input from **Percent without High Schools** is less than the threshold turn on the **White Class**, else turn off the **White Class**. Given that the average white census tract had significantly lower values for **Percent without High School**

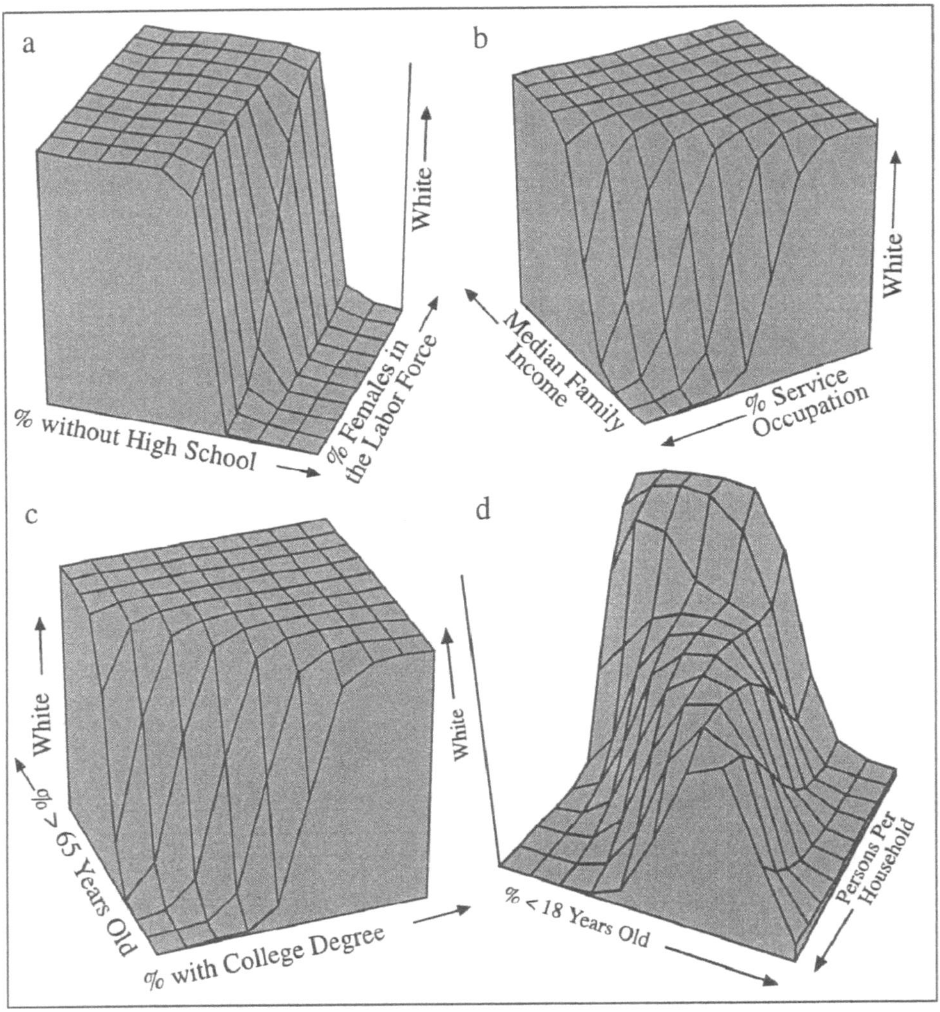

Figure 9.11: The response of the **White Class** to selected input neurons **Percent Without High School** and **Percent Females in the Labor Force** (a), **Median Family Income** and **Percent Service Occupation** (b), **Percent Greater Than 65 Years Old** and **Percent With College Degree** (c), and **Percent Less Than 18 Years Old** and **Persons Per Household** (d). After Lloyd (1996a).

(Table 9.2), this relationship makes intuitive sense. It also seems reasonable that the **Percent Females in the Labor Force** input neuron had virtually no impact on the activation of the **White** output neuron (Figure 9.11a). This was shown by the lack of change of the response surface along the **Percent Females in the Labor Force** axis. This might be stated as an *ignore rule*:

$$\text{If X then do not change Z} \tag{9.11}$$

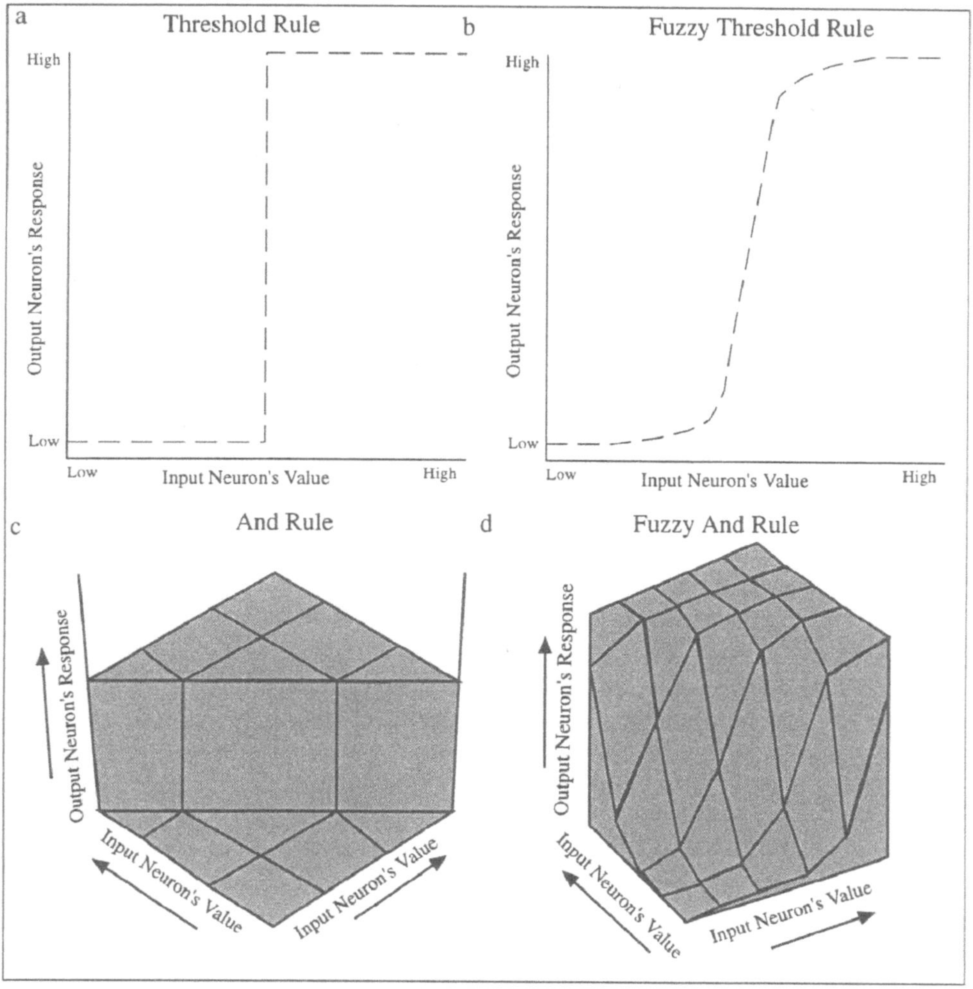

Figure 9.12: Simple rules that are Implicit in the response surfaces. After Lloyd (1996a).

An output neuron will not respond to the values from a particular input neuron if weights connecting the two units have a value of zero. A weight stabilizing at a value of zero indicates the network has learned that sender neuron can be ignored.

When two (or more) input neurons form a pattern that is repeated by the exemplars frequently then *threshold rules* can be combined into *and rules* (Figure 9.12c). These rules may combine information from any number of input neurons that form consistent patterns. *And rules* were illustrated by two of the response surfaces:

$$\text{If } X<T_x \text{ and } Y>T_y \text{ then increase Z else decrease Z} \qquad (9.12)$$

If **Median Family Income** is very low and **Percent Service Occupation** is very high then turn off the **White Class** output neuron (Figures 9.11b) was one rule. This would appear to agree with the finding that the **White Class** mean was significantly higher for **Median Family Income** and significantly lower for **Percent Service Occupation** (Table 9.2). In this case the threshold is more of a steep ramp than a cliff, but the idea is the same. The *and rule* can be stated in other forms:

$$\text{If } X<T_x \text{ and } Y<T_y \text{ then decrease Z else increase Z} \quad (9.13)$$

If both **Percent Greater Than 65 Years Old** and **Percent with College Degree** are low then turn off the **White Class** neuron, else turn it on, is the same rule in a different form (Figure 9.11c). The neural network, of course, did not use such logical rules. It only processed the input values through the network using the connection weights it had learned. The rules were implicit in the behavior of the network, however, just as people might appear the be using a rule to make decisions when processing information with their biological neural network even when they are not aware of using a rule. Figure 9.11d illustrated some more complex and interesting relationships that cannot be described as simple linear rules. The highest activations for the **White Class** neuron were associated with low values of **Percent less than 18 Years Old** and **Persons Per Household**. There was some negative slope to the surface toward the opposite corner, which represents high values for both input neurons. The most interesting result was the surface's steep drop to the other two corners that represent high values for one input and low for the other. When one input was high and the other was low the **White Class** neuron was turned off. This is because the two input variables both reflect family size and were naturally correlated. The learning process never presented a white neighborhood with either of these two patterns. In any city it is unlikely a census tract would have a large population under 18 year old and small families or large families and few children. The network appeared to be able to learn about the patterns it was presented during training and also learn to make inferences about patterns it was not presented. Similar results were reported by Lloyd (1994) for neural networks learning information about classes of maps.

The Response of the Black Class. The relationships between the patterns presented to the input neurons and the activation of the black output neuron suggested some relationships that can be stated as simple rules. If **Persons Per Household** was high and **Percent Less Than the Poverty Level** was low then turn on the **Black Class** is a fuzzy variation of the *and rule* (Figures 9.12b and 9.13a). If **Percent Service Occupation** was low then turn off the **Black Class**, else turn it on is another example of a *threshold rule* (Figure 9.13b). Input values from the **Percent Managerial and Professional Occupation** neuron apparently had little impact on the activation of the output neuron for the **Black Class** (Figure 9.13b). This was another example of an *ignore rule*. Figure 9.13c illustrated a response surface with subtle variations that were not linear. There was generally a moderate response by

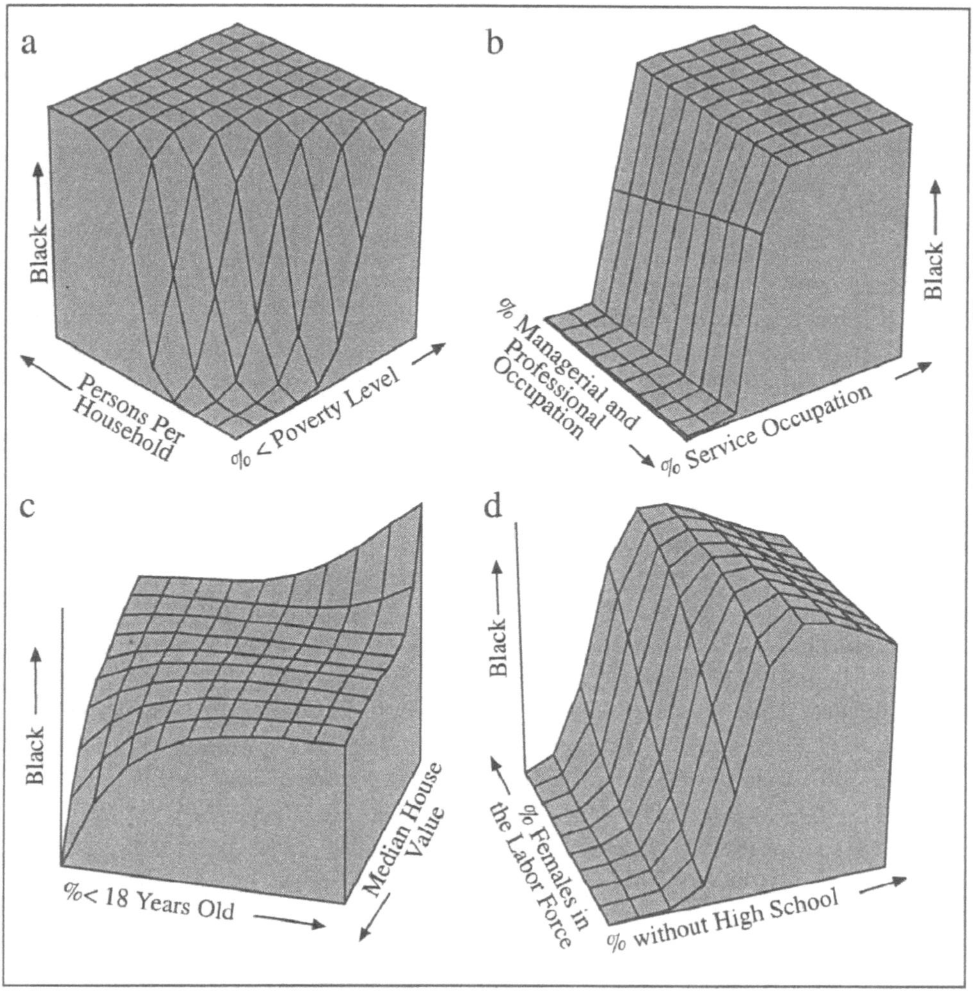

Figure 9.13: The response of the **Black Class** to selected input neurons **Persons Per Household** and **Percent Less Than the Poverty Level** (a), **Percent Managerial and Professional Occupation** and **Percent Service Occupation** (b), **Percent Less Than 18 Years Old** and **Median House Value** (c), and **Percent Females in the Labor Force** and **Percent Without High School** (d). After Lloyd (1996a).

the **Black Class** output neuron to inputs from the **Percent Less Than 18 Years Old** and **Median House Value** input neurons. When **Median House Value** presented a very high value and **Percent Less Than 18 Years Old** presented a very low value, however, the **Black Class** had a larger positive response. When the opposite was true, **Median House Value** presented a very low value and **Percent Less Than 18 Years Old** presented a very high value, the **Black Class** had a very negative response. Figure 9.13d illustrated versions of an *ignore rule* and a *fuzzy threshold rule* (Figures 9.12b and 9.12b). This response surface was similar to the one illustrated by Figure 9.11b, but has some subtle variation that was not linear. The

Black Class responded very little to the inputs from **Percent Female in the Labor Force**. It increased only slightly when combined with moderate values of **Percent Without High School**. The **Black Class** had and S-shaped response to **Percent Without High School.** Very low values from this input neuron turned the **Black Class** off. It turned on gradually as **Percent Without High School** increased, then increased greatly with moderate values, and then declined gradually with higher values.

The Response of the Integrated Class. Integrated neighborhoods were perhaps the most interesting and difficult to understand with simple logic. This complexity was reflected in the response surfaces that represented the relationships between selected input neurons and the **Integrated Class** output neuron. The relationship between the **Integrated Class** and **Percent Managerial and Professional Occupation** input neuron appeared to be an example of a simple *ignore rule.* Other relationships were not easily described with simple rules. For example, the **Integrated Class** neuron was moderately high with high values of **Percent Service Occupation** (Figure 9.14a). The activation for the **Integrated Class** declined as **Percent Service Occupation** decreased and was lowest for moderate values of this input neuron. As **Percent Service Occupation** increased from moderate to high values, however, there was a dramatic increase in the activation for the **Integrated Class** output neuron. Some integrated neighborhoods apparently had a low percentage of service workers and others had a high percentage of service workers, but none were associate with moderate values. The response surface for the **Integrated Class** over values for **Persons Per Household** and **Median Family Income** was also not easily described by a simple rule (Figure 9.14b). When **Persons Per Household** values were high, the response was lowest for low income values and increased to a moderate level as income increased. When **Persons Per Household** values were low, the response was extremely high when incomes were low and declined to moderate levels as income values increased. The strongest response was for moderately low income values and extremely low household values (Figure 9.14b). This suggested that at least some integrated neighborhoods tended to have moderate incomes and smaller family sizes, e.g., tracts # 20.02 (Figures 9.8 and 9.9).

The response surface for other input neurons suggested a different type of integrated neighborhood. A strong activation occurred for the **Integrated Class** output neuron when **Median House Value** and **Percent Less Than 18** were both high (Figure 9.14c). For both of these variables the **Integrated Class** mean was not different from the **White Class** mean but significantly different from the **Black Class** mean (Table 9.2). This suggested that at least some integrated neighborhoods had expensive houses and suburban locations, e.g., tract # 103.04, (Figures 9.8 and 9.9). The response surface for the **Integrated Class** related to **Percent Greater Than 65** appeared to be an *ignore rule* (Figure 9.14d). There was a subtle response for lower values for this neuron when combined with moderately high values for the **Percent Without High School** neuron. A more interesting feature of this surface was the **Integrated Class**'s bi-modal response to the **Percent Without High School** neuron. The **Integrated Class** turned on with both high and low values for this input neuron, but turned off for moderate values. This suggested that some integrated neighborhoods had relatively high levels of education, while others had relatively low levels. The notion that integrated neighborhoods were transition zones between black

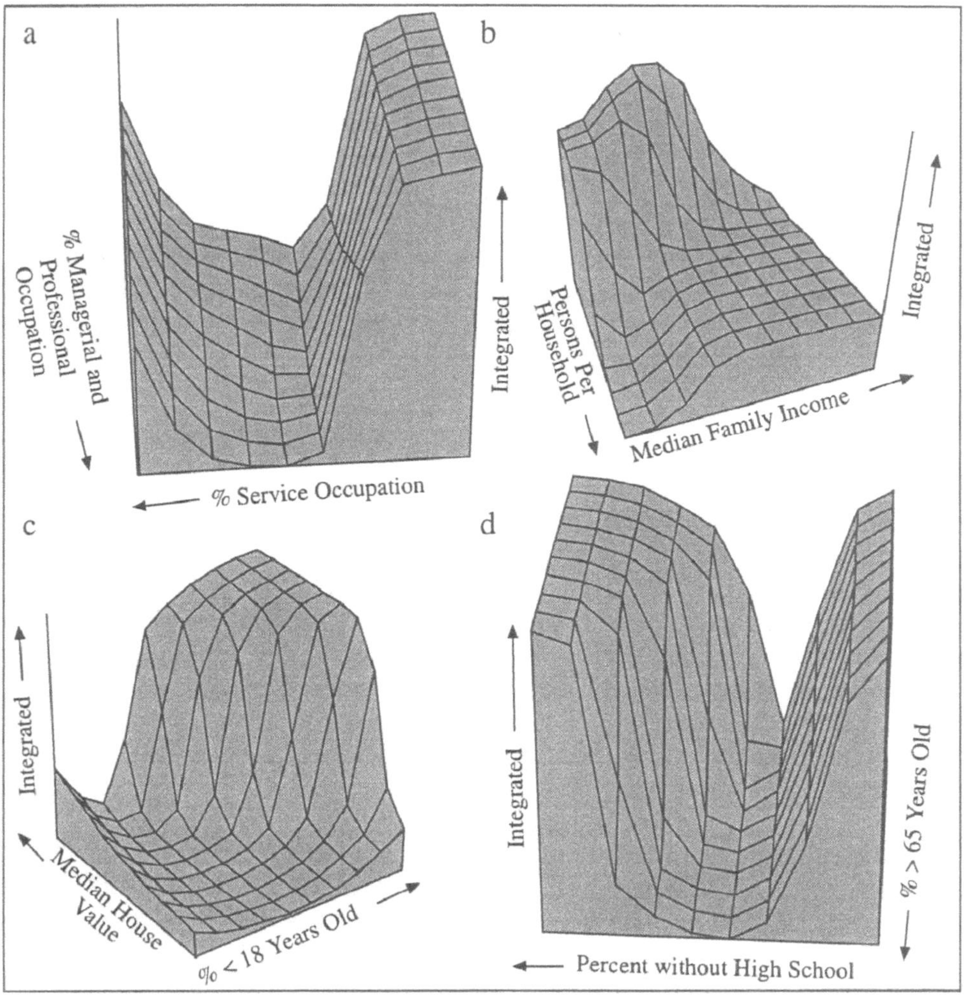

Figure 9.14: The response of the **White Class** to selected input neurons **Percent Managerial and Professional Occupation** and **Percent Service Occupation** (a), **Persons Per Household** and **Median Family Income** (b), **Median House Value** and **Percent Less Than 18 Years Old** (c), and **Percent Greater Than 65 Years Old** and **Percent Without High School** (d). After Lloyd (1996a).

and white neighborhoods may have some merit when viewed in terms of their spatial locations (Figure 9.8) or their average socio-economic status (Figure 9.9). The transition between black and white neighborhoods, however, was accomplished in different ways in different parts of the city. These differences were reflected by the complex relationships between the **Integrated Class** output neuron and the various characteristics represented by the input neurons.

9.5. Acquiring Spatial Information

The cognitive processes underlying visual search have been discussed in Chapter 5. One approach to understanding cognitive search processes is to simulate them with models (Cave and Wolfe 1990). A key to understanding visual search is measuring the attention that objects in a visual scene warrant as they compete to be noticed (Duncan and Humphreys 1989,1992). Neural networks have been shown to be useful for acquiring information from maps (Lloyd 1994). This example focuses on the acquisition of information being used to search a map for a target symbol that is defined by specific characteristics. It represents an unsupervised neural network producing a continuous estimate of the activation for map symbols at various locations on a map (Figure 9.1). Activation in this case is used as a measure of the visual attention that should be paid to a particular location on the map.

9.5.1. MODELING VISUAL SEARCH

Lloyd (1996b) used a hybrid model that combined a pattern associator and an auto-associator neural network for performing two tasks (McClelland and Rumelhart (1989). The first task was simply to learn the target that would be searched for on a given trial. Having learned the target, the pattern associator model could also be used to decide if a particular map symbol being considered during the search was the target. A second auto-associator neural network was used for simulating the acquisition of visual information related to a number of map symbols distributed on the map. In the following discussion k refers to 1 of 46 locations on a map of South Carolina, j refers to 1 of 4 input (sender) neurons, and i refers to 1 of 2 (pattern associator model) or 1 of 4 (auto-associator model) output (receiver) neurons.

Map symbols were defined in terms of four characteristics. These characteristics, color, shape, size, and orientation, were typical dimensions used by cartographers for producing effective symbols for maps. For a given search trial,two types of symbols were defined. A target symbol that could potentially be on the trial map and distractor symbols that always occupied some locations on the map. For example, the target might be a small, red, horizontal rectangle and distractors might be a small, blue, horizontal rectangle (Figure 9.15). Symbols were randomly assigned to a fixed number of locations within South Carolina associated with the centroids of the 46 counties. For a given trial, there could be 6, 16, 26, 36, or 46 locations randomly selected as a symbol location (Figure 9.16). Two types of simulations were conducted. One type was a feature search where the target had some unique feature. For example, if the target's features for a simulation were red, triangle, small, and horizontal, these characteristics could be assigned a digital code of 1 because they were target features (Figure 9.17). For a feature search ,the target would differ from the distractors on some unique dimension. If the unique dimension was color, then the distractor's characteristics would be blue, triangle, small, and horizontal. The unique color of the target should cause it to pop out of the display. A conjunctive search differs from a feature search because the target does not have a unique feature. Since the target shares all its characteristics with some distractors it is not as likely to pop out of the map display. For example, a conjunctive target also might be assigned red, triangle, small, and horizontal as characteristics and digital codes of 1. (Figure 9.18). Some distractors could then be defined to share color with the target and differ in shape while others could share shape and differ in color. Size and orientation are held constant for this example.

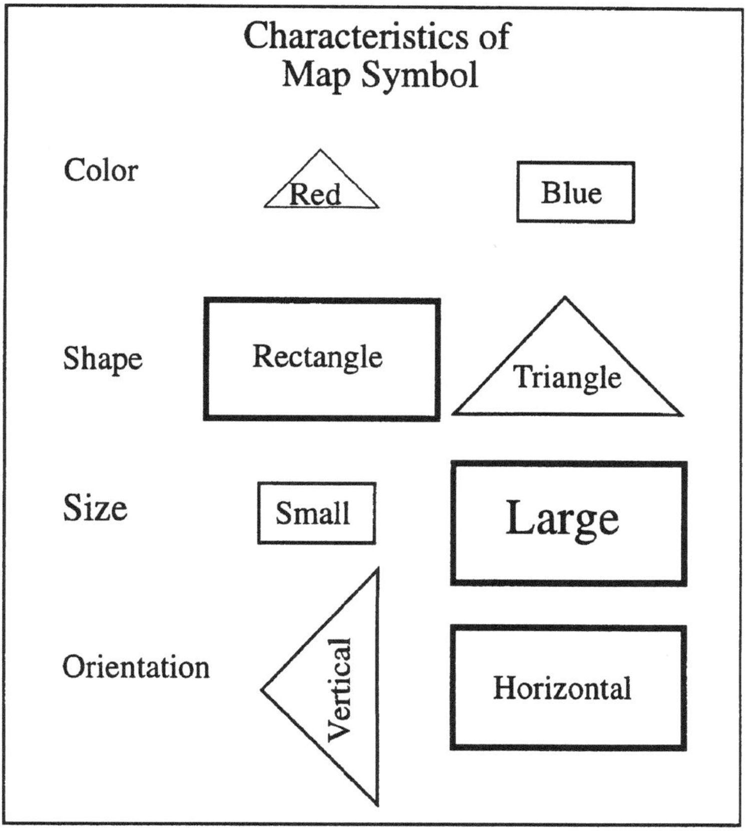

Figure 9.15: The dimensions used to encode map symbols for the target and distractor symbols. After Lloyd (1996b).

Learning the Target

The first stage in the simulation process was to learn to identify the target and store its memory for later use. This was done with a simple network that linked an input layer of neurons directly with an output layer (Figure 9.19). The four dimensions used to define targets and distractors were represented by the four input neurons and the class of **Target** or **Not Target** were represented by the output neurons. Each input neuron was connected to each output neuron producing eight connections. Digital examples of the target and distractor used for a particular trial were passed through the network and a net activation (Net_i) produced for each output neuron:

$$Net_i = \Sigma w_{i,j} A_j \tag{9.14}$$

The activation of each input neuron (A_j), or the digital code for the characteristic, is multiplied by the weight connecting the input and output neurons ($w_{i,j}$) and these are summed over the four connections. The output for the neuron is produced by processing

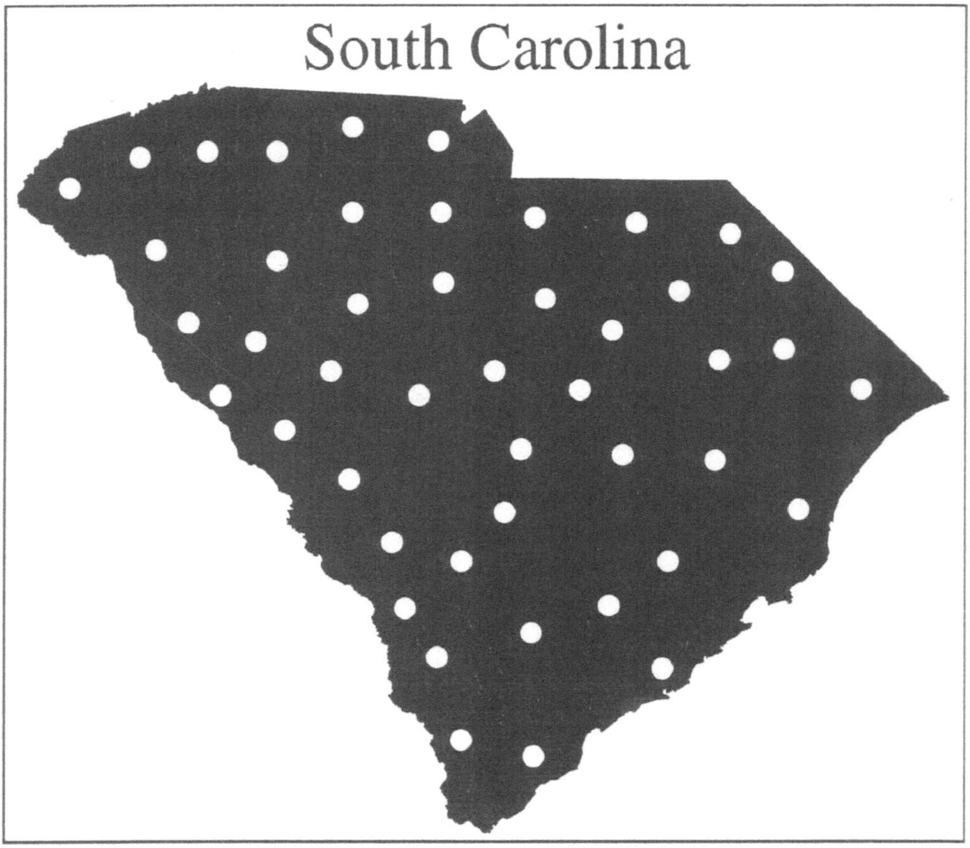

Figure 9.16: Map with the 46 locations that could have map symbols.

the net activation through a logistic function:

$$O_i = \frac{1}{1 + e^{-net_i/T}} \tag{9.15}$$

This produces an output between 0 (neuron turned off) and 1(neuron turned on). The constant T affects the slope of the logistic curve (Figure 9.2). The delta rule was used to update the weights after each pass:

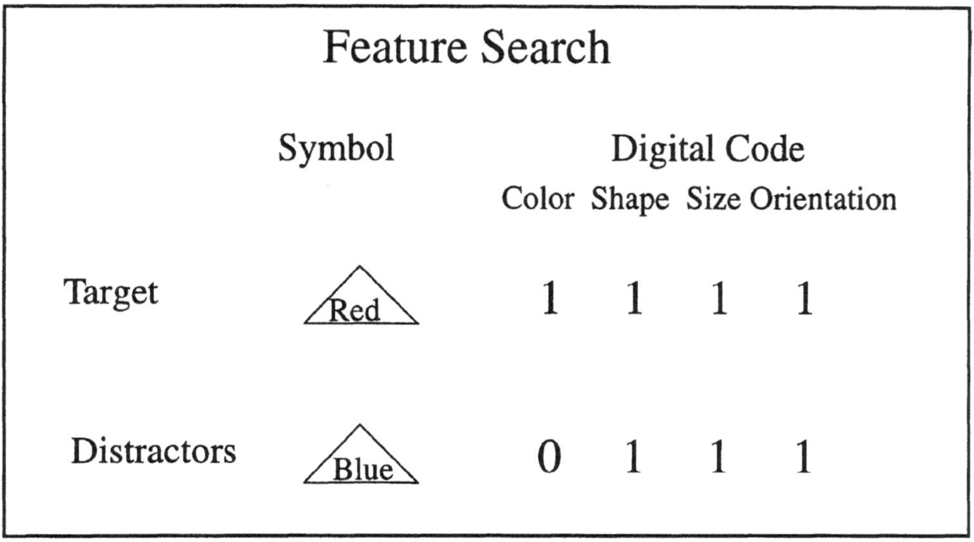

Figure 9.17: Example of a target and distractor map symbol used for a feature search.

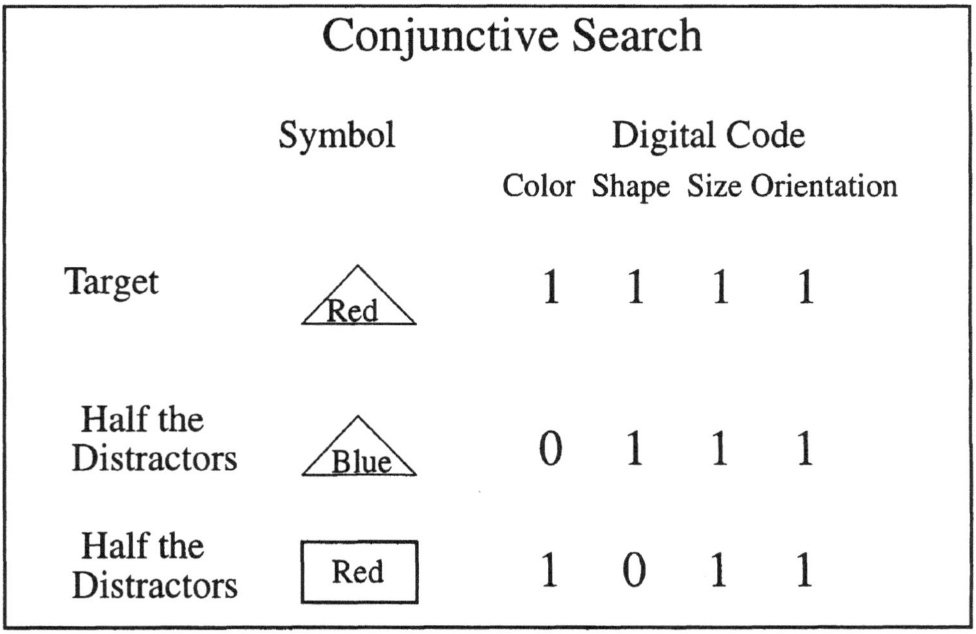

Figure 9.18: Example of a target map symbol and distractor map symbols used for a conjunctive search.

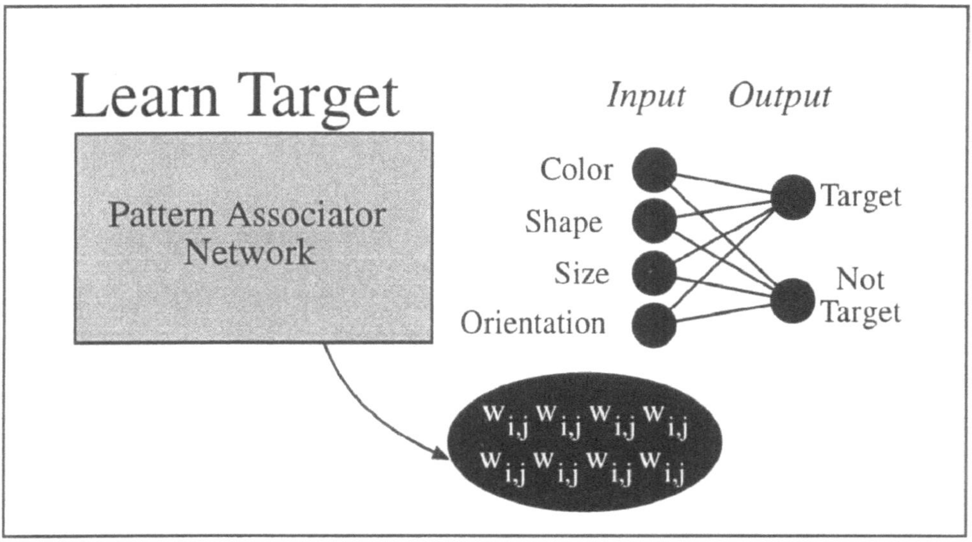

Figure 9.19: The pattern associator neural network used to learn the target map symbol.

$$\Delta w_{i,j} = \beta(t_i - O_i)A_j \tag{9.16}$$

The change made to the weight connecting input neuron *j* to output neuron *i* is proportional to the learning constant (β) times the error (target - output) times the activation of the input neuron. The weights were adjusted on each pass until their values stabilized. The network had then learned to distinguish the target from a distractor and this knowledge was stored in the weights.

Acquiring Information from the Map

The second stage of the simulation acquired an overall activation measure for the spatial locations that served as a yardstick for attention. The basic notion was the higher the activation a location was signaling then the greater the attention one should pay to that location. It was, therefore, necessary to measure the overall activation in such a way that locations with target symbols and locations with symbols similar to target symbols produced the highest overall activation. The initial task was to pass the digital codes for each location through an auto-associator network and cycle the network until the weights stabilized (Figure 9.20). Four neurons that represented the dimensions used to define the map symbols were used for the auto-associator network. In this model the neurons serve as both input and output neurons. The final result was an activation value for the Color, Shape, Size, and Orientation neurons that could be used to compute overall activation. Each neuron was connected to each other neuron, but not to itself. This produced 12 connection among the neurons (Figure 9.20). The activation for an individual neuron was produced from an internal and external signal. The external signal is just the digital codes that define the four characteristics of a map symbol at location k. The internal signal comes to a neuron from the other neurons:

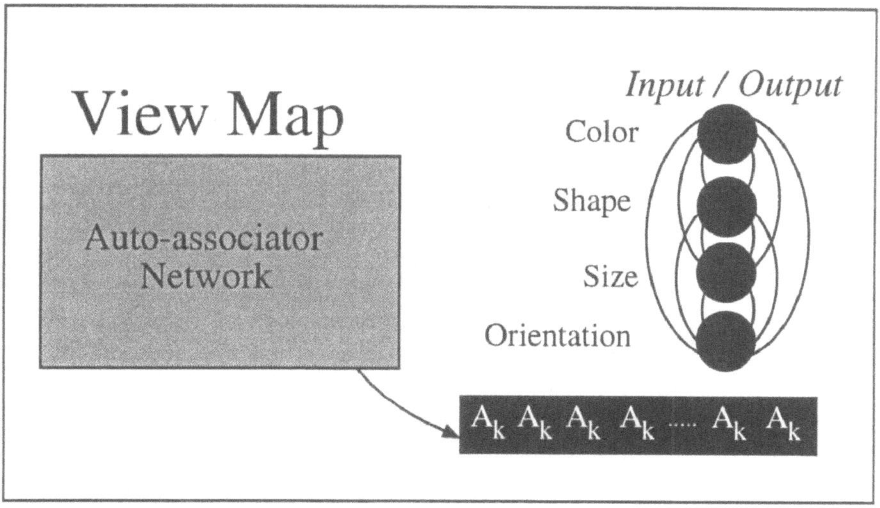

Figure 20: The auto-associator neural network used to acquire information from the map.

$$\text{IntInput}_i = \Sigma w_{i,j} A_j \tag{9.17}$$

The internal input for a neuron is the current activation of a sender neuron times the weight connecting the sender neuron to the receiver neuron summed over all the senders. This signal is combined with the external signal to produce a Netinput for each neuron:

$$\text{Net}_i = (\text{estr})\ \text{ExtInput}_i + (\text{istr})\text{IntInput}_i \tag{9.18}$$

The net input (Net_i) equals a scaling constant for the external input times the external input plus a scaling constant times the internal input time the internal input. Depending on the sign of the net input, the output for the neuron (A_i) is adjusted differently. For a positive net input the new activation is computed:

$$\text{If } (+) \qquad A_i = A_i + \text{Net}_i\ (\max - A_i) - (\text{decay})A_i \tag{9.19}$$

The new activation is computed by adding to the current activation the net activation times the difference between the maximum activation and the current activation and subtracting a decay constant times the current activation. The maximum activation was set to +1.0. Weighting the net input by the maximum minus the activation produces larger changes when the activation is near 0.5 and smaller changes as the activation approaches the maximum The decay constant is used to return the activation to a resting point if net inputs are weak.

For a negative net input the new activation is computed:

$$\text{If } (-) \quad A_i = A_i + Net_i\,(A_i - \min) - (decay)A_i \tag{9.19}$$

The new activation is computed by adding the current activation to the net activation times the current activation minus the minimum activation and subtracting a decay constant times the current activation. The minimum activation was set to 0.0. Multiplying the net input by the difference between the current activation and minimum activation produces larger changes when the activation is near 0.5 and smaller changes as the activation approaches zero.

Bottom-Up Information. Bottom-up information measures the contrast between each map symbol and the other symbols on the map. If there was a great contrast, as when the target symbol had a unique feature, the symbol should pop-out of the map display (Lloyd 1996c). Since bottom-up information was to measure the contrast between a symbol at a given location and the rest of the symbols on the map, a value was computed for each location:

$$\text{Bottom-up}_k = \Sigma_j \Sigma_i (A_{i,k} - A_{i,j})^2 \tag{9.20}$$

The squared difference between the current location and another location for each of the i neuron's activation is computed and summed. These j values are then summed to compute the bottom-up information for location k. For a feature search the location with the target should have activations different from other locations with distractors. This should produce a high level of bottom-up information and result in a visual contrast between the target symbol and other distractor symbols. Since a distractor symbol at any given location should have activations very similar to the other distractors, it should have a lower value for bottom-up information and less visual contrast with other locations. The same is not true for a conjunctive search. Because the location with the target symbols has no unique feature, its bottom-up information will not differ greatly from other locations and the result should not be a visual contrast between the target and distractor symbols.

Top-Down Information. Top-down information measures how similar a map symbol at a particular location is to the target map symbol and separate values can be computed for each location:

$$\text{Top-down}_k = \Sigma_i (A_{i,t} - A_{i,k})^2 \tag{9.21}$$

The squared difference between the activation learned for the target by the pattern associator neural network for neuron i and the activation for neuron i produced by the auto-associator network is computed and summed over the neurons for each of the k locations. For a feature search, the location with the target should produce a relatively small top-down value and a location with a distractor should produce a relatively large value. For a conjunctive search the differences between a target and distractor location would not be as large. Bottom-up and top-down information were standardized to equalize their variances and combined to produce a final attention signal for each location:

$$\text{Signal}_k = \text{Bottom-up}_k - \text{Top-down}_k \tag{9.22}$$

The vector representing the attention signal for all the locations with map symbols was then sorted to put the strongest signal first and the weakest last in the vector (Figure 9.21). The codes for the strongest to weakest location were then passed through the pattern associator model and it was cycled until the **Target** or **Not Target** neuron reached a threshold level to make a *Yes* or *No* decision (Figure 9.22). For a feature search producing a pop-out effect for a target on a trial map, the location with the target should be processed first, or nearly first, in the queue because that location should have the strongest attention signal. For a conjunction search not producing a pop-out effect for the target, the location with the target would not attract any more attention than any other location. If it had an average signal, the target location would not be processed until approximately half the other locations were processed. When the target pattern was processed, it would turn on the **Target** neuron and the decision would be made that the target was on the map. For trial maps without a target, the search would continue for both a feature or conjunctive search until it was hopeless. Cave and Wolfe (1990) suggested humans abandon searches before considering all possible locations based on

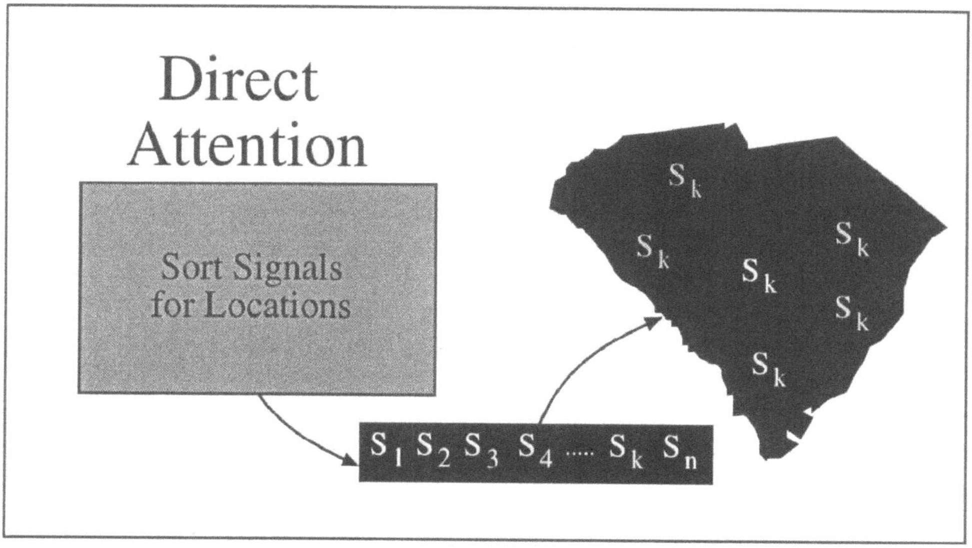

Figure 9.21: Attention is directed to the location with the strongest signal based on the combination of bottom-up and top-down information. Signals are sorted in an order from strongest to weakest.. After Lloyd (1996c).

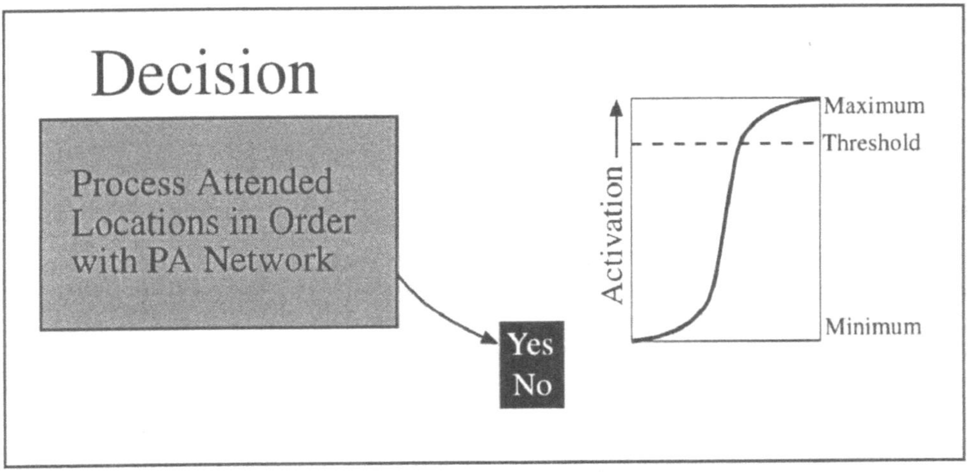

Figure 9.22: The final Yes or No decision regarding the presence of the target symbol on the map was made by processing the attended location through the pattern associator network that had learned the target. After Lloyd (1996c).

some internal expectation they have of their likely success given the effort they have already made. This notion was simulated by a random process that generated a number to be subtracted from the N locations to be considered.

Reaction Times

If the model successfully simulated the processes used to visually search for targets on maps, the times needed to make decisions should produce a predicted pattern. Each simulated trial was a search of a map having N symbols. On half the simulation trials the target was present at a randomly determined location and on the other half the target was not present. At the conclusion of each trial the decision regarding the presence of the target (*Yes* or *No*) was recorded as was the time needed to make the decision (Reaction Time). Feature integration theory argued that feature searches are the result of a parallel process that produce the same reaction times regardless of the number of symbols being considered (Treisman 1988). The reaction times for a feature search should be the same for maps with 6 symbols and 46 symbols. A conjunctive search should produce a different pattern of reaction times. Reaction times for *Yes* responses should be faster than *No* responses with the reaction time increasing as the number of map symbols increases. The rate of increase for the *No* responses should be approximately twice as fast as the *Yes* responses. Duncan and Humphreys (1989) reviewed many search experiments done with human subjects and reported the rates of change (slopes of regression lines) that were associated with parallel feature searches and serial conjunctive searches (Figure 9.23). They reported parallel searches with slopes

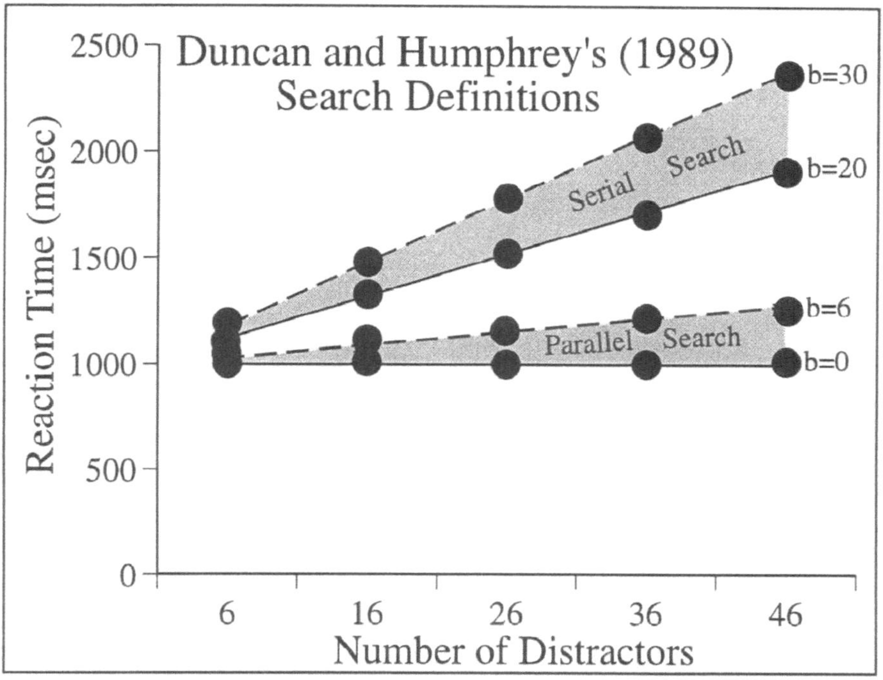

Figure 9.23: Summary of search results for experiments done with human subjects. After Duncan and Humphreys (989).

ranging from 0 to 6 seconds/distractor and serial searches with slopes ranging from 20 to 30 seconds/ distractor. Lloyd's (1996c) simulation results appeared to be within the ranges reported for human subjects. The feature searches for 10,000 trials were summarized with a reaction time versus number of distractors plot (Figure 9.24). The plot indicated some increase of reaction times with the number of distractors for the **No** answers, but the slopes appeared to be within the range reported for human subjects. The simulation results for 10,000 conjunctive search trials were summarized with a similar reaction time versus number of distractor plot (Figure 9.25). This plot indicated the expected increase of reaction times with number of distractors for both the *Yes* and *No* responses with the slope of the *No* responses being nearly twice as steep as the *Yes* responses. These results also appeared to be within the range of those reported for human subjects and suggested the hybrid neural network did a good job of simulating the visual search processes used by humans.

9.6. Pattern Recognition

The lower right corner of Figure 9.1 represents the fourth example that is an unsupervised classification problem. Being able to recognize patterns is a very basic

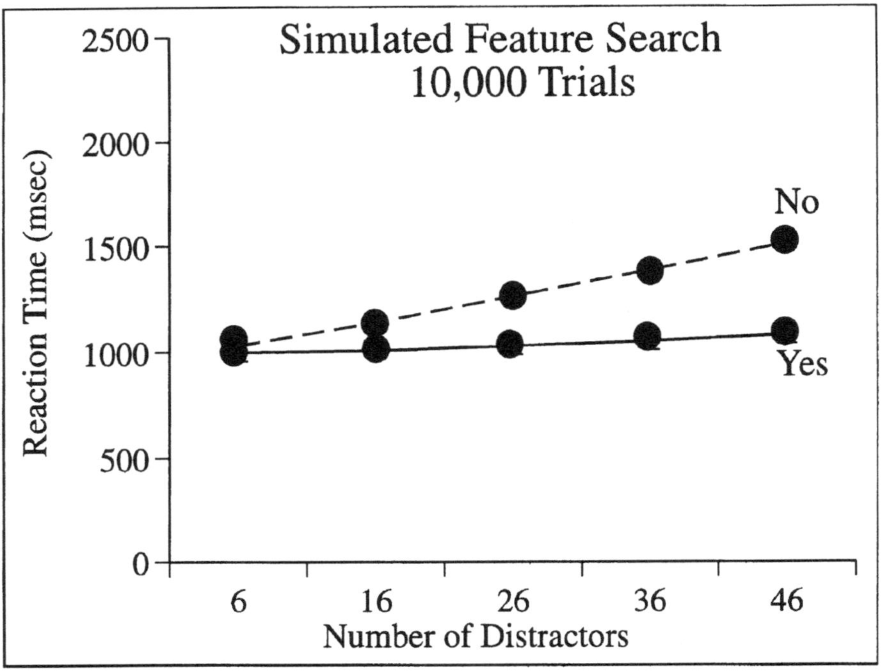

Figure 9.24: Plot of mean reactions times by number of distractors for the simulation of a feature search based on 10, 000 trials. (After Lloyd 1996c).

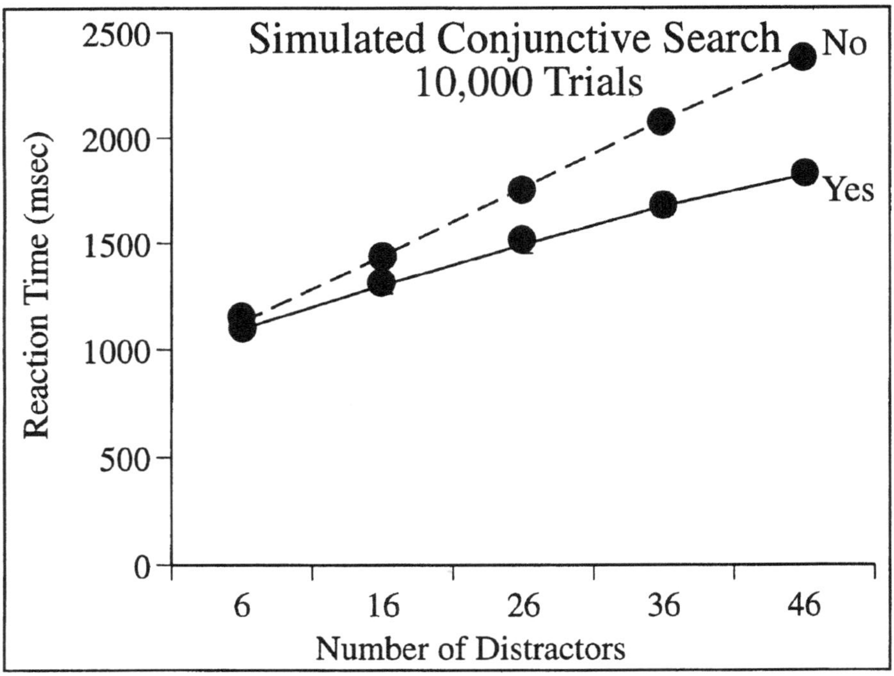

Figure 9.25: Plot of mean reaction time by number of distractors for the simulation of a conjunctive search based on 10,000 trials. After Lloyd (1996c).

visual ability. Humans learn to do this very early in life and continue to learn new patterns throughout life. Map symbols like those used in the previous section are familiar and simple, but useful for encoding information. The current example uses a Kohonen neural network to acquire categorical information (Caudill and Butler 1992a, 1992b; Clothiaux and Bachmann 1994; Kohonen 1989; Openshaw 1994; Winter and Hewitson 1994). This type of neural network is also called a self-organizing map. The network is self-organizing because it learns through a strictly bottom-up process that adjusts the connections in the network by considering only the patterns of the inputs. It is a map because learning takes place within a two-dimensional field of competing neurons frequently referred to as the Kohonen layer. The training procedure encourages neighbors in the space (Kohonen layer) to make similar connections with the input neurons. The self-organizing nature of the network is attractive because it simulates learning categories by detecting patterns in the data rather than by having an expert provide correct classifications for input patterns. Another good argument for this type of network structure is that the brain seems to operate in a similar way (Kosslyn and Koenig 1992). Neurons that are near each other in the Kohonen layer are encouraged by the learning process to make similar connections with neurons in the input layer. This produces a distance decay effect within the Kohonen layer and makes similar connection for neurons in the same region of the layer.

9.6.1. MAP SYMBOLS

The map symbols for this example were defined using color, shape, and orientation dimensions and the notion of a visual field. The term visual field is used here to refer to the area on a monitor containing a map symbol that a person might look at during a map-reading experience (Lloyd 1996c). The researched discussed in the previous section defined digital codes for map symbols in a simple way as just having a characteristic (1) or not having a characteristic (0). Lloyd (1996d) defined symbols as fields of visual inputs that implicitly defined characteristics such as color, shape, and orientation by activation levels of a field of neurons. The field was simply a five by five array of neurons that had two layers. This made a total of 5 X 5 X 2 or 50 input neurons available to define a map symbol. Decisions on the size of the field and number of layers were based on limitations of the software being used for the simulations (Caudill and Butler 1992a, 1992b) Each layer could by thought of as attached to a color gun in an RGB monitor. The two layers are used to define three colors for this example. Using three layers would allow the production of additional colors. Using a larger field size would also allow the manipulation of size as a basic characteristic. Lloyd (1996d) defined 12 symbols by activating some of the 50 neurons in the visual field and not activating other in the field (Figures 9.26 and 9.27). Half the neurons were designated the red layer and the other half of the neurons were designated the blue layer. If only neurons in the red layer were activated, then the map symbol had a red color (Figures 9.26a, 9.26b, 9.27a, and 9.27b). If only neurons in the blue layer were activated, then the map symbol had a blue color (Figures 9.26c, 9.26d, 9.27c, and 9.27d). If neurons in

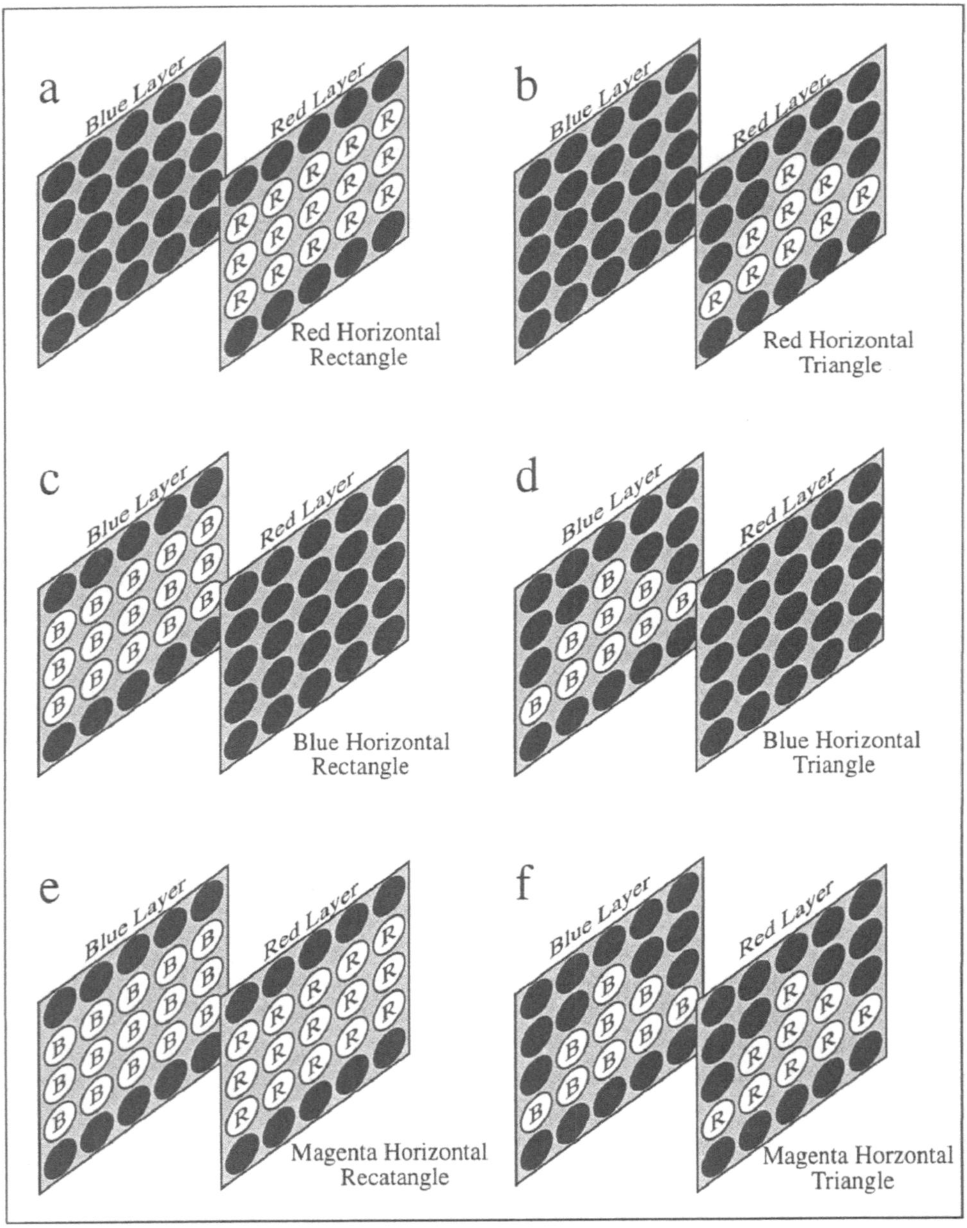

Figure 9.26: The 12 horizontal map symbols learned by the Kohonen neural network.

the same locations were turned on in both layers, then the map symbol had a magenta color (Figures 9.26e, 9.26f, 9.27e, and 9.27f). Shape and orientation were defined by the spatial arrangement of the activated neurons within the layers. Shape was defined within the visual field as a rectangle (Figures 9.26a. 9.26c, 9.26e, 9.27a, 9.27c, and 9.27e) or as a triangle (Figures 9.26b. 9.26d, 9.26f, 9.27b, 9.27d, and 9.27f). Orientation was defined within the visual field as horizontal (Figure 9.26) or vertical

(Figure 9.27). Lloyd (1988) reported that subjects who first viewed a map and then determined if target symbols had been on the map were able to use a parallel process to

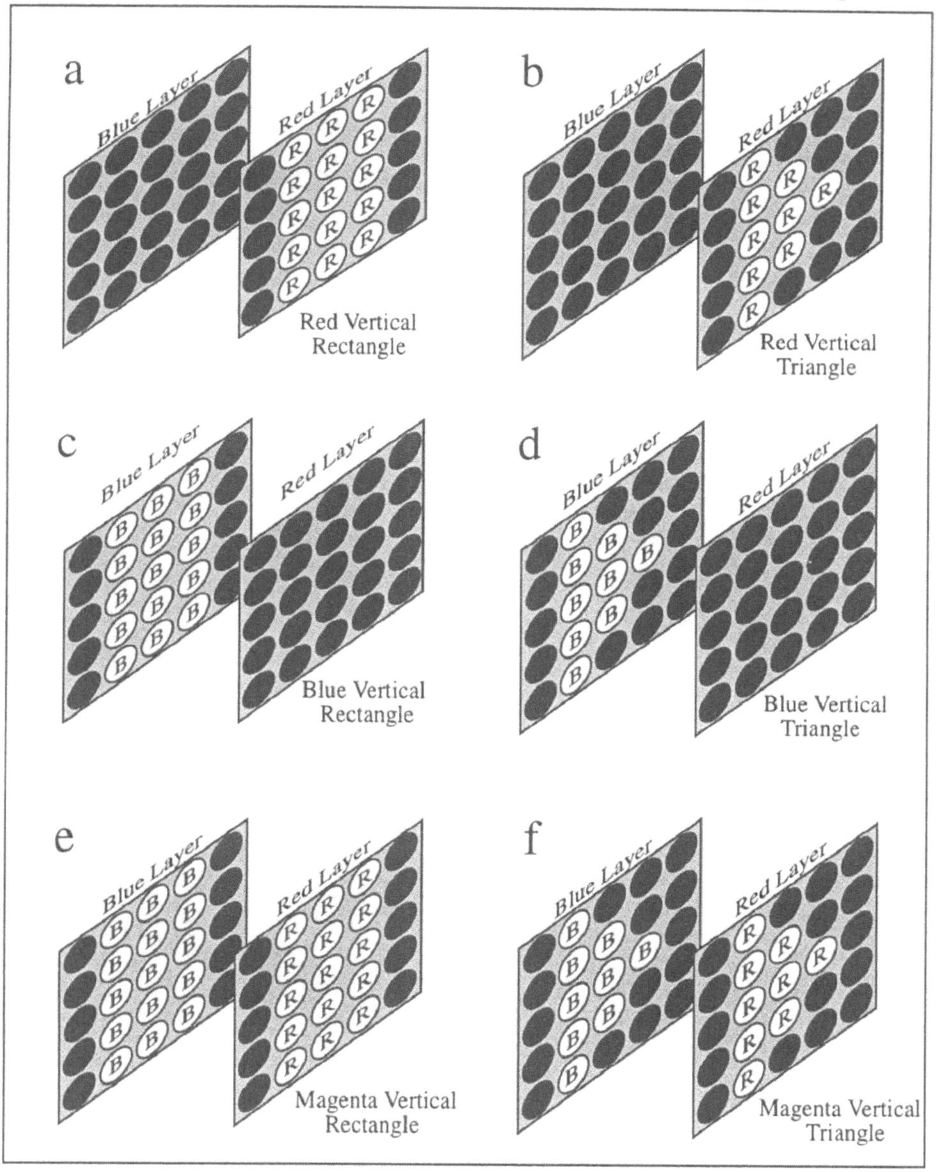

Figure 9.27: The 12 vertical map symbols learned by the Kohonen neural network.

make their decisions. This was indicated by the relationship between decision times and the number of symbols on trial maps. Subjects could answer in the same amount of time regardless of the number of symbols on the map. Processing times were equal when subjects had to remember maps with 4 or 24 symbols. The process used by the map readers in Lloyd's (1988) experiment can be simulated with a Kohonen neural network.

9.6.2. KOHONEN MODEL

Lloyd's (1996d) Kohonen model learned the 12 map symbols discussed above (Figures 9.26 and 9.27). Once the model had learned to classify the visual fields as separate map symbols, it could identify any of them with little effort or processing time. The input layer had 50 neurons organized in a 5 by 10 array (Figure 9.28). The 25 on the left side of the array represented the red layer of the visual field and the 25 on the right side represented the blue layer of the visual field. Digital codes for the 12 map symbols were presented to the input layer as binary data. Any of the input neurons could take on a value of 1 or 0. For example, the map symbol represented in Figure 9.26c is a blue horizontal rectangle. It is represented as binary digital codes for the input layer as 1's in the second, third, and fourth rows of the right side of the input layer with all other locations being set to 0 (Figure 9.28).

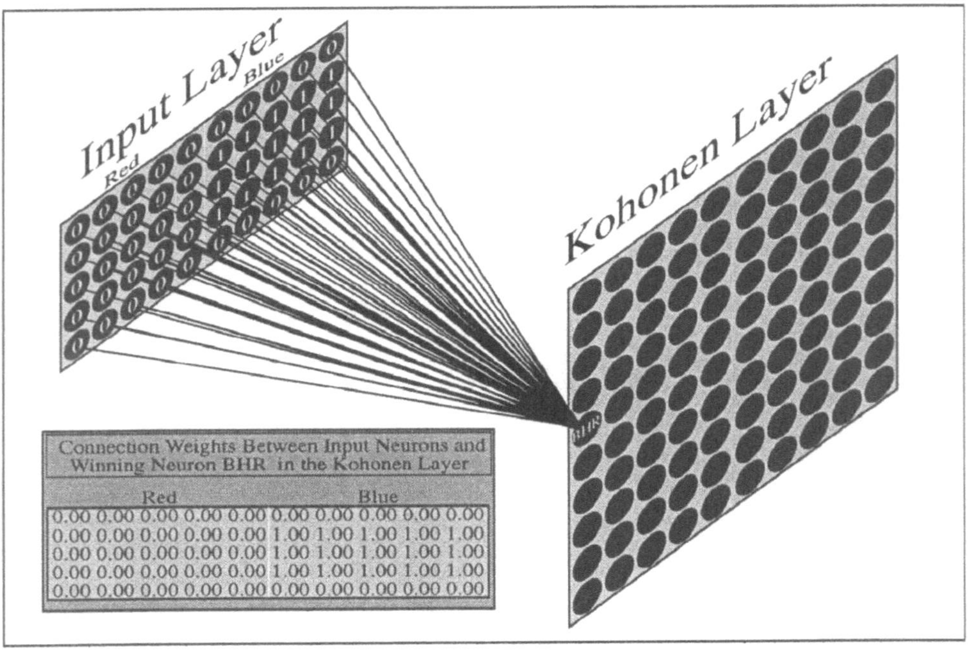

Figure 9.28: The Kohonen neural network model used to learn the map symbols. A blue horizontal rectangle is shown in the input layer. Connections from the input neurons to the neuron in the Kohonen layer that learned the map symbols are also shown.

The Kohonen layer was defined as 100 neurons organized as 10 X 10 two-dimensional space. The Kohonen model was very simple. Each of the *i* neuron in the Kohonen layer had a connection to each of the *j* neurons in the input layer. There were no other connections among neurons within either the input or Kohonen layers. There were, therefore, 50 X 100 connections that were initially set to some small positive random number. When a map symbol was presented as a pattern to the input layer each of the neurons in the Kohonen layer used these inputs to compute an activation:

$$A_i = \Sigma_j \, w_{i,j} \, I_j \tag{9.23}$$

The current weight connecting neuron *i* in the Kohonen layer to neuron j in the input layer was multiplied by the current value for the input neuron *j* and summed over all the connections. Since the weights were initially determined by a random process, any neuron could have the highest activation on the first pass. The winner-takes-all rule was used for adjusting the weights. For the winning neuron and its designated neighbors the weights were changed using a variation of the delta rule:

$$\Delta \, w_{i,j} = \beta \, (I_j - w_{i,j}) \tag{9.24}$$

The change made to the weight was the learning constant (0.2) times the difference between the value for in input neuron and the weight connecting it to the neuron in the Kohonen layer. Error in this model was defined as a difference between an input and a weight. When the model had learned the map symbol, there would be no difference and the weights connecting the input layer to the winning neuron in the Kohonen layer would be a virtual representation of the input pattern. Neighbors were defined initially as any neuron within five or fewer steps from the winning neuron. After every two epochs of training this definition was decreased by one step until only the winner was being updated. This procedure makes the neighbors of winners have weights with similar values. Any neuron not considered a winner or a neighbor on a pass was not updated:

$$\Delta w_{i,j} = 0.0 \tag{9.25}$$

The network was trained for 36 epochs until the average distance between the input values and the weights of winning neurons were less than a minimum threshold value (0.01). After training was completed 1 of the 100 neurons in the Kohonen layer would have the highest activation (turn on) when a map symbol was presented to the network. Within the Kohonen layer 12 of the neurons had become connected with the 12 map symbols through their weights that replicated the patterns. For example, the neuron at row 5 and column 1 learned the blue horizontal rectangle map symbol. Its weights have taken on values that suggest a virtual image of the map symbol (Table 9.3). When winners were mapped according to the characteristics of the pattern they had learned some regions can be detected within the Kohonen layer (Figure 9.29). The most obvious pattern was for orientation with vertical symbols in one corner of the space and horizontal symbols in the opposite corner. The specific pattern was of little interpretative value, however, because patterns in the Kohonen layer are partially a

Table 9.3: Weights for neuron at row 5 and column 1 that learned the blue horizontal rectangle map symbol.

Red					Blue				
0.000	0.000	0.000	0.000	0.000	0.000	0.000	0.000	0.000	0.000
0.000	0.000	0.000	0.000	0.000	0.999	0.999	1.000	0.999	0.999
0.000	0.000	0.000	0.000	0.000	0.999	1.000	1.000	1.000	0.999
0.000	0.000	0.000	0.000	0.000	0.999	1.000	1.000	0.999	0.999
0.000	0.000	0.000	0.000	0.000	0.000	0.000	0.000	0.000	0.000

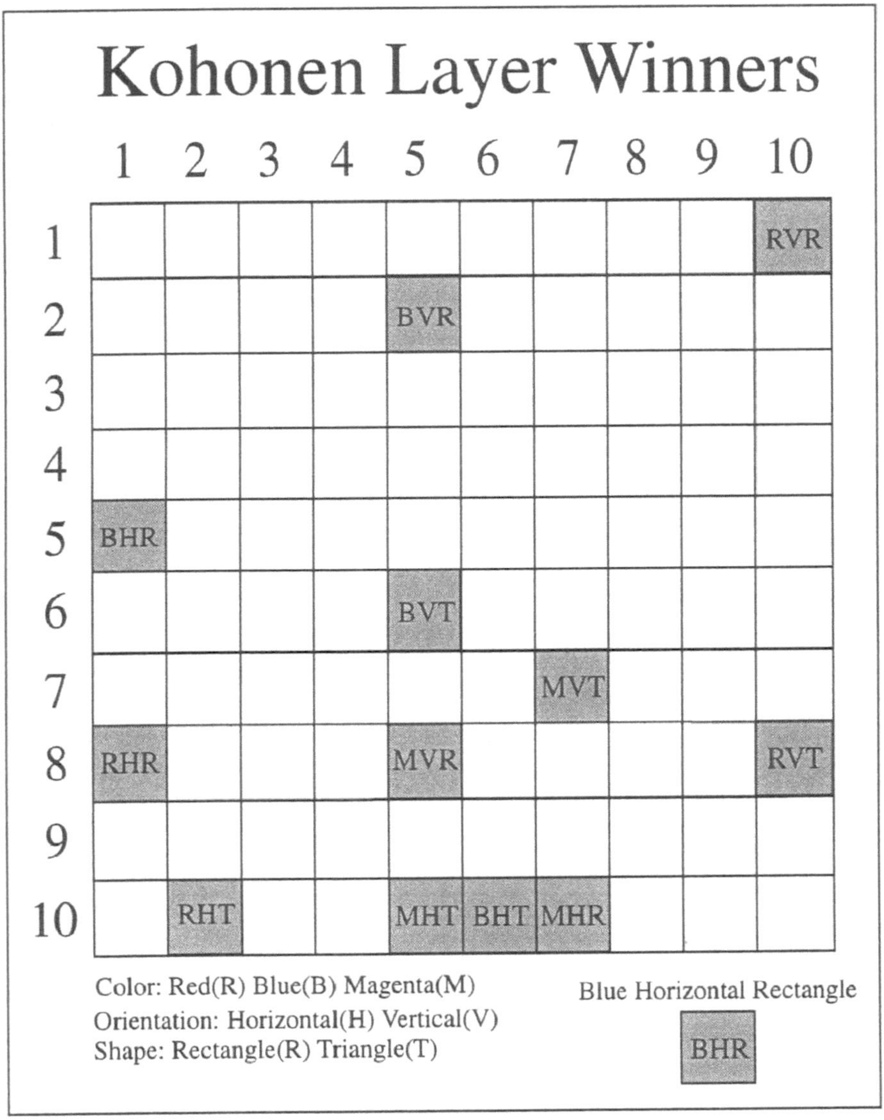

Figure 9.29: The winning neurons in the Kohonen layer after training was completed.

product of the initial random values assigned to the weights. Training the model with a new set of initial values is likely to result in 12 different winning neurons. This is similar to what would happen if a number of humans learned the map symbols. Which neuron is not as important as the fact that learning always occurs and some neuron learns to replicate each of the input patterns. Once the connections between the input and Kohonen layer are learned for a set of map symbols the network can later consider any symbol as input by processing its values through the network. If the new symbol resembles one of the learned symbols it will produce the highest activation for the symbol in the Kohonen layer associated with the learned symbol. This allows the identification of exemplars for a category that are not prototypical examples of the ccategory.

9.7. Conclusions

This chapter has demonstrated the construction of neural network models for a variety of geographic problems. Models that could be used for making predictions and performing classifications were illustrated as were models that learned through supervised and unsupervised training. The next and final chapter summarizes the central themes of the book and looks at future directions in spatial cognition.

CHAPTER 10

CONCLUSIONS

10.1 Introduction

The discipline of geography traditionally has been defined to include a wide range of interests in both the physical environment and the human activities taking place in the environment. A simple connectionist definition of geography might be the study of how people are connected to the geographic space serving as their environment. Although elements of physical and human geography can be studied in isolation, any approach that focuses on the connections between humans and their environments offers a holistic integrated perspective for geographic research. The theories and methods of cognitive science that have been discussed in the previous chapters offer such a strategy. Geographic researchers using this research strategy make the important connection between physical and human geography by focusing on how humans process information about environments to determine their behaviors. These cognitive explanations of human behavior in geographic environments should include statements of what environmental information was important for decision making, how the information was acquired, how the information was stored in human memory, and how it was processed to make the decisions that directed behavior.

10.2 Summary of Ideas

Martindale's (1991) model of the order and connections among cognitive processes that was used to explain map reading in Chapter 2 is redrawn in this chapter as the top part of Figure 10.1. This figure was designed to summarize some of the important ideas presented in this book. The model starts with input from a physical stimulus and ends with the output of some overt human response. The information from the environment is processed along the way, transformed, and passed on by sensory, perceptual, and conceptual analyzers. The information also interacts with previously stored information and may itself be abstracted and stored in some form in long-term memory. In this way the current information becomes part of our knowledge of the environment and can influence future decision making. There are six levels (a through f) below the model that illustrate some of the import ideas discussed in this book. These levels are organized similar to Martindale's (1991) model with the left (black) part of a level representing the processes used to acquire and store geographic information. These processes result in spatial information being stored in long-term memory and are done with systems that are sensory, perceptual, and conceptual analyzers. Each level represents a general (a) or specific task (b-f) that was discussed in previous chapters of this book. The middle (white) part of a level represents a decision being made using the information available from the earlier processes. Specific tasks represented in the levels have a miniature version of a figure used to illustrate either an experimental trial requiring a human subject's decision (c and f) or a neural network designed to simulate a decision for the particular task (b, d, and e). The right part (grey) of each level indicates the general (a) or particular (b-f) behavior in space that might be associated with the processes represented in that level's example. Viewing the model or

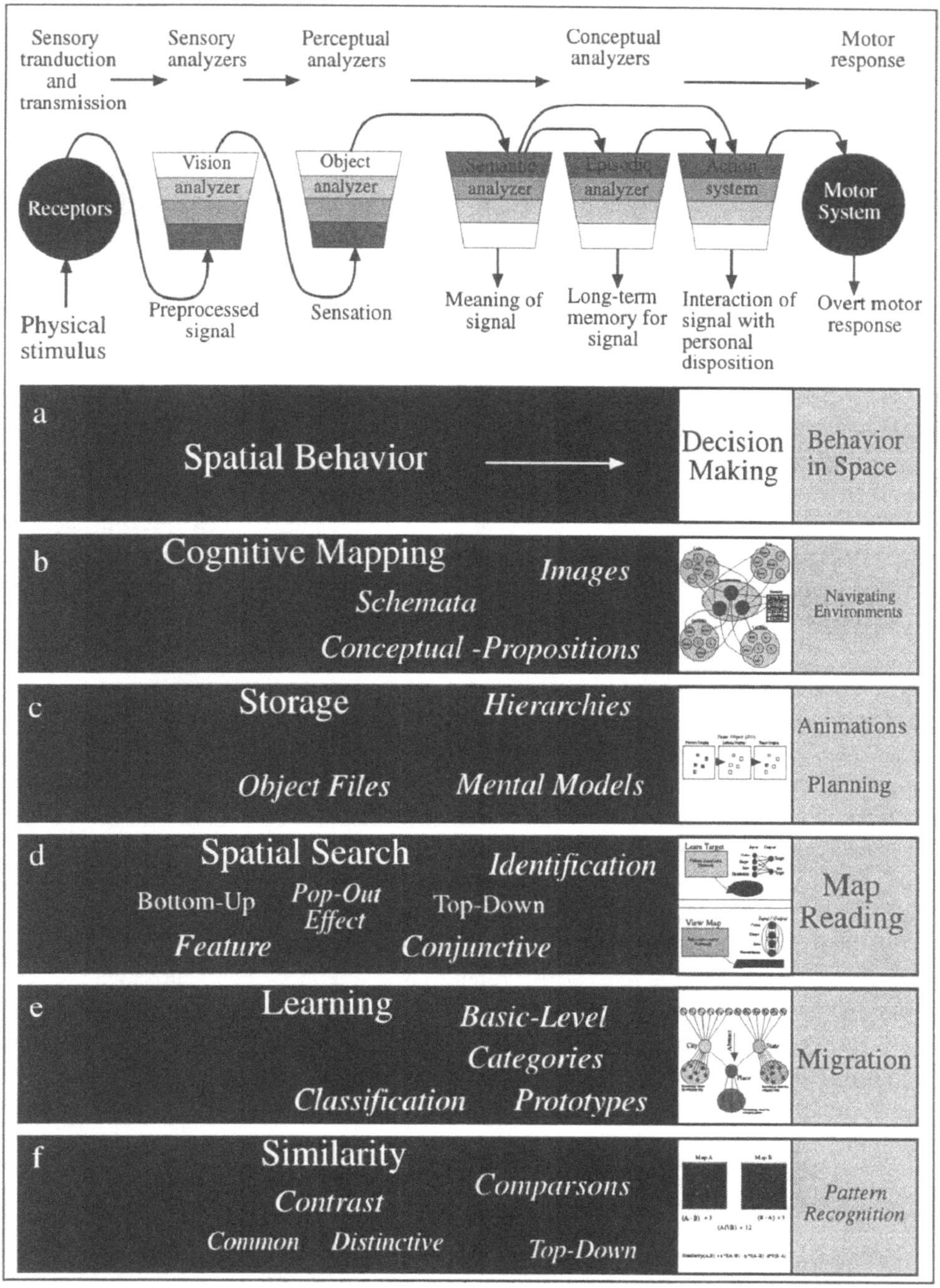

Figure 10.1: The top of the figure represents Martindale's information processing model and the levels below represent ideas discussed in previous chapters.

the levels in Figure 10.1 from left to right one can think of the order of processing over time. Information is initially received from the environment on the far left of the figure. The information is processed by a series of sensory, perceptual, and conceptual analyzers to be stored in long-term memory and/or used to make a decision causing some overt behavior on the far right. Most of the ideas discussed in this book have a locus in Figure 10.1 in the middle (perceptual analysis) or right (conceptual analysis) of the left side (black) of a level. For example, object files relate to perceptual processing and mental models relate to conceptual structures in long-term memory. Some processes, e.g., schemata, identification, or classification, combine information in long-term memory with information being currently perceived.

10.2.1 SPATIAL BEHAVIOR AS RULES FOR DECISION MAKING

The earliest behavioral geographers made arguments similar to those presented in the introduction of this chapter. One of the most influential arguments was made by Rushton (1969) and is represented here as the general example for the model (Figure 10.1a). Rushton argued for a distinction between *spatial behavior* and *behavior in space*. He argued that *spatial behavior* should be thought of as the *rules* that are used to make decisions and *behavior in space* as the overt expressions of those decisions, e.g., trips to the grocery store. A fair question might be, "How do people acquire such rules for decision making?" Although people may behave as though they are following some rule, a connectionist or wet mind version of Rushton's idea can support the same argument (Kosslyn and Koenig 1992). The general notion is that sensory, perceptual, and conceptual processes have prepared our biological neural networks to make decisions by processing relevant information. Experiences with environmental information have trained our neural networks to make these decisions. The collection of experiences that has trained someone's neural network is the *spatial behavior* that has prepared that person for decision making. The experiences could be primary experiences in the environment or secondary experiences such as map reading or getting advice from a friend or a book. Each decision produces some overt behavior and contributed to the further training of the neural network for making subsequent decisions. The *spatial behavior* that Rushton (1969) studied was the common decision-making rule for grocery shopping used by farmers in Iowa. Individual rules were produced by the farmers' interactions with grocery stores in the central places of Iowa. These primary and secondary grocery-shopping experiences trained the farmers' neural networks to make satisfactory decisions. The apparent rules used for decision making emerge because the connections between neurons in the farmers' neural networks stabilized after learning to make satisfactory decisions.

10.2.2. COGNITIVE MAPPING

With and without conscious effort we continuously encoded information about the environment into our memories (Figure 10.1b). Although the current information we have encoded in our memories may not be cartographically accurate, it should be accurate enough to produce successful navigation decisions after a reasonable number of environmental experiences. A connectionist view of cognitive mapping involves the training and use of a neural network structured in long-term memory that stores important environmental information. McClelland and Rumelhart's (1989) Interactive Activation Competition (IAC) neural networks (Figure 3.1) are an attractive starting point for modeling cognitive maps because they represent associative memory well. This allows a simulation to connect into the memory structure at any location in the network. Thinking about a landmark or a characteristic for a landmark initiates the spread of activation in the network to simulate the retrieval of important information. Navigation behavior frequently requires only a simple decision to determine the next overt act, e.g., should I turn left or right at the intersection? On-the-ground experiences like these are usually inspired by some landmark object that is currently being viewed. Seeing this landmark automatically activates neurons in the network associated with the object and, through the various connections, other related information is also activated. The more experiences you have with an environment the greater the number of connections and the more likely the information needed to make a satisfactory decision will be activated. You do not need to activate the entire environment to successfully navigate, but just enough to make the current decision. You are officially lost when you are viewing landmark objects that fail to activate any useful information for making a confident decision regarding your next movement.

Some advanced form of neural network might also be useful someday for thinking about environments from multiple perspectives and constructing generalized plans. Such advanced networks could allow multiple neurons to be activated and their influence to spread throughout the network. This could be useful for simulating making plans when you are not actually navigating in the environment, but just thinking about options that might be possible for some future experience. People can easily do this with their biological neural networks. Suppose you have to go to the airport on the other side of the city the next morning and are planning the best route the night before your trip. You can image the direct route through the center of the city and the route on the outerbelt to determine which is likely to be faster during morning traffic. You can think about places along both routes where you might stop to get money from a drive-through bank machine or where you might get breakfast. You can estimate the time you would need to depart home to travel the routes, make the stops, and get to the airport one half hour before the plane leaves. Although such planning is easily done by people familiar with their home cities, such high-level planning activity would be an impossible challenge for current artificial neural networks. Since it is not impossible to imagine a neural network with enough neurons and connections to encode all the significant landmarks in a city, a geographic information system that can process such planning problems without human direction might be constructed someday. Such a sophisticated GIS should be able to generate from its network of neurons and connections both images that depict the environment and conceptual-propositions that describe the environment to provide the traveler with the best route to the airport.

10.2.3. STORING INFORMATION

The connectionist view of the information processing done by the human brain is quite different from the software programs, hard drives, and central processing units that are included in most current discussions of geographic information processing (Figure 10.1c). The external view is that the human brain is very complicated because it has 100 billion neurons and quadrillions of connections all packed into three pounds of meat. The information in a neural network is distributed throughout the network in the connections between neurons and the processing is done in parallel. When the network processes information, the connections produce patterns of activations that have some particular meaning to the owner of the network. The internal view is that the human brain is relatively simple. Each neuron is an individual processing unit that collects information from other neurons sending it signals, produces an activation from the these inputs, generates its own output, and passes it to other neurons that receive its signal. The neural networks demonstrated in Chapter 9 and in other chapters of the book have been relatively simple and directed at very specific problems they were specifically designed to solve. The efficient storage of information is the greatest problems to be overcome before a GIS can be based on a neural network. The GIS would need to be able to generalize information from experiences and permanently store important knowledge it has learned. To be efficient, this knowledge could be stored as a model of the environment similar to a mental model created by a person and make use of categories and hierarchical structures. The GIS should be able to combine learned knowledge with new information being processed and spontaneously make decisions without the assistance of a human operator. Like humans, the GIS should know what tasks need done, when to do them, and how to do them.

10.2.4. SEARCHING

Multiple decisions are made during an active search for a target in the environment or on a map. The most important decision is identifying an object being viewed as the target or as a nontarget. This involves a comparison of information related to the target encoded in long-term memory with the information currently being viewed for an attended object. The information matches or it does not to produce a decision. Another important decision is related to *attention*. What should the search look at next? Some searches are almost effortless processes because the target pops out of the background. If the target has a unique feature, it demands our attention. A feature search is shown as a sensory analysis on the left side of Figure 10.1d because searchers are thought to be aware of unique feature before focused attention. Conjunctive searches are positions farther to the right in Figure 10.1d because attention is directed by a combination of bottom-up information from the environment or map and top-down information in long-term memory when the target does not have unique features. Another decision in a search process is the decision to abandon the search. This decision can be conceived as completing the analyses directed at identifying the target, but never making a satisfactory match between information related to an object currently being perceived and information encoded in long-term memory related to the target. Having eliminated all the objects that have potential for being the target or having exhausted the time you are willing to devote to the search, the search is abandoned for more interesting activities that deserve your attention.

10.2.5. LEARNING

Everyone has had a variety of learning experiences and probably would say they know what the term *learning* means (Figure 10.1e). Almost everyone has been part of formal learning processes as students (and sometimes teachers) in education systems. Certainly we all have been part of informal learning processes as children and adults. The connectionist view of learning is very simple. It is the changes in the connections between neurons. When the strengths of these connections change, learning occurs. Changes in connections can occur that result in information processing that is more or less successful. Usually changes in connections (learning) are directed by some goal that marks its success. With this additional constraint, learning can be defined as changes in the strength of connections between neurons that produce a positive outcome for the process. We constantly process information. We can have a significant amount of processing without significant learning. Once the connections between neurons have changed to produce a successful result consistently, the connections should stabilize because there is little incentive to change, i.e., there is nothing more to learn. We learn to classify exemplars and create prototypes for categories by experiences with exemplars of the categories. We generalize abstract patterns in the information and create basic-level categories that can be used efficiently for communication. We learn to make good decision or to execute a skilled behavior through experience and practice.

10.2.6. JUDGING SIMILARITY

A cognitive science approach can be used to address questions traditionally answered using other methods. For example, the similarity of patterns on maps can be measured with statistical methods such as the standard produce-moment correlation coefficient (Figure 10.1f). As discussed in Chapter 8, this approach works well provided the goal is not to explain human behavior. An explanation of the visual similarity experienced by human map readers requires a different approach that explicitly includes knowledge of how humans process visual information to judge similarity (Goldstone 1994; Lloyd, Rostkowska-Covington, and Steinke 1996; Tversky 1977). When we experience an object, our awareness of the object automatically activates other similar objects and related information stored in the network. This happens because similar objects have similar connections in the network. Two geographic objects like Alabama and Mississippi should have similar connections in most people's neural networks. This is because they are connected in space as neighbors with a common boundary, connected to the same region (The South), connected to the same historical events, economy, culture, and basic-level category (state). The same is true to a degree for any other geographic objects closely located in space (Tobler 1970). Similar objects produce similar activations in the neural network (Cammack and Lloyd 1993).

10.2.7 NEURAL GEOGRAPHIC INFORMATION SCIENCE (NGIS)

Speculations about the future are expected at the closing of a book like this one. This final section of the book continues that tradition. Research based on cognitive science and research based on traditional geographic information processing, as done with a GIS, have some common and distinct features. Both are focused on processing information. Cognitive scientists consider subjectively experienced information processed by humans and a GIS usually considers objectively measured information processed by computer programs. The processing and storage of information are distinctly different. Humans acquire information (learn) by experiences that guide changes to connections between neurons in neural networks. A GIS program acquires information by reading number out of a file stored on a hard drive. Humans tend not to store the details of their experiences, but generalize large volumes of information by creating abstract categories in hierarchical structures, e.g., basic-level categories, that store the essence of what has been learned. A GIS does not learn in the same sense; it is provided new raw data from a different file. Humans make intuitive decisions that are not based on equations, objective information, or sound theoretical principles. A neural network trained by experiences has produced connections that reflect those experiences. When later asked to provide a decision, the network produces activations that also reflect those same experiences.

What advantages might there be in having a neural geographic information system? One big advantage is that the NGIS could learn from its experiences. This would also mean the NGIS could automatically store what it has learned for future use. This idea has come up in a number of contexts throughout the book. One example is the distinction between top-down and bottom-up information to process a visual search. Top-down information (the nature of the target) has been previously learned and influences the visual search for a target on a map that provides the bottom-up information. A better example might be Kahneman and Treisman's (1984) distinction between *types* and *tokens*. It would be an advantage for an NGIS to store common objects in it's long-term memory as *types* rather than storing *tokens* (Figure 10.2). *Types* would be permanent categories of environmental objects that have defined characteristics. The descriptions would function like prototypes of the category. The definitions should be defined by the experiences of the system with that particular category of object and could evolve with experience. *Tokens* are all objects being processes in an input scene. They would be treated like object files as temporary entities that are not stored in the system. No information needs to be permanently stored about them except that they are objects. As it is encountered, a particular *token* automatically could be compared to long-term memories of *types* to identify it a as a member of a category . For example, the *tokens* in Figure 10.2a have their characteristics stored as layers of information. Each *token* could be processed with a neural network to learn *types* that could be permanently stored for future use (Figure 10.2b).

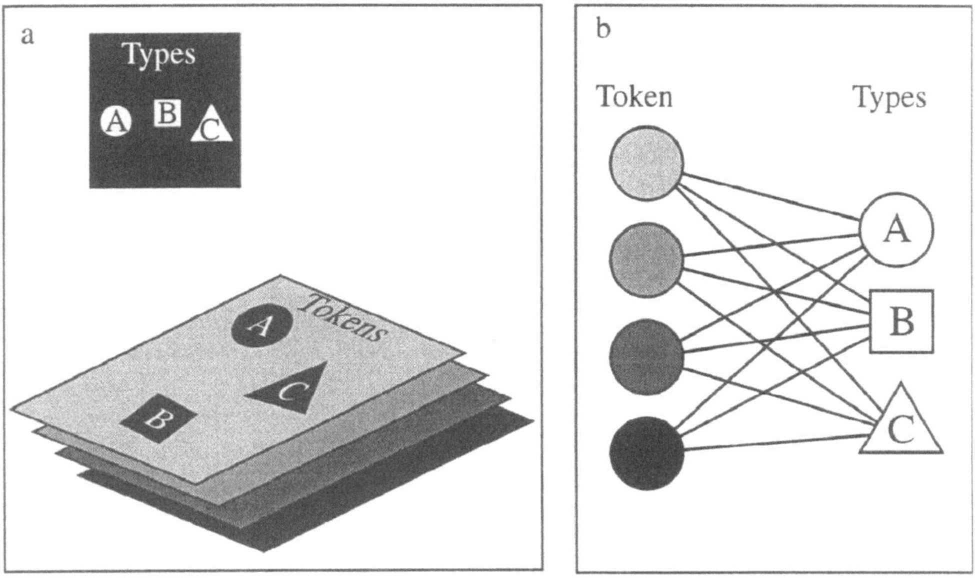

Figure 10.2: An example of objects in a display (*Tokens*) represented by multiple layers of information (a) being used to learn object categories (*Types*) by a neural network (b).

Another advantage of storing learned knowledge in an NGIS is that creativity would be possible. Computer programs that reside in a GIS execute decisions made by the operator of the system on data files also selected by the operator. If spatial knowledge were learned and stored in an NGIS, this generalized knowledge could be accessed from a variety of locations within the system to generate activity within the network. This would provide many different perspectives for the same problem and possibly create new solutions.

Another useful advantage of storing generalized information learned from experiences is that a particular NGIS could become skilled at solving particular problems. Like human experts, the more experiences it could have with a problem the more it could learn and the more skilled it would become over time. For example, suppose the NGIS was to become an urban landuse specialist. It could analyze patterns for objects in many different urban areas and store this knowledge in its long-term memory. This is quite different from just having the data available for analysis stored on the hard drive.

Finally, intuitive decisions could be made by the NGIS for small tasks that usually have to be made by the human operator. Simple examples are decisions that cartographers might consider artistic skills. For example, objects in the environment can be included or deleted when environments are represented on maps at different scales. The details in the lines used to represent rivers and political boundaries also have to be adjusted to accommodate scale changes. These types of decision-making skills, that cartographers learn by making maps, could also be learned by the NGIS as it produces maps for viewing.

Note that it has not been argued that any of the ideas discussed above would be easily implemented. It may be naive even to think an NGIS can learn to do anything a human can do. It also has not been argued when these ideas might be implemented. If enough geographers come to appreciate the possibilities of cognitive science and work toward establishing and reaching common goals, the connections between humans and their environments should have an interesting future.

References

Anderson, J. (1978) Arguments concerning representations for mental imagery, *Psychological Review* **85**, 249-277.

Anderson, J. (1985) *Cognitive Psychology and Its Implications*, W. H. Freeman, New York.

Anderson, J. and Bower, G. (1973) *Human Associative Memory*, V.H. Winston and Sons, New York.

Ashby, F. and Gott, R. (1988) Decision rules in the perception and categorization of multidimensional stimuli, *Journal of Experimental Psychology: Learning, Memory, and Cognition* **14,** 33-53.

Ashby, F. and Perrin, N. (1988) Toward a unified theory of similarity and recognition, *Psychological Review* **95,** 124-50.

Baddeley, A. (1986) *Working memory*, Clarendon Press, Oxford.

Ben-Shakhar, G. and Gati, I. (1987) Common and distinctive features of verbal and pictorial stimuli as determinants of psychophysiological responsivity, *Journal of Experimental Psychology: General* **116**, 91-105.

Berlin, B. and Kay, P. (1969) *Basic color terms: Their universality and evolution*, University of California Press, Berkeley.

Biederman, I. (1987) Recognition-by-components: A theory of human image understanding, *Psychological Review* **94**, 115-147.

Blakemore, C. (1970) The representation of three-dimensional visual space in the cat's striate cortex, *Journal of Physiology* **209**, 155-178.

Brennan, N. and Lloyd, R. (1993) Searching for boundaries on maps: The cognitive process, *Cartography and Geographic Information Systems* **20**, 222-236.

Briggs, R. (1973) Urban cognitive distance, in R. Downs and D. Stea (eds.), *Image and Environment: Cognitive Mapping and Spatial Behavior*, Aldine, Chicago, pp. 361-388.

Bryant, D. and Tversky, B. (1992) Assessing spatial frameworks with object and direction probes, *Bulletin of the Psychonomic Society* **30**, 29-32.

Bryant, D., Tversky, B., and Franklin, N. (1992) Internal and external spatial frameworks for representing described scenes, *Journal of Memory and Language* **31**, 74-98.

Bundesen, C. and Pedersen, L. (1983) Color segregation and visual search, *Perception and Psychophysics* **33**, 487-493.

Bunting, T. and Guelke, L. (1979) Behavioral and perception geography: A critical appraisal, *Annals of the Association of American Geographers* **69**, 448-462.

Busemeyer, J. and Myung, I. (1988) A new method for investigating prototype learning, *Journal of Experimental Psychology: Learning, Memory, and Cognition* **14,** 3-11.

Byrne, R. (1979) Memory for urban geography, *Quarterly Journal of Experimental Psychology* **31**, 147-154.

Byrne, R. and Johnson-Laird, P. (1989) Spatial reasoning, *Journal of Memory and Language* **28**, 564-75.

Cadwallader, M. (1996) *Urban geography: An analytical approach*, Prentice Hall, Upper Saddle River, N. J.

Cammack, R. (1995) The cognition of cartographic animation with object files, unpublished Ph.D. dissertation, Department of Geography, University of South Carolina.

Cammack, R. and Lloyd, R. (1993) Connected space: Regional neural networks. Paper presented at the Annual Meeting of the Association of American Geographers, Atlanta, Georgia.

Canter, D. and Tagg, S. (1975) Distance estimations in cities, *Environment and Behavior* **7**, 59-80.

Carpenter, P. and Just, M. (1986) Spatial ability: An information processing approach to psychometrics, in R. Sternberg (ed.), *Advances in the Psychology of Human Intelligence,* Erlbaum, Hillsdale, NJ, pp. 221-252.

Carter, R. (1982) Visual search with color, *Journal of Experimental Psychology: Human Perception and Performance* **8**, 127-136.

Caudill, M. and Butler, C. (1992a) *Understanding Neural Networks, Computer Explorations, Volume 1: Basic Networks*, MIT Press, Cambridge, MA.

Caudill, M. and Butler, C. (1992b) *Understanding Neural Networks, Computer Explorations, Volume 2: Advanced Networks*, MIT Press, Cambridge, MA.

Cave, K. and Wolfe, J. (1990) Modeling the role of parallel processing in visual search, *Cognitive Psychology* **22**, 225-71.

Chernoff, H (1973) The use of faces to represent points in k-dimensional space graphically, *Journal of the American Statistical Association* **68**, 361-368.

Civco, D. (1993) Artificial neural networks for land-cover classification and mapping, International *Journal of Geographic Information Systems* **7**, 173-186.

Clothiaux, E. and Bachmann, C. (1994) Neural networks and their applications, in B. Hewitson and R. Crane (eds.), *Neural Nets: Applications in Geography*, Kluwer, Dordrecht.

Cohen, R. (1977) *The Psychology of Cognition,* Academic Press, New York.

Cohen, R., Baldwin, L., and Sherman, R. (1978) Cognitive maps of naturalistic settings, *Child Development* **49**, 1216-1218.

Cohen, A. and Rafael, R. (1991) Attention and feature integration: Illusory conjunctions in a patient with a parietal lobe lesion, *Psychological Science* **2**, 106-110.

Conerway, V. (1991) *The Effects of Complexity on the Mental Rotation of Map Images,* MA Thesis, University of South Carolina.

Couclelis, H., Golledge, R., Gale, N., and Tobler, W. (1987) Exploring the anchor point hypothesis of spatial cognition, *Environmental Psychology* **7**, 99-122.

Cox, K. and Golledge, K. (1969) *Behavioral Problems in Geography: A symposium,* Northwestern University Press, Evanston, IL.

Craik, F. and Lockhart, R. (1972) Levels of processing: A framework for memory research, *Journal of Verbal Learning and Verbal Behavior* **11**, 671-684.

Dent, B. (1972) Visual organization and thematic map communication, *Annals of the Association of American Geographers* **62**, 79-93.

DeSarbo, W., Johnson, M., Manrai, A., Manrai, L., and Edwards, E. (1992) Tscale: A New Multidimensional Scaling Procedure Based on Tversky's Contrast Model, *Psychometrika* **57**, 43-70.

Dobson, M. (1973) Choropleth maps without class intervals?: A comment, *Geographical Analysis* **5**, 358-560.

Dobson, M. (1980) Comment: perception of continuously shaded maps, *Annals of the Association of American Geographers* **70**, 106-108.

Dobson, M. (1983) Visual information processing and cartographic communication: The utility of redundant stimulus dimensions, in D. Taylor (ed.), *Graphic Communication and Design in Contemporary Cartography,* Wiley, New York.

Downs, R. (1979) Critical appraisal or determined philosophical skepticism? *Annals of the Association of American Geographers* **69**, 468-471.

Downs, R. (1981) Maps and metaphors, *Professional Geographer* **33**, 287-293.

Downs, R. and Stea, D. (1973) *Image and Environment: Cognitive Mapping and Spatial Behavior,* Aldine, Chicago.

Downs, R. and Stea, D. (1977) *Maps in minds: Reflections on cognitive mapping*, Harper and Row, New York.

Duncan, J. and Humphreys, G. (1989) Visual search and stimulus similarity, *Psychological Review* **96**, 433-458.

Duncan, J. and Humphreys, G. (1992) Beyond the search surface: Visual search and attentional engagement, *Journal of Experimental Psychology: Human Perception and Performance* **18**, 578-588.

D'Zmura, M. (1991) Color in visual search, *Vision Research* **31**, 951-966.

Eastman, R. (1985) Graphic organization and memory structures for map learning, *Cartographica* **22**, 1-20.

Egbert, S. and Slocum, T. (1992) EXPLOREMAP: An exploration system for choropleth maps, *Annals of the Association of American Geographers* **82**, 275-88.

Egeth, H., Virzi, R., and Garbart, H. (1984) Searching for conjunctively defined targets, *Journal of Experimental Psychology: Human Perception and Performance* **10**, 32-39.

Estes, W. (1994) *Classification and Cognition,* Oxford University Press, Oxford.

Evans, G. and Pezdek, K. (1980) Cognitive mapping: Knowledge of realworld distance and location information, *Journal of Experimental Psychology: Human Learning and Memory* **6**, 13-24.

Ferguson, E. and Hegarty, M. (1994) Properties of cognitive maps constructed from texts, *Memory and Cognition* **22**, 455-473.

Finke, R. (1979) The functional equivalence of mental images and error movement, *Cognitive Psychology* **11**, 235-264.

Finke, R. and Kosslyn, S. (1980) Mental imagery acuity in the peripheral vision field, *Journal of Experimental Psychology: Human Perception and Performance* **6**, 126-139.

Finke, R. and Schmidt, M. (1977) Orientation-specific color aftereffects following imagination, *Journal of Experimental Psychology: Human Perception and Performance* 3, 599-606.

Franklin, N. and Tversky, B. (1990) Searching imagined environments, *Journal of Experimental Psychology: General* **119**, 63-76.

Franklin, N., Tversky, B., and Coon, V. (1992) Switching points of view in spatial mental models, *Memory and Cognition* **20**, 507-518.

Gabriel, M. and Moore, J. (1990) *Learning and computational neuroscience: Foundations and adaptive networks*, The MIT Press, Cambridge, MA.

Gärling, T. and Golledge, R. (1993) *Behavior and Environment: Psychological and Geographical Approaches,* Elsevier Science Publishers, North-Holland.

Garner, W. (1974) *The Processing of Information and Structure,* Erlbaum, Potomac, MD.

Gati, I. and Tversky, A. (1982) Representations of qualitative and quantitative dimensions, *Journal of Experimental Psychology: Perception and Performance* **8**, 325-340.

Gati, I. and Tversky, A. (1984) Weighting common and distinctive features in perceptual and conceptual judgments, *Cognitive Psychology* **16**, 341-370.

Gati, I. and Tversky, A. (1987) Recall of common and distinctive features of verbal and pictorial stimuli, *Memory and Cognition* **15**, 97-100.

Gelman, S. (1988) The development of induction within natural kind and artifact categories, *Cognitive Psychology* **20,** 65-95.

Gentner, D. (1983) Structure-mapping: A theoretical framework of analogy, *Cognitive Science* **7**, 155-170.

Getis, A., Getis, J., and Feldman, J. (1994) *Introduction to Geography,* Wm. C. Brown Publishers, Dubuque, Iowa.

Gibson, J. (1950) *The Perception of the Visual World,* The Riverside Press, Cambridge, MA.

Glenberg, A. and Langston, W. (1992). Comprehension of illustrated text: Pictures help to build mental models, *Journal of Memory and Language* **31**, 127-151.

Glenberg, A and McDaniel, M. (1992) Mental models, picture, and text: Integration of spatial and verbal information, *Memory and Cognition* **20**, 458-460.

Gluck, M. (1991) Stimulus generalization and representation in adaptive network models of category learning, *Psychological Science* **2**, 50-55.

Gold, J. (1980) *An Introduction to Behavioural Geography,* Oxford University Press, Oxford.

Goldberg, J., MacEachren, A., and Kotval, X. (1992) Mental image transformation in terrain map comparisons. unpublished manuscript, Department of Geography, The Pennsylvania State University.

Goldstone, R. (1994) Similarity, interactivity, activation, and mapping, *Journal of Experimental Psychology: Learning, Memory, and Cognition* **20**, 3-28.

Goldstone, R., Medin, D., and Gentner, D. (1991) Relational similarity and nonindependence of features in similarity judgments, *Cognitive Psychology* **23**, 222-262.

Golledge, R. G. (1978) Learning about urban environments, in T. Carlstein, D. Parks and N. Thrift (eds.), *Timing Space and Spacing Time,* Arnold, London, pp. 76-98.

Golledge, R., Gale, N., Pellegrino, J., and Doherty, S. (1992) Spatial knowledge acquisition by children: Route learning and relational distances, *Annals of the Association of American Geographers* **82**, 223-244.

Golledge, R. (1993) Geographical perspectives on spatial cognition, in T. Garling and R. Golledge (eds.), *Behavior and Environment: Psychological and Geographic Approaches,* Elsevier Science Publishers, North-Holland, pp. 16-46.

Golledge, R. Dougherty, V., and Bell, S. (1995) Acquiring spatial knowledge: Survey versus route-based knowledge in unfamiliar environments, *Annals of the Association of American Geographers* **85**, 134-158.

Golledge, R. and Stimson, R. (1987) *Analytical Behavioural Geography,* Croom Helm, London.

Gould, P. (1966) On Mental Maps, Michigan Inter-University Community of Mathematical Geographers, Discussion Paper #9.

Gould, P. (1975) Acquiring spatial information, *Economic Geography* **51**, 87-99.

Graham, E. (1976) What is a mental map?, *Area* **9**, 259-262.

Gulliver, F. (1908) Orientation of maps, *Journal of Geography* **7**, 55-58.

Hampton, J. (1988) Overextension of conjunctive concepts: Evidence for a unitary model of concept typicality and class inclusion, *Journal of Experimental Psychology: Learning, Memory, and Cognition* **14**, 12-32.

Hartshorne, R. (1959) *Perspective on the nature of geography*, Rand McNally Co, Chicago.

Hewiston, B. and Crane, R. (1994) *Neural nets: Applications in geography*, Kluwer Academic Publishers, Dordrecht.

Hebb, D. (1949) *The Organization of Behavior*, Wiley, New York.

Heermann, P. and Khazenie, N. (1992) Classification of multispectral remote sensing data using a back-propagation neural network, *IEEE Transactions on Geoscience and Remote Sensing* **30**, 81-88.

Heider, E. (1971) "Focal" color areas and the development of color names, *Developmental Psychology* **4**, 447-55.

Heider, E. (1972) Universals in color naming and memory, *Journal of Experimental Psychology* **93,** 10-20.

Hintzman, D, O'Dell, C. and Arndt, D. (1981) Orientation in cognitive maps, *Cognitive Psychology* **13**, 149-206.

Hirtle, S. and Jonides, J. (1985) Evidence of hierarchies in cognitive maps, *Memory and Cognition* **13**, 208-217.

Hoffman, J. (1986) The psychology of perception, in J. LeDoux and W. Hirst (eds.), Mind and Brain: Dialogues in Cognitive Neuroscience, Cambridge University Press, Cambridge, p 12.

Holmes, J. (1984) *Cognitive Processes Used to Recognize Perspective Three-Dimensional Map Surfaces,* MA Thesis, University of South Carolina.

Holmes, E. and Gross, C. (1984) Effects of temporal lesions on discrimination of stimuli differing in orientation, *Journal of Neuroscience* **4**, 3063-3068.

Holyoak, K. and Mah, W. (1982) Cognitive reference points in judgments of symbolic magnitude, *Cognitive Psychology* **14**, 328-352.

Holyoak, K. and Thagard, P. (1989) Analogical mapping by constraint satisfaction, *Cognitive Science* **13**, 295-355.

Hubel, D. (1988) *Eye, Brain, and Vision*, W. H. Freeman, New York.

Huttenlocher, J., Hedges, L., and Duncan, S. (1991) Categories and particulars: Prototype effects in estimating spatial location, *Psychological Review* **98,** 352-376.

Ittelson, W. (1973) *Environment and Cognition,* Seminar Press, New York.

Jenks, G. and Caspall, F. (1971) Error on choropleth maps: Definition, measurement, reduction, *Annals of the Association of American Geographers* **61**, 217-244.

Jensen, J. (1986) *Introductory digital image processing: A remote sensing perspective*, Prentice-Hall, Englewood Cliffs, NJ

John, O., Hampson, S. And Goldberg, L. (1990) The basic level in personality-trait hierarchies: Studies of trait use and accessibility in different contexts, Journal of Personality and Social Psychology **60** , 348-361.

Johnson, M. (1987) *The body in the mind,* University of Chicago Press, Chicago.

Johnson-Laird, P. (1983) *Mental Models: Towards a Cognitive Science of Language, Inference, and Consciousness,* Cambridge University Press, Cambridge.

Johnson-Laird, P. (1993) *Human and machine thinking*, Lawrence Erlbaum Associates, Hillsdale, NJ.

Johnston, R. (1978) *Multivariate statistical analysis in geography*, Longman, London.

Kahneman, D. and Treisman, A. (1984) Changing views of attention and automaticity, in R. Parasuraman and D. Davies (eds.), *Varieties of attention,* Academic Press, New York, pp. 29-61.

Kahneman, D, Treisman, A. and Gibbs, B. (1992) The Reviewing of object files: object-specific integration of information, *Cognitive Psychology* **24**, 175-219.

Kilcoyne, J. (1974) Pictographic symbols in cartography, *Proceedings of the Association of American Geographers* **6**, 87-90.

Kirk, W. (1951) Historical geography and the concept of the behavioral environment, in G. Kuriyan (ed.), *Indian Geographical Journal, Silver Jubilee Edition,* Indian Geographical Society, Madras, pp. 152-160.

Köppen, W. (1931) *Grundriss der Klimakunde*, Walter de Gruyter Company, Berlin.

Kohonen, T. (1989) *Self-organization and associative memory*, Springer-Verlag, Berlin.

Konorski, J. (1967) *The Integrative Activity of the Brain*, University of Chicago Press, Chicago.

Kosko, B. (1992) *Neural networks and fuzzy systems: A dynamical systems approach to machine intelligence* Prentice-Hall, Upper Saddle River, NJ.

Kosko, B. 1993. Fuzzy *Thinking: The New Science of Fuzzy Logic*, Hyperion, New York.

Koysslyn, S. (1976) Using imagery to retrieve semantic information: A developmental study, *Child Development* **47**, 434-444.

Kosslyn, S. 1980. Image and Mind, Harvard University Press, Cambridge.

Kosslyn, S., Flynn, R., Amsterdam, J., and Wang, G. (1990) Components of high-level vision: A cognitive neuroscience analysis and accounts of neurological syndromes, *Cognition* **34**, 203-277.

Kosslyn S. and Koenig, O. (1992) *Wet Mind: The New Cognitive Neuroscience*, Free Press, New York.

Kosslyn, S., Koenig, O., Barrett, A., Cave, C., Tang, J., and Gabrieli, J. (1989) Evidence for two types of spatial representations: Hemispheric specialization for categorical and coordinate relations, *Journal of Experimental Psychology: Human Perception and Performance* **15**, 723-735.

Kosslyn, S., Murphy, G., Bemesderfer, M., and Feinstein, K. (1977) Category and continuum in mental comparisons, *Journal of Experimental Psychology: General* **106**, 341-375.

Kosslyn, S., Pick, H., and Fariello, G. (1974) Cognitive maps in children and men, *Child Development* **45**, 707-716.

Kosslyn, S. and Pomerantz, J. (1977) Imagery, propositions, and the form of internal representations, *Cognitive Psychology* **9**, 52-76.

Kosslyn, S., Reiser, B., Farah, M., and Fliegel, S. (1983) Generating visual images: Units and relations, *Journal of Experimental Psychology: General* **112**, 278-303.

Kosslyn, S., Segar, C., Pani, J., and Hillger, L. (1990) When is imagery used? A diary Study, *Journal of Mental Imagery* **14**, 131-152.

Kosslyn, S. and Swartz, S. (1978) Visual images as spatial representations in active memory, in A. Hanson and E. Riseman (eds.), *Computer Vision Systems*, Academic Press, New York, pp. 223-241.

Kosslyn, S., Holtzman, J., Farah, M., and Gazzangia, M. (1985) A computational analysis of mental image generation: Evidence from functional dissociations in split-brain patients, *Journal of Experimental Psychology: General* **114**, 311-341.

Kumler, M. and. Groop, R. (1990) Continuous-tone mapping of smooth surfaces, *Cartography and Geographic Information Systems* **17**, 279-291.

Lakoff, G. (1986) *Women, fire, and dangerous things*, University of Chicago Press, Chicago.

Lawrence, J. (1994) *Introduction to neural networks*, California Scientific Software Press, Nevada City, CA.

Lee, T. (1976) *Psychology and the Environment*, Methuen, London.

Levine, M., Jankovic, I., and Palij, M. (1982) Principles of spatial problems solving, *Journal of Experimental Psychology: General* **111**, 157-175.

Lewis, P. (1976) *New Orleans: The making of an urban landscape*, Ballinger, Cambridge.

Lindberg, E. and Gärling, T. (1981) Acquisition of locational information about reference points during locomotion with and without a concurrent task: Effects of number of reference points, *Scandinavian Journal of Psychology* **22**, 109-115.

Livingston, M. and Hubel, D. (1984) Anatomy and physiology of a color system in the primate visual cortex, Journal *of Neuroscience* **4**, 309-386.

Lloyd, R. (1982) A look at images, *Annals of the Association of American Geographers* **72**, 532-548.

Lloyd, R. (1988) Searching for map symbols: The Cognitive Process, *The American Cartographer* **15**, 363-78.

Lloyd, R. (1989a) Cognitive mapping: Encoding anu uecoding information, *Annals of the Association of American Geographers* **79**, 101-124.

Lloyd, R. (1989b) The estimation of distance and direction from cognitive maps, *The American Cartographer* **16**, 109-122.

Lloyd, R. (1992) Comparing map symbols to prototypes, unpublished research paper, Department of Geography, University of South Carolina.

Lloyd, R. (1993) Cognitive processes and cartographic maps, in T. Garling and R. Golledge (eds.), *Behavior and Environment: Psychological and Geographic Approaches,* Elsevier Science Publishers, North-Holland, pp. 141-169.

Lloyd, R. (1994) Learning spatial prototypes, *Annals of the Association of American Geographers* **84**, 418-40.

Lloyd, R. (1996a) Learning racial classes for urban neighborhoods, unpublished research paper, Department of Geography, University of South Carolina.

Lloyd, R. (1996b) Simulating visual search processes, Paper presented at the annual meeting of the Association of American Geographers, Charlotte, NC.

Lloyd, R. (1996c) Cognitive Theories of Spatial Search, unpublished paper, Department of Geography, University of South Carolina.

Lloyd, R (1996d) Learning visual fields with self-organizing maps, unpublished paper, Department of Geography, University of South Carolina.

Lloyd, R. and Cammack, R. (1996) Constructing cognitive maps with orientation biases, in J. Portugali (ed.), *The Construction of Cognitive Maps,* Kluwer Academic Publishers, Dordrecht, pp. 187-214.

Lloyd, R., Cammack, R., and Holliday, W. (1995) Learning environments and switching perspectives, *Cartographica* **32**, 5-17.

Lloyd, R. and Carbone, G. (1995) Comparing human and neural network learning of climate categories, *Professional Geographer* **47**, 237-250.

Lloyd, R. and Heivly, C. (1987) Systematic distortions in urban cognitive maps, *Annals of the Association of American Geographers* **77**, 191-207.

Lloyd, R. and Gilmartin, P. (1991) Cognitive maps of the world: Distortions and individual differences, *The National Geographical Journal of India.* **37**, 118-29.

Lloyd, R. and Hooper, H. (1991) Urban cognitive maps: Computation and structure, *Professional Geographer* **43**, 15-27.

Lloyd, R. and Jennings, D. (1978) Shopping behavior and income: Comparisons in an urban environment, Economic Geography **54**, 157-167.

Lloyd, R., Patton, D., and Cammack, R. (1995) Cognitive Structure of Spatial Knowledge, paper presented at the annual meeting of the Southeastern Division of the Association of American Geographers, Knoxville, TN.

Lloyd, R., Patton, D., and Cammack, R. (1996) Basic-level geographic categories, *Professional Geographer* **48**, 181-194.

Lloyd, R., Rostkowska-Covington, E., and Steinke, T. (1996) Feature matching and the similarity of maps, in C. Wood and C. Keller (eds.), Cartographic Design: Theoretical and practical perspectives, John Wiley & Sons, Chichester, pp. 211-236.

Lloyd, R, and Steinke, T. (1976) The Decision-making process for judging the similarity of choropleth maps, *The American Cartographer* **3**, 177-184.

Lloyd, R. and Steinke T. (1977) Visual and Statistical Comparison of Choropleth Maps, *Annals of the Association of American Geographers* **67**, 429-36.

Lloyd, R. and Steinke, T. (1984) Recognition of disoriented maps: The cognitive process, *The Cartographic Journal* **21**, 55-59.

Lowe, D. (1985) *Perceptual Organization and Visual Recognition,* Kluwer Academic Publishers, Boston.

Lowe, D. (1987a) Three-dimensional object recognition from single two-dimensional images, *Artificial Intelligence* **31**, 355-395.

Lowe, D. (1987b) The viewpoint consistency constraint, *International Journal of Computer Vision* **1**, 57-72.

Lowenthal, D. (1961) Geography, experience, and imagination: Toward a geographical epistemology, *Annals of the Association of American Geographers* **51**, 241-261.

Lutgens, F. and E. Tarbuck (1995) *The Atmosphere: An Introduction to Meteorology,* Prentice Hall, Englewood Cliffs, NJ.

Lynch, K. (1960). *Image of the City,* MIT Press, Cambridge, MA.

MacEachren, A. (1982) The role of complexity and symbolization method in thematic map effectiveness, *Annals of the Association of American Geographers* **72**, 495-513.

MacEachren, A. (1992a) Application of environmental learning theory to spatial knowledge acquisition from maps, *Annals of the Association of American Geographers* **82**, 245-274.

MacEachren, A. (1992b) Learning spatial information from maps: Can orientation-specificity be overcome?, *Professional Geographer* **44**, 431-443.

Malt, B. (1989) An on-line investigation of prototype and exemplar strategies in classification, *Journal of Experimental Psychology: Learning, Memory and Cognition* **15,** 539-55

Mani, R. and Johnson-Laird, P. (1982) The mental representation of spatial descriptions, *Memory and Cognition* **10**, 181-187.

Marr, D. (1982) *Vision: A computational investigation into the human representation and processing of visual information*, W. H. Freeman, San Francisco.

Martindale, C. (1980) Subselves: The internal representation of situational and personal dispositions, in L Wheeler (ed.), *Annual Review of Personality and Psychology, Vol. 1*, Sage Publications: Newbury Park, CA.

Martindale, C. (1991) Cognitive Psychology: A Neural-network Approach, Brooks/Cole Publishing Company, Pacific Grove, CA.

Maunsell, J. and Newsome, W. (1987) Visual processing in monkey extrastriate corex, *Annual Review of Neuroscience* **10**, 363-401

McCarty, H. and Salisbury, N. (1961) Visual comparison of isopleth maps as a means of determining correlations between spatially distributed phenomena, University of Iowa Studies in Geography No. 3, Iowa City.

McClelland, J. (1981) Retrieving general and specific knowledge from stored knowledge of specifics, *Proceedings of the Third Annual Conference of the Cognitive Science Society*, 170-172.

McCelland, J. and Rumelhart, D. (1986) *Parallel Distributed Processing: Explorations in the Microstructure of Cognition, Volume 2: Psychological and Biological Models*, MIT Press, Cambridge, MA.

McClelland, J. and Rumelhart, D. (1989) *Explorations in Parallel Distributed Processing: A Handbook of Models, Programs, and Exercises*, MIT Press, Cambridge, MA.

McNamara, T., Halpin, J., and Hardy, J. (1992) The representation and integration in memory of spatial and nonspatial information, *Memory and Cognition* **20**, 519-532

McNaughton, B. and Nadel, L. (1990) Hebb-Marr Networks and the neurobiological representation of action in space, in M. Gluck and D. Rumelhart (eds.), *Neuroscience and Connectionist Theory,* Erlbaum, Hillsdale, NJ, pp. 1- 64.

McNiff, M. (1991) Memory for land use categories on maps: A comparison of focal, associative, and false color schemes, MA Thesis, University of South Carolina.

Medin, D., Altom, M., and Murphy, T. (1984) Given versus inducted category representations: Use of prototype and exemplar information in classification, *Journal of Experimental Psychology: Learning, Memory, and Cognition* **10**, 333-52.

Medin, D. and Schwanenflugel, P. (1981) Lenearn separability in classification learning, *Journal of Experimental Psychology: Learning, Memory, and Cognition* **7**, 355-368.

Milner, B. (1971) Interhemispheric differences and psychological processes, *British Medical Bulletin Supplement* **27**, 272-277.

Moar, , I. and Bower, G. (1983) Inconsistency in spatial knowledge, *Memory and Cognition* **11**, 107-113.

Monmonier, M. (1992) Authoring graphic scripts: Experiences and principles, *Cartography and Geographic Information Systems* **19,** 247-260.

Montgomery, R. Weyel, P. , and Petersen, J. (1988) The gang of fourteen: A game for learning about world climates, *Journal of Geography* **87,** 174-78.

Moraglia, G, Maloney, K., Fekete, E., and Al-Basi, K. (1989) Visual search along the colour dimension, *Canadian Journal of Psychology* **43**, 1-12.

Morrill, R. (1987) A Theoretical Imperative, *Annals of the Association of American Geographers* **77**, 535-41.

Morris, M. and Murphy, G. (1990) Converging operations on a basic level in event taxonomies, Memory and Cognition **18**, 407-418.

Muehrcke, P. (1973)Visual pattern comparison in map reading, *Proceedings of the Association of American Geographers* **5**, 190-194.

Muehrcke, P. (1986) *Map use: Reading, analysis, and interpretation*, JP Publications, Madison.

Muehrcke, P. and Muehrcke, J. (1992) *Map use: Reading, analysis, and interpretation*, JP Publications, Madison, WI.

Müller, H. and Rabbitt, P. (1989) Spatial cueing and the relation between the accuracy of "where" and "what" decisions in visual search, *The Quarterly Journal of Experimental Psychology* **41**, 747-773.

Muller, J. (1976) Visual association in choroplethic mapping, *Proceeding of the Association of American Geographers* **8**, 160-164.

Muller, J. (1979) Perception of continuously shaded maps, *Annals of the Association of American Geographers* **69,** 240-249.

Muller, J. (1980) Visual comparison of continuously shaded maps, *Cartographica* **17**, 40-52.

Murphy, D.(1976) A diagrammatic key for teaching climate classification, *Journal of Geography* **75,** 333-338.

Murphy, G. (1991a) Parts in object concepts: Experiments with artificial categories *Memory and Cognition* **19**, 423-38.

Murphy, G. (1991b) More on parts in object concepts: Response to Tversky and Hemenway Memory and Cognition **19**, 443-447.

Murphy, G. and Wisniewski, E. (1989) Categorizing objects in isolation and in scenes: What a superordinate is good for, *Journal of Experimental Psychology: Learning, Memory, and Cognition* **15**, 572-86.

Murphy, G. and Smith, E. (1982) Basic-level superiority in picture categorization, *Journal of Verbal Learning and Verbal Behavior* **21**, 1-20.

Myung, I. and Busemeyer, J. (1992) Measurement-free tests of a general state-space model of prototype learning, *Journal of Mathematical Psychology* **36,** 32-67.

Nagy, A., Sanchez, R., and Hughes, T. (1990) Visual search for color differences with foveal and peripheral vision, *Journal of the Optical Society of America A* **7**, 1995-2001.

Nakayama, K. and Silverman, G. (1986a) Serial and parallel encoding of visual feature conjunctions, *Investigative Ophthalmology and Visual Science* **27**, 182.

Nakayama, K. and Silverman, G. (1986b) Serial and parallel processing of visual feature conjunctions, *Nature* **320**, 264-65.

Neisser, U. (1976) *Cognition and Reality: Principles and Implications of Cognitive Psychology,* Freeman, San Francisco .

Nelson, E. (1995) Colour detection on bivariate choropleth maps: The visual search process *Cartographica* ***32***, 33-43.

Nelson, E. and Gilmartin, P. (1996) An evaluation of multivariate, quantitative point symbols for maps, in C. Wood and P. Keller (eds) *Cartographic design: Theoretical and practical perspectives*, John Wiley & Sons, London, pp. 191-210.

Nigrin, A. (1993) *Neural networks for pattern recognition,* MIT press, Cambridge, MA.

Nosofsky, R. (1987) Attention and learning processes in the identification and categorization of integral stimuli, *Journal of Experimental Psychology: Learning, Memory, and Cognition* **13**, 87-108.

Nosofsky, R. (1988) Similarity, frequency, and category representations, *Journal of Experimental Psychology: Learning, Memory, and Cognition* **14,** 54-65.

O'Keefe, J. and Nadel, L. (1978) *The Hippocampus as a Cognitive Map,* Clarendon Press, Oxford.

O'Keefe, J. and Nadel, L. (1979) Precis of O'Keefe and Nadel's The Hippocampus as a Cognitive Map, *The Behavioral and Brain Sciences* **2**, 487-494.

Olson, J. (1970) The effects of class interval systems on the visual correlation of choropleth maps, Ph.D. dissertation, University of Wisconsin.

Openshaw, S. (1994) Neuroclassification of spatial data, in B. Hewitson and R. Crane (eds.), *Neural Nets: Applications in Geography,* Kluwer, Dordrecht, pp. 53-69.

Orleans, P. (1973). Differential Cognition of Urban Residents: Effects of Social Scale on Mapping, in R. Downs and D. Stea (eds.), *Image and Environment: Cognitive Mapping and Spatial Behavior,* Aldine, Chicago, pp. 115-130.

Parker, R. (1973) *The Godwuf Manuscript,* Houghton Mifflin, Boston.

Patton, D. (1995) Learning geographic categories within a map-reading environment. Ph.D. Dissertation, University of South Carolina.

Patton, D. and Cammack, R. (1996) An examination of the effects of task type and map complexity on sequenced and static choropleth maps, in C. Wood and P. Keller (eds.), *Cartographic design: Theoretical and practical perspectives*, John Wiley & Sons, London, pp. 237-252.

Pavio, A. (1969) Mental imagery in associative learning and memory, *Psychological Review* **76**, 241-263.

Pavio, A. (1971) *Imagery and Verbal Processes,* Holt, Rinehart, and Winston, Inc., New York.

Perrig, W. and Kintsch, W. (1985) Propositional and situational representations of text, *Journal of Memory and Language* **24**, 503-518.

Pinker, S. (1979) Mental maps, mental images, and intuitions about space, *The Behavioral and Brain Sciences* **2**, 513.

Posner, M. (1988) Structures and functions of selective attention, in T. Boll and B. Bryant (eds.), *Clinical Neuropsychology and Brain Function: Research , Measurement, and Practice,* American Psychological Association, Washington, DC, pp. 173-202.

Posner, M. and Keele, S. (1968) On the genesis of abstract ideas, *Journal of Experimental Psychology* **77**, 353-363.

Posner, M. and Keele, S. (1970) Retention of abstract ideas, *Journal of Experimental Psychology* **83**, 304-308.

Presson, C. and Hazelrigg, M. (1984) Building spatial representations through primary and secondary learning, *Journal of Experimental Psychology: Learning, Memory, and Cognition* **10**, 716-722.

Presson, C., DeLange, N, and Hazelrigg, M. (1989) Orientation-specificity in Spatial Memory: What Makes a Path Different from a Map of a Path?, *Journal of Experimental Psychology: Learning, Memory, and Cognition* **15**, 887-897.

Pylyshyn, Z. (1973) What the mind's eye tells the mind's brain: A critique of mental imagery, *Psychological Bulletin* **80**, 1-24.

Pylyshyn, Z. (1984) *Computation and Cognition: Toward a Foundation for Cognitive Science,* MIT Press, Cambridge, MA.

Quinlan, P. and Humphreys, G. (1987) Visual search for targets defined by combinations of color, shape, and size: An examination of the constraints on feature and conjunction searches, *Perception and Psychophysics* **37**, 455-472.

Raaijamakers, J. and Shiffrin, R. (1981) Search of associative memory, *Psychological Review* **88**, 16-45.

Rice, K. (1989) Disoriented prism maps: A recognition experiment. *Cartographic Perspectives,* **4**, 32.

Ritter, N. and Hepner, G. (1990) Application of an artificial neural network to land-cover classification of thematic mapper imagery, *Computers and Geoscience* **16**, 873-880.

Robinson, A., Morrison, J., Muehrcke, P., Kimmerling, J., and Guptill, S. (1995) *Elements of cartography,* John Wiley & Sons, New York

Rosch, E. (1973) Natural categories. *Cognitive Psychology* **4**, 328-350.

Rosch, E. (1975a) Cognitive representations of semantic categories, *Journal of Experimental Psychology: General* **104**, 192-233.

Rosch, E. (1975b) Cognitive reference points, *Cognitive Psychology* **7**, 532-547.

Rosch, E. (1975c) The nature of mental codes for color categories, *Journal of Experimental Psychology: Human Perception and Performance,* **1**, 303-322.

Rosch, E. and Lloyd, B. (1978) *Cognition and categorization,* Earlbaum, Hillsdale, NJ

Rosch, E. and Mervis, C. (1975) Family resemblances: Studies in the internal structure of categories, *Cognitive Psychology* **7**, 573-605.

Rosch, E., Mervis, C., Gray, C., Johnson, D., and Boyes-Braem, P. (1976) Basic objects in natural categories, *Cognitive Psychology* **8** , 382-439.

Rueckl, J., Cave, K., and Kosslyn, S. (1989) Why are "what" and "where" processed by separate cortical visual systems? A computational investigation, *Journal of Cognitive Neuroscience* **1**, 171-186.

Rumelhart, D., Lindsay, P, and Norman, D. (1972) A process model for long-term memory, in E. Turving and W. Donaldson (eds.), *Organization of Memory,* Academic Press, New York, pp. 197-246.

Rumelhart, D. and McClelland, J. (1986) *Parallel Distributed Processing: Explorations in the Microstructure of Cognition, Volume 1 : Foundations*, MIT Press, Cambridge, MA.

Rumelhart, D., Hinton, G., and McClelland, J. (1986) A general framework for parallel distributed processing, in D. Rumelhart and J. McClelland (eds.), *Parallel Distributed Processing: Explorations in the Microstructure of Cognition, Volume 1: Foundation,* MIT Press, Cambridge, MA, pp. 45-76.

Rushton, G. (1969) Analysis of spatial behavior by revealed space preference, *Annals of the Association of American Geographers* **59**, 391-400.

Rushton, G. (1979) On behavioral and perception geography, *Annals of the Association of American Geographers* **69**, 463-464.

Saarinen, T. (1979) Commentary-critique of Bunting-Guelke paper, *Annals of the Association of American Geographer* **69**, 464-468.

Sadalla, E., Burroughs, W., and Staplin, L. (1980) Reference points in spatial cognition, *Journal of Experimental Psychology: Human Learning and Memory* **5**, 516-528.

Sadalla, E. and Magel, S. (1980) The perception of traversed distance, *Environment and Behavior* **12**, 65-79.

Sadalla, E. and Staplin, L. (1980) An information storage model for distance cognition, *Environment and Behavior* **12**, 183-193.

Sagi, D. and Julesz, B. (1985) "Where" and "what" in vision, *Science* **228**, 1217-1219.

Sasanuma, S. (1974) Kanji versus kana processing alexia and transient agraphia, *Cortex* **10**, 89-97.

Sattath, S. and Tversky, A. (1987) On the relation between common and distinctive feature models, *Psychological Review* **94**, 16-22.

Shallice, T. (1978) The dominant action system: An information-processing approach to consciousness, in K. Pope and J. Singer (eds.), *Scientific investigations into the flow of human experience*, Plenum, New York.

Shank, R. and Abelson, R. (1977) *Scripts, plans, goals, and understanding*, Erlbaum, Hillsdale, NJ.

Shaver, P., Schwartz, J., Kirson, D., and O'Conner, C. 1987. Emotional knowledge.: Further explanation of a prototype approach, Journal of Personality and Social Psychology, **52**, 1061-1086.

Shepard, R. (1978) The mental image, *American Psychologist* **33**, 125-137.

Shepard, R. and Cooper, L. (1983) *Mental Images and Their Transformations*, MIT Press, Cambridge, MA.

Shepard, R. and Hurwitz, S. (1984) Upward direction, mental rotation, and discrimination of left and right turns in maps, *Cognition* **18**, 161-193.

Shepard, R. and Podgorny, P. (1978) Cognitive processes that resemble perceptual processes, in W. Estes (ed.), *Handbook of Learning and Cognitive Processes*, Erlbaum, Hillsdale, NJ, pp. 189-237.

Sholl, M. (1987) Cognitive maps as orienting schemata, *Journal of Experimental Psychology: Learning, Memory, and Cognition* **13**, 615-628.

Shortridge, B. (1982) Stimulus processing models from psychology: Can we use them in cartography, *The American Cartographer* **9**, 69-80.

Shortridge, B. and Welch, R. (1982) The effect of stimulus redundancy on the discrimination of town size on maps, *The American Cartographer* **9**, 69-80.

Slocum, T. and Egbert, S. (1993) Knowledge Acquisition from choropleth maps, *Cartography and Geographic Information Systems* **20**, 83-95.

Solso, R. (1979) *Cognitive Psychology*, Harcourt Brace Jovanovich, Inc., New York.

Smith E., and Zarate, M. (1992) Exemplar-based model of social judgment, *Psychological Review* **99**, 3-21.

Steinke, T. and Lloyd, R. (1981) Cognitive integration of objective map attribute information, *Cartographica* **18**, 13-23.

Steinke, T. and Lloyd, R. (1983a) Images of maps: A rotation experiment, *Professional Geographer* **35**, 455-461.

Steinke, T. and Lloyd, R. (1983b) Judging the similarity of choropleth map images. *Cartographica* **20**, 35-42.

Stevens, A. and Coupe, P. (1978) Distortions in judged spatial relations, *Cognitive Psychology* **10**, 422-437.

Taber, R. (1991) Knowledge processing and fuzzy cognitive maps, *Expert Systems with Applications* **2**, 83-87.

Tanaka and Taylor (1991) Object categories and expertise: Is the basic level in the eye of the beholder?, *Cognitive Psychology* **23**, 457-82.

Taylor, H. and Tversky, B. (1992a) Descriptions and depictions of environments, *Memory and Cognition* **20**, 483-496.

Taylor, H. and Tversky, B. (1992b) Spatial mental models derived from survey and route descriptions, *Journal of Memory and Language* **31**, 261-292.

Thorndyke, P. and Hayes-Roth, B. (1982) Differences in spatial knowledge acquired from maps and navigation, *Cognitive Psychology* **14**, 560-581.

Thorndyke, P. (1981) Distance estimations from cognitive maps, *Cognitive Psychology* **13**, 526-550.

Tobler, W. (1970) A computer movie simulating urban growth in the Detroit region, *Economic Geography Supplement* **46**, 234-240.

Tobler, W. (1973) Choropleth maps without class intervals?, *Geographical Analysis* **5**, 262-265.

Tolman, E. (1948) Cognitive maps in rats and men, *Psychological Review* **55**, 189-208.

Tootell, R., Silverman, M., and DeValois, R. (1981) Spatial frequency columns in primary visual coretx, *Science* **214**, 813-815.

Treisman, A. (1982) Perceptual grouping and attention in visual search for features and for objects, *Journal of Experimental Psychology: Human Perception and Performance* **8**, 194-214.

Treisman, A. (1988) Features and objects: The fourteenth Bartlett Memorial Lecture, *The Quarterly Journal of Experimental Psychology* **40A**, 201-237.

Treisman, A. (1991) Search, similarity, and integration of features between and within dimensions, *Journal of Experimental Psychology: Human Perception and Performance* **17**, 652-76.

Treisman, A. and Gelade, G. (1980) A feature integration theory of attention, *Cognitive Psychology* **12**, 97-136.

Treisman, A. and Gormican, S. (1988) Feature analysis in early vision: Evidence from search asymmetries, *Psychological Review* **95**, 15-48.

Treisman, A. and Sato, (1990) Conjunction search revisited, *Journal of Experimental Psychology: Human Perception and Performance* **16**, 459-78.

Treisman, A. and Souther, J. (1985) Search asymmetry: A diagnostic for preattentive processing of separable features, *Journal of Experimental Psychology: General* **114**, 285-310.

Trowbridge, C. (1913) On fundamental methods of orientation and imaginary maps, *Science* **38**, 888-897.

Tuan, Y. (1974). Topophilia: A study of environmental perception, attitudes, and values, Prentice-Hall Inc., Englewood Cliffs, NJ.

Tuan, Y. (1975) Images and mental maps, *Annals of the Association of American Geographers* **65**, 205-213.

Turving, E. (1985) How many memory systems are there? *American Scientist* **40**, 385-398.

Tversky, A. (1977) Features of similarity, Psychological Review **84**, 327-352.

Tversky, B. (1981) Distortions in memory for maps, *Cognitive Psychology* **13**, 407-433.

Tversky, B. (1992) Distortions in cognitive maps, *Geoforum* **23**, 131-138.

Tversky, B. and Hemenway, K. (1983) Categories of environmental scenes, *Cognitive Psychology* **15**, 121-49.

Tversky, B. and Hemenway, K. (1984) Objects, Parts, and Categories. *Journal of Experimental Psychology: General* **113**, 169-93.

Tversky, B. And Hemenway, K. (1991) Parts and the basic level in natural categories and artificial stimuli: Comments on Murphy (1991), Memory and Cognition **19**, 439-442.

Ungerleider, L and Mishkin, M. (1982) Two cortical visual systems, in D. Ingle, M. Goodale and R. Mansfield (eds.), *Analysis of Visual Behavior,* MIT Press, Cambridge, MA.

Usery, E. (1993) Category theory and the structure of features in geographic information systems, Cartography and Geographic Information Systems **20**, 5-12.

Wallace, (1989) Cognitive mapping and the origin of language and mind, *Current Anthropology* **30**, 518-526.

West, R. and Morris, C. (1985) Spatial cognition on nonspatial tasks: Finding spatial knowledge when you're not looking for it, in R. Cohen (ed.), *The Development of Spatial Cognition*, Erlbaum, Hillsdale, NJ, pp. 13-39.

White, G. (1945) Human adjustment to floods: A geographical approach to the flood problem in the United States, Department of Geography, University of Chicago, Research paper # 29.

Waddill, P. and McDaniel, M. (1992) Pictorial enhancement of text memory: Limitations imposed by picture type and comprehension skill, *Memory and Cognition* **20**, 472-82.

Widrow, G. and Hoff, M. (1960) Adaptive switching circuits, *Institute of Radio Engineers, Western Electronic Show and Convention Record, Part 4*, 96-104.

Wilson, S., Rinck, M., McNamara, T., Bower, G., and Morrow, D. (1993) Mental models and narrative comprehension: Some qualifications, *Journal of Memory and Language* **32,** 141-54.

Wilton, R. and File, P. (1975) Knowledge of spatial relations: A preliminary investigation, *Quarterly Journal of Experimental Psychology* **27**, 251-258.

Winter, K. and Hewitson, B.(1994) Self-organizing maps - Application to census data, in B. Hewitson and R. Crane (eds.), *Neural Nets: Applications in Geography*, Kluwer, Dordrecht, pp. 71-77.

Wittgenstein, L. (1953) *Philosophical investigations*, MacMillan, New York.

Wolfe, J., Cave, K., and Franzel, S. (1989) A modified feature integration model for visual search, *Journal of Experimental Psychology: Human Perception and Performance* **15** ,419-433.

Wolpert, J. (1964) The decision process in a spatial context, *Annals of the Association of American Geographers* **54**, 537-558.

Woodworth, R. (1948) *Contemporary Schools of Psychology,* Ronald Press, New York.

Wright, J. (1947) Terrae Incognitae: The place of imagination in geography, *Annals of the Association of American Geographers* **37** , 1-15.

—A—

—B—

—C—

—L—

—M—

Funding from the National Science Foundation that supported research discussed in Chapter 8 (Grant SES-8107126), Chapter 3 (Grant SES-8508676), and Chapters 5 and 9 (Grant SES-9224362) is gratefully acknowledged. Additional funding for research discussed in Chapters 7 and 8 was granted by the Carolina Venture Fund of the University of South Carolina.

Figure 1.2 was based on a map originally appearing in the June 30, 1991 edition of *The Times-Picayune*. Figure 1.3 was redrawn from *Cognition,* Vol. 34, p. 314, by S. Kosslyn, R. Flynn, J. Amsterdam, and G. Wang (1991) with kind permission of Elsevier Science - NL, Sara Burgerharstraat 25, 1055 KV Amsterdam, The Netherlands. Figure 2.8 was redrawn from *Cognitive psychology: A neural-network approach*, p. 60, by C. Martindale. Copyright © 1991 Brooks/Cole Publishing Company, Pacific Grove, CA 93950, a division of International Thomson Publishing Inc. Redrawn with permission of the publisher. Figure 2.9 was redrawn from *Neural networks and fuzzy systems: A dynamical systems approach to machine intelligence*, p. 153, by B. Kosko, © 1992. Redrawn with permission of Prentice-Hall, Upper Saddle River, NJ. Figure 2.10 was redrawn from *Expert systems with applications*, Vol. 2, p. 85, Knowledge processing and fuzzy cognitive maps, by R. Taber, © 1991, with kind permission from Elsevier Science Ltd, The Boulevard, Langford Lane, Kidlington Ox5 1GB, UK. Figure 2.11 was redrawn from the *Proceedings of the Third Annual Conference of the Cognitive Science Society*, © 1981, with the permission from J. McClelland. Figure 4.8 was redrawn from *Cartographic Design: Theoretical and Practical Perspectives*, p. 249, eds. C. Wood and C. Keller, An examination of the effects of task type and map complexity on sequenced and static choropleth maps, by D. Patton and R. Cammack, © 1996, redrawn with permission of John Wiley & Sons, Ltd. Figure 5.1 was redrawn from *The Quarterly Journal of Experimental Psychology*, Vol. 40A, p. 202, Features and objects: The fourteenth Bartlet Memorial Lecture, by A. Treisman, © 1988, with permission from Erlbaum (UK) Taylor & Francis, 27 Church Road , Hove, East Sussex, BN3 2FA, UK. Figure 5.10 was redrawn from *Psychological Review*, Vol. 96, p. 442, Visual search and stimulus similarity, by J. Duncan and G. Humphreys, copyright © 1989 by the American Psychological Association. Adapted with permission. Figure 5.11 and Figure 5.12 were redrawn from *Cognitive Psychology*, Vol. 22, pp. 234 and 236, Modeling the role of parallel processing in visual search, by K. Cave and J. Wolfe, © 1990, with permission from Academic Press, Inc. , Orlando, FL. Figure 6.15 was redrawn from *Cognitive Psychology*, Vol. 10, p. 435, Distortions in judged spatial relations, by A. Stevens and P. Coupe, © 1978, with permission from Academic Press, Inc., Orlando, FL. Figure 7.23 was redrawn from the *Journal of Experimental Psychology: Learning, Memory, and Cognition*, Vol. 14, p. 35, Categorization of multidimensional stimuli, by F. Ashby and R. Gott, copyright © 1988 by the American Psychological Association. Adapted with permission.

SPRINGER NATURE

GPSR Compliance

The European Union's (EU) General Product Safety Regulation (GPSR) is a set of rules that requires consumer products to be safe and our obligations to ensure this.

If you have any concerns about our products, you can contact us on ProductSafety@springernature.com

In case Publisher is established outside the EU, the EU authorized representative is:

Springer Nature Customer Service Center GmbH
Europaplatz 3
69115 Heidelberg, Germany

FSC
www.fsc.org
MIX
Papier | Fördert gute Waldnutzung
FSC® C083411

Kolibri 360 GmbH
Ferdinand-Jühlke-Straße 7
99095 Erfurt, Deutschland
produktsicherheit@kolibri360.de